The IMA Volumes in Mathematics and Its Applications

Volume 15

Series Editors
Hans Weinberger   Willard Miller, Jr.

Institute for Mathematics and Its Applications
IMA

The **Institute for Mathematics and its Applications** was established by a grant from the National Science Foundation to the University of Minnesota in 1982. The IMA seeks to encourage the development and study of fresh mathematical concepts and questions of concern to the other sciences by bringing together mathematicians and scientists from diverse fields in an atmosphere that will stimulate discussion and collaboration.

The IMA Volumes are intended to involve the broader scientific community in this process.

Hans Weinberger, Director
Willard Miller, Jr., Associate Director

IMA PROGRAMS

1982-1983	**Statistical and Continuum Approaches to Phase Transition**
1983-1984	**Mathematical Models for the Economics of Decentralized Resource Allocation**
1984-1985	**Continuum Physics and Partial Differential Equations**
1985-1986	**Stochastic Differential Equations and Their Applications**
1986-1987	**Scientific Computation**
1987-1988	**Applied Combinatorics**
1988-1989	**Nonlinear Waves**
1989-1990	**Dynamical Systems and Their Applications**

SPRINGER LECTURE NOTES FROM THE IMA:

The Mathematics and Physics of Disordered Media
Editors: Barry Hughes and Barry Ninham
(Lecture Notes in Math., Volume 1035, 1983)

Orienting Polymers
Editor: J.L. Ericksen
(Lecture Notes in Math., Volume 1063, 1984)

New Perspectives in Thermodynamics
Editor: James Serrin
(Springer-Verlag, 1986)

Models of Economic Dynamics
Editor: Hugo Sonnenschein
(Lecture Notes in Econ., Volume 264, 1986)

Donald G. Truhlar
Editor

Mathematical Frontiers in Computational Chemical Physics

With 89 Illustrations

Springer-Verlag
New York Berlin Heidelberg
London Paris Tokyo

Donald G. Truhlar
Department of Chemistry,
Chemical Physics Program,
and Supercomputer Institute
University of Minnesota
Minneapolis, MN 55455
USA

Mathematics Subject Classification (1980): 81-XX, 80A50

Library of Congress Cataloging-in-Publication Data
Mathematical frontiers in computational chemical physics / Donald G.
Truhlar, editor.
p. cm. — (The IMA volumes in mathematics and its
applications ; v. 15)
Proceedings of the Workshop on Atomic and Molecular Structure and
Dynamics, held June 15–July 24, 1987, at the Institute for
Mathematics and Its Applications, University of Minnesota.
Bibliography: p.
ISBN 0-387-96782-6
1. Chemistry, Physical and theoretical—Mathematics—Congresses.
I. Truhlar, Donald G., 1944– . II. Workshop on Atomic and
Molecular and Dynamics (1987 : University of Minnesota)
III. Institute for Mathematics and Its Applications. IV. Series.
QD455.3.M3M38 1988
541—dc19 88-20117

Camera-ready copy provided by the authors.
Printed and bound by R.R. Donnelley and Sons, Harrisonburg, Virginia.
Printed in the United States of America.

9 8 7 6 5 4 3 2 1

ISBN 0-387-96782-6 Springer-Verlag New York Berlin Heidelberg
ISBN 3-540-96782-6 Springer-Verlag Berlin Heidelberg New York

The IMA Volumes in Mathematics and Its Applications

Current Volumes:

Volume 1: Homogenization and Effective Moduli of Materials and Media
Editors: Jerry Ericksen, David Kinderlehrer, Robert Kohn, J.-L. Lions

Volume 2: Oscillation Theory, Computation, and Methods of Compensated Compactness
Editors: Constantine Dafermos, Jerry Ericksen, David Kinderlehrer, Marshall Slemrod

Volume 3: Metastability and Incompletely Posed Problems
Editors: Stuart Antman, Jerry Ericksen, David Kinderlehrer, Ingo Müller

Volume 4: Dynamical Problems in Continuum Physics
Editors: Jerry Bona, Constantine Dafermos, Jerry Ericksen, David Kinderlehrer

Volume 5: Theory and Applications of Liquid Crystals
Editors: Jerry Ericksen and David Kinderlehrer

Volume 6: Amorphous Polymers and Non-Newtonian Fluids
Editors: Constantine Dafermos, Jerry Ericksen, David Kinderlehrer

Volume 7: Random Media
Editor: George Papanicolaou

Volume 8: Percolation Theory and Ergodic Theory of Infinite Particle Systems
Editor: Harry Kesten

Volume 9: Hydrodynamic Behavior and Interacting Particle Systems
Editor: George Papanicolaou

Volume 10: Stochastic Differential Systems, Stochastic Control Theory and Applications
Editors: Wendell Fleming and Pierre-Louis Lions

Volume 11: Numerical Simulation in Oil Recovery
Editor: Mary Fanett Wheeler

Volume 12: Computational Fluid Dynamics and Reacting Gas Flows
Editors: Bjorn Engquist, M. Luskin, Andrew Majda

Volume 13: Numerical Algorithms for Parallel Computer Architectures
Editor: Martin H. Schultz

Volume 14: Mathematical Aspects of Scientific Software
Editor: J.R. Rice

Volume 15: Mathematical Frontiers in Computational Chemical Physics
Editor: Donald G. Truhlar

Forthcoming Volumes:

1987-1988: *Applied Combinatorics*

Applications of Combinatorics and Graph Theory to the Biological and Social Sciences

CONTENTS

DYNAMICAL GROUPS

FOREWORD

This IMA Volume in Mathematics and its Applications

MATHEMATICAL FRONTIERS IN COMPUTATIONAL CHEMICAL PHYSICS

is in part the proceedings of a workshop which was an integral part of the 1986-87 IMA program on SCIENTIFIC COMPUTATION. We are grateful to the Scientific Committee: Bjorn Engquist (Chairman), Roland Glowinski, Mitchell Luskin and Andrew Majda for planning and implementing an exciting and stimulating year-long program. We especially thank the Workshop Organizer, Donald Truhlar, for organizing a workshop which brought together many of the major figures in a variety of research fields connected with atomic and molecular structure for a fruitful exchange of ideas.

Willard Miller, Jr.

Hans Weinberger

MATHEMATICAL FRONTIERS IN COMPUTATIONAL CHEMICAL PHYSICS

PREFACE

The Workshop on Atomic and Molecular Structure and Dynamics was held June 15–July 24, 1987 at the Institute for Mathematics and Its Applications on the University of Minnesota Twin Cities campus as part of the 1986–87 I.M.A. Program on Scientific Computation. There were over 70 participants, including the eleven plenary lecturers whose contributions form the chapters of this volume. The chapters discuss a wide variety of topics in the subject area of the Workshop. Each chapter includes expository material that is especially prepared to introduce the subject to a mathematical audience interested in studying frontier areas in mathematical chemical physics, and in addition each chapter also discusses challenging problem areas where additional mathematical progress is necessary and desirable for the advancement of the field. In many cases, the lecturers have emphasized unsolved aspects of their subject field that provide fertile ground for future research. The mathematical problems involve the theory of partial differential equations, integral equations, analytic continuation, asymptotic expansions, group theory, Lie algebra, molecular dynamics, simulations, statistical mechanics, and of course quantum mechanics. Taken as a whole it is my hope that the collection of topics will provide the interested mathematician–or mathematically oriented physical scientist–with an entry into some of the most promising areas of current research in chemical physics and physical chemistry.

The proceedings open with an introduction to the Born-Oppenheimer approximation by which the electronic problem can be solved with fixed nuclei to provide an effective potential for internuclear motion. Every other chapter in the book is concerned with either the electronic problem (the chapters by Almlöf, Reinhardt, Langhoff, Paldus, and Shavitt) or the motion of atoms on a Born-Oppenheimer potential surface (the other chapters); although in a few cases the methods are general enough to be used for both kinds of problem (e.g., the techniques presented for electronic auto-ionization may also be applied to vibrational predissociation). The second introductory lecture is concerned with the basic orbital and configurational techniques used in modern treatments of electronic structure.

The next two chapters deal with the spectral theory of electronic Schroedinger operators. They provide an interesting contrast in the way a theorist may approach the calculation of a cross section function. In the first approach attention is centered on the quasibound states (variously called metastable states, finite-lifetime states, autoionizing states, or resonances) that are responsible for most of the structure in such curves, and one attempts to calculate the properties of these states directly. In the second approach one attempts to calculate the whole cross section function, thereby including any structure-producing states implicitly. In particular, the chapter included here focuses on the Stieltjes imaging moment technique for extracting the cross section curve from a finite-matrix representation of the Hamiltonian and on symmetry aspects of the continuum-state approximations that arise in this method. These two chapters are concerned primarily with few-body systems, e.g., the two electrons of H^- or the ten electrons of H_2O.

The next two chapters focus on systems involving many (up to Avogadro's number or more) atoms or molecules under conditions where their motions are well approximated by classical mechanics. The first chapter is concerned with the techniques and algorithms used to carry out such many-body simulations. The other chapter is concerned with collective phenomena in statistical mechanics and the geometry of potential energy hypersurfaces.

The two chapters on quantum mechanical dynamics include one on the statistical mechanics of many-body systems and another on the quantum theory of isolated collisions of an atom with a diatom.

Finally I grouped together three chapters under the heading of Dynamical Groups. In the first of these, compact dynamical groups in the strong sense are applied to the problem of vibrational energies. The final two chapters present an introduction to the very successful way in which unitary groups are used to express conveniently the Hamiltonian operator for the calculation of electronic structure problems with millions of antisymmetric, spin-adapted basis functions.

I would like to express my thanks and that of all the participants in the Workshop to Hans Weinberger, Willard Miller, Jr., George Sell, Mitch Luskin, and Avner Friedman of I.M.A. for their help and hospitality in planning and hosting the Workshop, with a special personal thanks to Hans Weinberger who helped me with the organization at every stage from the initial planning to finally serving as a very helpful Series Editor for this volume. It is also a pleasure to thank the I.M.A. staff, especially Pat Kurth, Kaye Smith, and Robert Copeland for their very special assistance in the running of the Workshop and Patricia Brick and George Swan for their expert assistance in the production of these proceedings.

Donald G. Truhlar
Minneapolis
July, 1987

INTRODUCTORY CHAPTERS

THE BORN-OPPENHEIMER APPROXIMATION IN MOLECULAR QUANTUM MECHANICS

C. ALDEN MEAD†

Abstract. This chapter introduces the reader to the formulation of the Born-Oppenheimer adiabatic aproximation, and its significance in molecular quantum mechanics. The Hamiltonian operator for a molecular system is set up, and it is shown in some detail how different contributions to the energy and wave function can be classified according to the way they depend on the ratio of electronic to nuclear mass. There is some discussion of the significance of these results. Finally, a brief introduction is presented to some topics of current research, including properties of the coupling between different electronic levels, and the properties of topological phase factors experienced by electronic wave functions when the nuclear coordinates traverse a closed path.

1. Introduction. The purpose of these introductory lectures is to acquaint the audience with the very fundamental adiabatic approximation in molecular quantum mechanics, which was first formulated in a classic paper by Born and Oppenheimer [1]. Another milestone in the development of the theory was the somewhat different treatment by Born and Huang [2]. A brief but clear summary of the different approaches has been given by Ballhausen and Hansen [3]. This approach, in which the large mass of a nucleus compared to that of an election permits an approximate separation of electronic and nuclear motion, greatly simplifies the quantum mechanical treatment of molecular systems, and also provides the essential link between quantum mechanics and traditional chemistry, in that it justifies the "ball-and-stick" picture of simple molecules which had been so successful over the years in accounting for a large body of chemical knowledge. Of course, if quantum mechanics had *not* been reconcilable with that picture, it would constitute a serious defect of the quantum theory. Thus, it would not be an exaggeration to say that the quantum mechanical theory of molecules was founded by the publication of reference [1]; certainly, the approximation that bears the names of its authors remains basic to all molecular quantum mechanics, so it is important to understand its meaning, its implications, and also its limitations.

After these brief words of introduction, we now proceed to the formulation of the approximation and the lowest-order corrections to it, followed by a brief overview of some selected topics of current research, which will show an unabashed bias toward the author's own interests.

2. Formulation. A molecular system consists of n nuclei, which we will label with Latin indices $j = 0, 1, \ldots, (n-1)$, and N electrons, with Greek indices $\alpha = 1, 2, \ldots, N$. These interact with one another via electrostatic Coulomb forces. The nonrelativistic quantum mechanical treatment of such a system involves the following constants: The electronic mass m_e; the masses of the nuclei M_j; the Dirac constant $\hbar = h/2\pi$, where h = Planck's constant; and the electronic charge e_0. In

†Chemistry Department, University of Minnesota, Minneapolis, Minnesota, 55455, USA

ordinary units, all of these are numerically very small numbers. The treatment becomes much more transparent, however, if we use atomic units, in which the units of length, mass, and time are chosen so that m_e, e_0, and $\hbar$ are all equal to unity. The unit of length is the first Bohr radius, $\hbar^2/m_e e_0^2$ in ordinary units; the unit of time is the period of the lowest Bohr orbit divided by 2π, $\hbar^3/m_e e_0^4$ in ordinary units; and the unit of mass is the electron mass. In these units, distances characteristic of molecules, such as chemical bond lengths, are of order of magnitude unity, as are electronic energies. The only parameters remaining are the nuclear masses M_j, which are now numerically *large*: all are greater than 10^3, all but the masses of the two lightest elements H and He are greater than 10^4, and the masses of the heavier elements are greater than 10^5. However, the nuclear masses do not differ from each other by large factors, so we can select some average nuclear mass M for our problem, and denote each M_j by $b_j M$. The problem will now contain *one* numerically large parameter M, while all other parameters are of order unity. In what follows, we will discuss the relative magnitudes of various contributions to molecular properties such as energy in terms of their dependence on M. Thus, a term which is independent of M will be referred to as of order unity. In practice, of course, such a term may turn out to be numerically rather large or small because of accumulation of factors such as 2π; but this will depend on peculiarities of a particular problem, and cannot be included in a discussion of general theory.

2.1 Hamiltonian. We denote the coordinates of nuclei and electrons relative to the laboratory by vectors $\mathbf{R}_j$ and $\mathbf{R}_\alpha$ respectively. The classical Lagrangian for the system consists of a kinetic energy T and potential energy V:

$$L = T - V, \tag{1}$$

with

$$T = \frac{1}{2}\Big\{\sum_{j=0}^{n-1} M_j \dot{\mathbf{R}}_j^2 + \sum_{\alpha=1}^{N} \dot{\mathbf{R}}_\alpha^2\Big\} \tag{2a}$$

$$V = \sum_{j<k} Z_j Z_k / R_{jk} - \sum_{j\alpha} Z_j / R_{j\alpha} + \sum_{\alpha<\beta} 1/R_{\alpha\beta} \tag{2b}$$

where Z_j denotes the charge on nucleus j in units of the electronic charge, a dot above a letter denotes time differentiation, and R_{jk}, $R_{j\alpha}$, and $R_{\alpha\beta}$ are interparticle distances.

The system has $3(n+N)$ degrees of freedom, but three of these are just the trivial motion of the center of mass, so it is convenient to define new coordinates in which the center-of-mass motion is separated from the relative motion. We do this in the following way (not the only possible way): We introduce script letters

$$\mathcal{M}_j = \sum_{k=0}^{j-1} M_k, \tag{3a}$$

(so that $\mathcal{M}_n$ = total mass of nuclei) and

$$\mathcal{M} = \text{ total mass of electrons and nuclei,} \tag{3b}$$

and define a center-of-mass coordinate $\mathbf{r}_0$, nuclear relative coordinates $\mathbf{r}_j$ ($j = 1, 2, \ldots, n-1$), and electronic relative coordinates $\mathbf{r}_\alpha$ as follows:

$$\text{(3c)} \qquad \mathbf{r}_0 = \mathcal{M}^{-1}\Big\{\sum_j M_j \mathbf{R}_j + \sum_\alpha \mathbf{R}_\alpha\Big\};$$

$$\text{(3d)} \qquad \mathbf{r}_1 = \mathbf{R}_1 - \mathbf{R}_0;$$

$$\text{(3e)} \qquad \mathbf{r}_j = \mathbf{R}_j - \mathcal{M}_j^{-1} \sum_{k=0}^{j-1} M_k \mathbf{R}_k, \qquad j = 1, 2, \ldots, (n-1);$$

$$\text{(4)} \qquad \mathbf{r}_\alpha = R_\alpha - \mathcal{M}_n^{-1} \sum_{k=0}^{n-1} M_k \mathbf{R}_k.$$

In (3c) and (3d), after introducing the center-of-mass coordinate, we have defined relative coordinates for the nuclei one at a time, with a coordinate for each nucleus being defined relative to the center of mass of the nuclei already introduced. We could have done the same for the electrons, but we prefer (4), which treats all electrons on the same basis. In terms of the new coordinates, the Lagrangian becomes:

$$\text{(5)} \qquad L = \frac{1}{2}\mathcal{M}\dot{\mathbf{r}}_0^2 + \frac{1}{2}\sum_{j=1}^{n-1} \mu_j \dot{\mathbf{r}}_j^2 + \frac{1}{2}\sum_\alpha \dot{\mathbf{r}}_\alpha^2 - \frac{1}{2}\mathcal{M}^{-1}\Big(\sum_\alpha \dot{\mathbf{r}}_\alpha\Big)^2 - V(\mathbf{r}_1, \ldots, \mathbf{r}_{n-1}, \mathbf{r}_\alpha)$$

where we have introduced reduced masses

$$\text{(6a)} \qquad \mu_j = M_j \mathcal{M}_j / \mathcal{M}_{j+1}.$$

The first term in (5) is the kinetic energy of the center of mass, and all other terms are independent of $\mathbf{r}_0$. The center of mass thus behaves as an independent free particle, and it will be omitted from now on. We are thus left with a system of $3(n+N-1)$ degrees of freedom representing the relative motion.

We now introduce an average nuclear mass M, define coefficients (of order of magnitude unity) b_j by

$$\text{(6b)} \qquad \mu_j = M b_j,$$

and define mass-scaled coordinates $\mathbf{u}_j$ by

$$\text{(6c)} \qquad \mathbf{u}_j = b_j^{\frac{1}{2}} \mathbf{r}_j.$$

In terms of these, (5) becomes:

$$\text{(7)} \qquad L = \frac{1}{2} M \sum\nolimits_j \dot{\mathbf{u}}_j^2 + \frac{1}{2}\sum \dot{\mathbf{r}}_\alpha^2 - \frac{1}{2}\mathcal{M}^{-1}\Big(\sum_\alpha \dot{\mathbf{r}}_\alpha\Big)^2 - V$$

We now go over to the classical Hamiltonian in the usual way be defining conjugate momenta

(8a) $$\mathbf{P}_j = \partial L/\partial \dot{\mathbf{u}}_j; \qquad \mathbf{P}_\alpha = \partial L/\partial \dot{\mathbf{r}}_\alpha;$$

and Hamiltonian

(8b) $$H = \sum_j \mathbf{P}_j \cdot \dot{\mathbf{u}}_j + \sum_\alpha \mathbf{P}_\alpha \cdot \dot{\mathbf{r}}_\alpha - L,$$

yielding:

(8c) $$H = \frac{1}{2M}\sum_{j=1}^{n-1} \mathbf{P}_j^2 + \frac{1}{2}\sum_\alpha \mathbf{p}_\alpha^2 + \frac{1}{2bM}\Big(\sum_\alpha \mathbf{p}_\alpha\Big)^2 + V(\mathbf{u}_j, \mathbf{r}_\alpha),$$

where the parameter b is defined by $bM = \mathcal{M}_n$.

When we quantize, the quantities in (8c) will be replaced by quantum mechanical operators, denoted by a hat, $\hat{}$. As is well known $\hat{\mathbf{u}}_j = \mathbf{u}_j$, $\hat{\mathbf{P}}_j = -i\nabla_j$, etc. We see from (8c) that the Hamiltonian contains two terms independent of M, and two others proportional to the small parameter $(1/M)$. Following Born and Oppenheimier [1], we define the parameter $\kappa = (1/M)^{\frac{1}{4}}$. The quantized version of (8c) can now be written as follows:

(9a) $$\hat{H} = \hat{T}_n + \hat{H}_{el}(\mathbf{u}) + \hat{H}_{mp},$$

where the first term is the nuclear kinetic energy

(9b) $$\hat{T}_n = \frac{\kappa^4}{2}\sum_j \hat{\mathbf{P}}_j^2 = -\frac{\kappa^4}{2}\sum \nabla_j^2;$$

the second term is called the electronic Hamiltonian,

(9c) $$\hat{H}_{el}(\mathbf{u}) = \frac{1}{2}\sum_\alpha \hat{\mathbf{P}}_\alpha^2 + \hat{V}$$

and the last term is called the mass polarization,

(9d) $$\hat{H}_{mp} = \frac{\kappa^4}{2b}\Big(\sum_\alpha \hat{\mathbf{P}}_\alpha\Big)^2.$$

The notation emphasizes that $H_{el}(\mathbf{u})$ can be thought of as an operator in the electronic degrees of freedom which depends parametrically on the nuclear coordinates: for any specified value of the nuclear coordinates $\mathbf{u}$, there is a well-defined operator $H_{el}(\mathbf{u})$ operating on functions of the electronic coordinates. Here, of course, $\mathbf{u}$ is a shorthand for the set of all nuclear coordinates $\mathbf{u}_j$.

For future use, we will also find it convenient to transform the nuclear kinetic energy (9b) in such a way as to separate the rotational degrees of freedom from the vibrational ones. We can define a molecule-fixed coordinate system in some convenient way, for example with the z-axis perpendicular to the plane of $\mathbf{u}_1$ and $\mathbf{u}_2$ and the x-axis in the direction of $\mathbf{u}_1$. Three degrees of freedom, denoted collectively by Ω, are angles giving the orientation of this frame relative to the laboratory, and the variables conjugate to these are related to the three components of angular momentum J_σ, where σ is an index taking on three values, one for each of the directions in space. The components of angular momentum may be defined relative to the laboratory or to the molecule-fixed frame, but the latter is more convenient for our purposes. We only sketch the way in which the other $(3n-6)$ nuclear degrees of freedom are introduced, since the procedure is not unique and the details are in any case only incidentally relevant to our topic. Our method is appropriate to our purposes, and is not necessarily the most convenient for the more usual purposes of analyzing vibrational spectra. For more detailed treatments of the conventional methods, the reader is referred to the *Handbuch* article by Nielsen [4], or the more recent book by Califano [5]. In our approach, the $\mathbf{u}_j$ are introduced one at a time. At each stage, a total angular momentum $\mathbf{J}_j$ can be defined for the degrees of freedom already introduced, and a corresponding angular velocity ω_j by applying the inverse of the inertia tensor to $\mathbf{J}_j$. Introducing $\mathbf{u}_{j+1}$, we can define its angular momentum relative to the frame moving with angular velocity ω_j, and a new total angular momentum and angular velocity with respect to which $\mathbf{u}_{j+2}$ rotates, etc. When this is finished, we have three overall rotational degrees of freedom Ω, and $(3n-6)$ "vibrational" or internal degrees of freedom Q_ℓ, with conjugate momenta P_ℓ. The kinetic energy, in quantized notation, becomes

$$\hat{T}_n = \frac{\kappa^4}{2}\Big\{\sum_{\sigma\tau} \hat{J}_\sigma \Gamma_{\sigma\tau}(Q)\hat{J}_\tau + \sum_{\ell m} \hat{P}_\ell \Lambda_{\ell m}(Q)\hat{P}_m\Big\} \tag{10}$$

or in shorthand:

$$T_n = \frac{\kappa^4}{2}\{\hat{\mathbf{J}}\cdot\overline{\Gamma}(Q)\hat{\mathbf{J}} + \hat{\mathbf{P}}\cdot\overline{\Lambda}(Q)\hat{\mathbf{P}}\}, \tag{11}$$

where $\overline{\Gamma}$ and $\overline{\Lambda}$ are matrices of order 3 and $(3n-6)$, respectively. The bar on top denotes a matrix. For $\overline{\Gamma}$, we have $\overline{\Gamma} = \overline{I}^{-1}$, where $\overline{I}$ is the nuclear inertia tensor given by:

$$I_{\sigma\tau} = \sum_{j=1}^{n-1}\{\mathbf{u}_j^2\delta_{\sigma\tau} - u_{j\sigma}u_{j=\tau}\}. \tag{12}$$

$\overline{\Lambda}$ involves inertia tensors for relative rotations, as well as the expression of angular momenta in terms of linear momenta and coordinates.

2.2 Quantization and Expansion. We want to formulate the eigenvalue problem for the Hamiltonian (9) in such a way that its solution can be approached by successive approximations utilizing the smallness of the parameter κ. To do this, we note that $\hat{H}_{el}$ now depends on the variables collectively denoted by Ω and Q, and that for any values of these it presumably possesses a complete set of eigenfunctions:

$$\hat{H}_{el}(\Omega, Q)\phi_\mu(\Omega, Q, r) = W_\mu(Q)\phi_\mu(\Omega, Q, r), \tag{13}$$

where the eigenvalue clearly depends only on Q because it is unchanged by a rigid rotation of all the nuclei together. The eigenfunctions can be chosen to be normalized for all values of the nuclear coordinates:

$$\int \phi_\mu^*(\Omega, Q, r)\phi_\nu(\Omega, Q, r)dr = \langle\phi_\mu|\phi_\nu\rangle(\Omega, Q) = \delta_{\mu\nu}. \tag{14}$$

For simplicity, we have assumed that the eigenvalues are all discrete. If, as in practice is usually the case, there is also a continuous spectrum, its presence would complicate the notation without affecting any of our basic conclusions.

We now expand the full wave function $\Psi(\Omega, Q, r)$ for our system as

$$\Psi(\Omega, Q, r) = \sum_\mu \psi_\mu(\Omega, Q)\phi_\mu(\Omega, Q, r), \tag{15}$$

and substitute this into the Schrödinger equation $\hat{H}\Psi = E\Psi$, making use of (9) and (11). In doing this, we recall that the operators $\hat{J}_\sigma$ and $\hat{P}_j$ are differential operators, so that the rules for applying them to products are the same as those for differentiation of a product. For example, we have

$$\hat{J}_\tau(\psi_\mu\phi_\mu) = (\hat{J}_\tau\psi_\mu)\phi_\mu + \psi_\mu(\hat{J}_\tau\phi_\mu). \tag{16}$$

We also note that the effect of $\hat{J}_\tau$ is to rotate the nuclei about the τ axis while holding the electrons fixed. Since the electronic eigenfunctions just move with the nuclei under such rotations, the effect of this is just the same as rotating the electrons in the opposite direction with nuclei fixed, which corresponds to the electronic angular momentum operator $\hat{L}_\tau$. Thus, (16) becomes

$$\hat{J}_\tau(\psi_\mu\phi_\mu) = (\hat{J}_\tau\psi_\mu)\phi_\mu - \psi_\mu(\hat{L}_\tau\phi_\mu). \tag{17}$$

Applying this twice, we find

$$\hat{\mathbf{J}} \cdot \overline{\Gamma}\hat{\mathbf{J}}(\psi_\mu\phi_\mu) = (\hat{\mathbf{J}} \cdot \overline{\Gamma}\hat{\mathbf{J}}\psi_\mu)\phi_\mu - 2(\hat{\mathbf{L}}\phi_\mu) \cdot \overline{\Gamma}\hat{\mathbf{J}}\psi_\mu + (\hat{\mathbf{L}} \cdot \overline{\Gamma}\hat{\mathbf{L}}\phi_n)\psi_\mu \tag{18}$$

The other term in (11) is similar, except that there is no analog of (17). We find:

$$\hat{\mathbf{P}} \cdot \overline{\Lambda}\hat{\mathbf{P}}(\psi_\mu\phi_\mu) = (\hat{\mathbf{P}} \cdot \overline{\Lambda}\hat{\mathbf{P}}\psi_\mu)\phi_\mu + 2(\hat{\mathbf{P}}\phi_\mu) \cdot \overline{\Lambda}\hat{\mathbf{P}}\psi_\mu + (\hat{\mathbf{P}} \cdot \overline{\Lambda}\hat{\mathbf{P}}\phi_\mu)\psi_\mu. \tag{19}$$

Inserting (9), (11), (13), (15), (18), and (19) into the Schrödinger equation, multiplying from the left by one of the ϕ_μ^*, integrating, and making use of (14), we obtain:

$$\begin{aligned} &\Big\{\frac{\kappa^4}{2}[\hat{\mathbf{J}} \cdot \overline{\Gamma}\hat{\mathbf{J}} + \hat{\mathbf{P}} \cdot \overline{\Lambda}\hat{\mathbf{P}}] + W_\mu\Big\}\psi_\mu \\ &\frac{\kappa^4}{2}\sum_\nu\Big\{ - 2\mathbf{L}_{\mu\nu} \cdot \overline{\Gamma}\hat{\mathbf{J}} + 2\mathbf{F}_{\mu\nu} \cdot \overline{\Lambda}\hat{\mathbf{P}} \\ &+ K_{\mu\nu} + G_{\mu\nu} + D_{\mu\nu}\Big\}\psi_\nu = E\psi_\mu, \end{aligned} \tag{20}$$

where

$$\mathbf{L}_{\mu\nu}(Q) = \int \phi^*_\mu(\Omega, Q, r)\hat{\mathbf{L}}\phi_\nu(\Omega, Q, r)dr = \langle\phi_\mu|\mathbf{L}|\phi_\nu\rangle(Q) \tag{21}$$

$$\mathbf{F}_{\mu\nu}(Q) = \langle\phi_\mu|\hat{\mathbf{P}}|\phi_\nu\rangle(Q) = \langle\phi_\nu| - i\nabla|\phi_\nu\rangle(Q); \tag{22}$$

$$K_{\mu\nu}(Q) = \langle\phi_\mu|\hat{\mathbf{L}}\cdot\overline{\Gamma}\hat{\mathbf{L}}|\phi_\nu\rangle(Q); \tag{23}$$

$$G_{\mu\nu}(Q) = \langle\phi_\mu|\hat{\mathbf{P}}\cdot\overline{\Lambda}\hat{\mathbf{P}}|\phi_\nu\rangle(Q); \tag{24}$$

$$D_{\mu\nu}(Q) = b^{-1}\langle\phi_\mu|(\sum_\alpha \hat{\mathbf{P}}_\alpha)^2|\phi_\nu\rangle(Q). \tag{25}$$

The matrix elements defined in eqs. (21-25) are all independent of κ, hence in general of order of magnitude unity. If the number of electrons is even, it can be shown by time reversal arguments [6] that the electronic wave functions ϕ_μ can be chosen real, so that diagonal elements of $\mathbf{L}$ and $\mathbf{F}$ are all zero. (But cf. Section 4).

It is convenient to think of the ψ_μ collectively as a column of functions denoted by the single letter ψ, each component of which is one of the functions $\psi_\mu(\Omega, Q)$. The Hamiltonian is then a combination of ordinary operators and matrices operating on the column. Equation (20) can now be rewritten as

$$(\hat{H}_{BO} + \hat{H}_{cp})\psi = E\psi, \tag{26}$$

where $\hat{H}_{BO)}$ is the diagonal Born-Oppenheimer Hamiltonian

$$\hat{H}_{BO} = \frac{1}{2}\kappa^4\{\hat{\mathbf{J}}\cdot\overline{\Gamma}\hat{\mathbf{J}} + \hat{\mathbf{P}}\cdot\overline{\Lambda}\hat{\mathbf{P}}\} + \mathcal{W}, \tag{27}$$

and $\hat{H}_{cp}$, in addition to a small diagonal part, gives the coupling between different electronic levels:

$$\hat{H}_{cp} = \frac{1}{2}\kappa^4\{-2\mathcal{L}\cdot\overline{\Gamma}\hat{\mathbf{J}} + 2\mathcal{F}\cdot\overline{\Lambda}\hat{\mathbf{P}} + \tilde{\mathcal{G}}\}, \tag{28}$$

where script letters now denote matrices operating on the column and

$$\tilde{\mathcal{G}} = \mathcal{K} + \mathcal{G} + \mathcal{D}. \tag{29}$$

Since $\hat{H}_{BO}$ is diagonal in the electronic states, neglect of $\hat{H}_{cp}$ leads to an approximate separation of electronic and nuclear motion, with eigenfunctions having only one nonzero entry in the column.

We now proceed to investigate the magnitudes of the various contributions to wave function and energy.

3. Hierarchy of terms. As a first approximation, we will neglect $\hat{H}_{cp}$, taking as a provisional approximate eigenfunction an eigenfunction of $\hat{H}_{BO}$ with

$$\psi_\mu = \psi_0\delta_{\mu 0}, \tag{30}$$

for which the approximate Schrödinger equation is

$$\{\frac{1}{2}\kappa^4[\hat{\mathbf{J}}\cdot\overline{\Gamma}\hat{\mathbf{J}}+\hat{\mathbf{P}}\cdot\overline{\Lambda}\hat{\mathbf{P}}]+W_0\}\psi_0=E\psi_0, \tag{31}$$

corresponding to motion of the nuclei in an effective potential energy given by $W_0(Q)$.

The approximate reduction of the molecular problem to a problem of nuclear motion in an effective potential, embodied in eq. (31), is the essence of the Born-Oppenheimer approximation.

It is not entirely obvious that this is the correct eigenfunction in the limit of small κ, since both $\hat{H}_{B0}$ and $\hat{H}_{cp}$ contain terms proportional to κ^4, but the analysis will show that this is true.

We concentrate on the simplest case, that of a quasi-rigid molecule; the modifications necessary for other cases will be fairly easy to see, and will be discussed later. For a quasi-rigid molecule, $W_0(Q)$ has a minimum value W_e at some equilibrium configuration Q_e. In the immediate vicinity of the minimum, the potential energy is approximately quadratic, corresponding to a multidimensional harmonic oscillator, with force constants of order unity since the potential energy does not depend on κ. The mass is essentially κ^{-4}. According to the standard theory of the harmonic oscillator, this leads to frequencies of order κ^2, and root-mean-square displacements from the Q_e of order κ. In the approach of Born and Oppenheimer [1], this is exploited by defining new coordinates for the displacement from equilibrium:

$$Q_j=Q_{ej}+\kappa q_j. \tag{32}$$

The momentum conjugate to q_j is

$$\hat{p}_j=\kappa^{-1}\hat{P}_j. \tag{33}$$

Root-mean-square values of the q_j are now expected to be of order unity.

Quantities appearing in the Hamiltonian which depend on Q can now be (formally at least) expanded in power series in the q_j in the vicinity of Q_e. For the potential energy W_0 we have:

$$W_0=W_e+\frac{1}{2}\kappa^2\sum_{jk}W_{jk}q_jq_k+(1/6)\kappa^3\sum_{jkl}W_{jkl}q_jq_kq_l+\dots \tag{34}$$

In (34), there is no linear term because we are expanding about the minimum. The coefficients W_{jk}, etc., are all independent of κ, thus of order unity.

We obtain similar expansions for the matrices $\overline{\Gamma}$ and $\overline{\Lambda}$:

$$\overline{\Gamma}=\overline{\Gamma}_e+\kappa\sum_j\overline{\Gamma}_jq_j+\frac{1}{2}\kappa^2\sum_{jk}\overline{\Gamma}_{jk}q_jq_k+\dots \tag{35a}$$

$$\overline{\Lambda}=\overline{\Lambda}_e+\kappa\sum_j\overline{\Lambda}_jq_j+\frac{1}{2}\kappa^2\sum_{jk}\overline{\Lambda}_{jk}q_jq_k+\dots \tag{35b}$$

The various terms in $\hat{H}_{B0}$ can now be classified according to their physical significance and according to their dependence on κ. We do this in the following way:

$$\hat{H}_{B0} = W_e + \kappa^2 \hat{h}_{vib} + \kappa^4 (\hat{h}_{rot} + \hat{h}_{vr}). \tag{36}$$

In eq. (36), W_e is the electronic energy, the effective potential evaluated at the equilibrium position; it is a constant as far as the nuclear motion is concerned. The vibrational Hamiltonian $\hat{h}_{vib}$ can be further broken up into parts as follows:

$$\hat{h}_{vib} = \hat{h}_{harm} + \hat{h}_{anh}, \tag{37}$$

where $\hat{h}_{harm}$ is the harmonic part, consisting of quadratic terms in the p's and q's, which can be transformed by the standard theory of small vibrations into a sum of independent harmonic oscillators:

$$\hat{h}_{harm} = \frac{1}{2}\{\hat{\mathbf{p}} \cdot \overline{\Lambda}_e \hat{\mathbf{p}} + \sum_{jk} W_{jk} q_j q_k\} \tag{38}$$

The contribution of $\hat{h}_{harm}$ to the energy will obviously be of order κ^2, since $\hat{h}_{harm}$ itself is independent of κ, and it appears in the full Hamiltonian multiplied by κ^2. The higher-order anharmonic corrections are included in $\hat{h}_{anh}$:

$$\begin{aligned} \hat{h}_{anh} = \frac{1}{2}\hat{\mathbf{p}} \cdot \Big\{ &\kappa \sum_j \overline{\Lambda}_j q_j \\ &+ \frac{1}{2}\kappa^2 \sum_{jk} \overline{\Lambda}_{jk} q_j q_k + \dots \Big\} \mathbf{p} \\ &+ (1/6)\kappa \sum_{jkl} W_{jkl} q_j q_k q_\ell \\ &+ (1/24)\kappa^2 \sum_{jklm} W_{jklm} q_j q_k q_l q_m + \dots \end{aligned} \tag{39}$$

The rotational energy of rigid rotation with the nuclei held at the equilibrium position is represented by $\hat{h}_{rot}$:

$$\hat{h}_{rot} = \frac{1}{2}\hat{\mathbf{J}} \cdot \overline{\Gamma}_e \hat{\mathbf{J}}. \tag{40}$$

Finally, $\hat{h}_{rv}$ represents the interaction between rotation and vibration due to the variation of the inertia tensor with nuclear coordinates:

$$\hat{h}_{vr} = \frac{1}{2}\hat{\mathbf{J}} \cdot \Big\{ \kappa \sum_j \overline{\Gamma}_j q_j + \frac{1}{2}\kappa^2 \sum_{jk} \overline{\Gamma}_{jk} q_j q_k + \dots \Big\} \hat{\mathbf{J}} \tag{41}$$

If $\hat{h}_{vr}$ is neglected, there is no coupling between the rotational degrees of freedom Ω and the vibrational ones q. Moreover, this neglect is justified if κ is small enough, since the lowest-order term in $\hat{h}_{vr}$ is of order κ, and it is multiplied by κ^4, meaning that its first contribution is at least of order κ^5. Through the fourth order in κ, therefore, we can separate the energy of the molecule (apart from the contribution of $\hat{H}_{cp}$, which we will consider presently) into independent electronic, vibrational, and rotational parts:

$$E = E_{el} + E_{vib} + E_{rot}, \tag{42}$$

where $E_{el} = W_{el}$, and E_{vib} and E_{rot} are the eigenvalues of $\hat{\kappa}^2 h_{vib}$ and $\hat{\kappa}^4 h_{rot}$, respectively. We now consider the κ-dependence of the various terms:

The electronic energy E_{el} is independent of κ, and is a constant as far as nuclear motion is concerned, being characteristic only of the electronic eigenfunction.

The lowest-order term in E_{vib} comes from the eigenvalues of $\hat{h}_{harm}$. These are independent of κ, and contribute terms to the total energy of order κ^2. The anharmonic corrections can be ascertained by perturbation theory. The lowest-order corrections are just the diagonal matrix elements of the perturbation, or, where these are zero, the square of an off-diagonal element divided by an energy denominator. Considering $\hat{h}_{anh}$ as a perturbation on $\hat{h}_{harm}$, we first note that, since $\hat{h}_{harm}$ is quadratic in the p's and q's, diagonal elements of terms of odd order in combined p's and q's will be zero. The cubic terms, proportional to κ, thus contribute only in second order, giving an eigenvalue correction to $\hat{h}_{vib}$ of order κ^2, leading to a correction to the total energy of order κ^4. The quartic terms in first order contribute corrections of the same magnitude. It is easily seen that higher-order anharmonic corrections are of order κ^6 or higher.

The eigenvalues of $\hat{h}_{rot}$ contribute to the energy in 4'th order in κ. The leading term in $\hat{h}_{rv}$, linear in κ but also in q, contributes only in second order, so the lowest-order contributions of vibration-rotation interaction are of order κ^6.

We must now consider the effect of $\hat{H}_{cp}$, which couples different electronic states, and whose off-diagonal part represents what one normally calls corrections to the Born-Oppenheimer approximation. Inserting (32) and (33) into (28), we find:

$$\hat{H}_{cp} = \kappa^3 \mathcal{F} \cdot \overline{\Lambda}\hat{\mathbf{p}} + \frac{1}{2}\kappa^4 \{\tilde{\mathcal{G}} - 2\mathcal{L} \cdot \overline{\Gamma}\hat{\mathbf{J}}\}. \tag{43}$$

Again, the energy correction is estimated by perturbation theory. The only diagonal term comes from $\tilde{\mathcal{G}}$, which in turn can be expanded about its value at the equilibrium position. The equilibrium value of $\tilde{G}_{00}$ contributes a constant term of order κ^4, with further corrections being of order κ^6 or higher. This diagonal element, however, does not break down the separation of electronic and nuclear motion. It can be, and sometimes is, simply absorbed into the effective potential W_0. It has been suggested by Ballhausen and Hansen [3] that the approximation of replacing $\hat{H}$ by $\hat{H}_{BO}$ be called the "Born-Oppenheimer adiabatic approximation": if $\tilde{G}_{00}$ is included as part of W_0, they suggest the term "Born-Huang adiabatic approximation" for the resulting treatment.

The true breakdown of the electronic-nuclear separation comes from the off-diagonal elements in (43), and it is easy to see that the lowest-order energy correction coming from these is of order κ^6, since they can contribute only in second order, the lowest-order matrix elements contain κ^3, and energy denominators are of order unity (differences between electronic energies).

Summarizing, the lowest-order contributions to the energy are all proportional to even powers of the Born-Oppenheimer parameter κ. Listed in order of the power of κ that they contain, the contributions are:

κ^0 : Electronic energy

κ^2 : Harmonic vibrational energy

κ^4 : Rotational energy; lowest anharmonic contributions to vibrational energy; constant correction due to $\tilde{G}_{00}$

κ^6 : Higher anharmonic corrections; rotation-vibration interaction; coupling between electronic states.

The molecular wave function itself can similarly be expanded in powers of κ. The lowest-order corrections are again estimated by perturbation theory, and are proportional to off-diagonal matrix elements divided by energy denominators. To zero order in κ, the wave function is a product of electronic, rotational, and vibrational parts, with the vibrational part being harmonic. The lowest anharmonic corrections are of order κ, while the vibration-rotation interaction and coupling between electronic states give corrections of order κ^3.

Several remarks are in order about this classification of contributions to energy and wave function:

(i) There is no assurance that the power series in κ will converge. In cases of interest, it is probably an asymptotic expansion. This is no disadvantage, however, since no one would be interested in trying to sum the entire series. The expansion is useful only if κ is small enough so that one gets a good approximation by keeping only a small number of terms.

(ii) The smallness of κ justifies not only the separation of electronic and nuclear motion, but also the separation of vibration and rotation and the harmonic approximation for the vibration. If anharmonic and/or vibration-rotation corrections are carried too far without at the same time considering Born-Oppenheimer corrections, one may be in danger of committing the inconsistency of leaving out terms that are as large as the terms one is including, or larger. The decision on which terms need to be included depends not only on the magnitudes involved, but also on what one is trying to calculate. For example, higher-order anharmonicities, rotation-vibration interaction, and coupling between electronic states all contribute to the energy at the 6'th order in κ, but the three contributions depend in different ways on the vibrational and rotational quantum numbers. If one wishes to calculate the absolute value of the energy through the 6'th order, all three contributions must be included; but if one is interested only in the *difference* in energy between two vibration-rotational levels, the coupling to other electronic states can be omitted through this order.

(iii) The smallness of κ also guarantees that the displacements of the nuclei from their equilibrium positions will be small on the average, thus justifying the traditional chemist's picture of the molecule.

(iv) If the molecule is not quasi-rigid, some of the $(3n-6)$ degrees of freedom that we have classified as vibrational may behave more like additional rotational degrees of freedom (in the case of floppy molecules with more or less free rotation about some of the bonds), or like translations (for molecular systems undergoing scattering processes, including chemical reactions). When this happens, it does not alter our general qualitative conclusions, but it does mean that each degree of freedom must be treated in an appropriate way. The replacement (32) and the subsequent expansion is appropriate only for genuine vibrational degrees of freedom.

(v) Our analysis only covers the κ-dependence of the various contributions. For finite values of κ, there can be special cases where the higher coefficients in the power series expansion are sufficiently larger than the lower ones to make a higher-order contribution actually larger than a lower-order one. Only in the limit of infinitesimally small κ does an analysis of the κ-dependence rigorously correspond to an analysis of relative sizes of the contributions. Such exceptions, of course, depend on the peculiarities of particular systems, and are not subject to general analysis.

We have now presented a brief account of the nature of the Born-Oppenheimer approximation. In the remaining two sections, we turn to a brief discussion of some current problems connected with it.

4. Derivative coupling between electronic states. As we have seen, the lowest-order Born-Oppenheimer correction to a molecular wave function is of order κ^3 (κ^4 if the degree of freedom involved is a rotation or translation). The correction, however, is made up of a matrix element divided by an energy denominator, and its magnitude must be reconsidered if the energy denominator is capable of becoming small. This can happen if the nuclear kinetic energy is very high, comparable with electronic energies, or if two electronic levels intersect for some Q, so that in the vicinity of the intersection the energy denominator becomes arbitrarily small.

In this chapter, we continue to assume that the number of electrons is even and that time-reversal invariance holds, so that electronic wave functions can be chosen real. In this case, it is known [6,7] that two conditions must be satisfied by the nuclear coordinates in order for an intersection to occur. For two nuclei, there is only one such coordinate apart from translation and rotation, so one does not get intersections except between states of different symmetry. For $n > 2$ nuclei, however, the number of relevant degrees of freedom is $(3n - 6)$, so there can be intersections on a submanifold of dimension $(3n - 8)$. For three nuclei, for example, the Q's form a three-dimensional space, within which an intersection can occur along a curve. One can always make the expansion (15), but near the intersection it is no longer a good approximation to keep only one term. It may still be a good approximation to keep only two terms, however, corresponding to the two electronic levels that intersect, and to neglect the mixing with other levels. To avoid nonessential excess baggage in the notation, we ignore the rotation, assume that the other degrees of freedom span a three-dimensional space (so that we can use ordinary vector notation), and also assume that $\overline{\Lambda}$ is just the unit matrix. This will make the essential points easier to see without affecting the main results. Treating our wave function as a two-dimensional column vector, we obtain the following eigenvalue problem:

$$\frac{\kappa^4}{2}\Big\{-\nabla^2 - 2\begin{pmatrix} 0 & \mathbf{v} \\ -\mathbf{v} & 0 \end{pmatrix}\cdot\nabla + \mathcal{G}\Big\}\begin{pmatrix}\psi_1\\ \psi_2\end{pmatrix} + \begin{pmatrix} W_1 & 0 \\ 0 & W_2 \end{pmatrix}\begin{pmatrix}\psi_1\\ \psi_2\end{pmatrix} = E\begin{pmatrix}\psi_1\\ \psi_2\end{pmatrix} \tag{44}$$

In arriving at (44), we have made use of the fact that $\mathbf{F}_{\mu\nu}$ in (22) is pure imaginary and antisymmetric in our case, and have defined

$$\mathbf{v} = i\mathbf{F}_{12} = \langle\phi_1|\nabla\phi_2\rangle = -\langle\phi_2|\nabla\phi_1\rangle. \tag{45}$$

The "derivative coupling" term involving $\mathbf{v}$ in (44) is troublesome, since it couples each component of the wave function to the derivative of the other. It can be modified by choosing new electronic basis functions which are linear combinations of the old ones. The new electronic functions are no longer eigenfunctions of $\hat{H}_{el}$, so the matrix $\mathcal{W}$ will no longer be diagonal, but this is of relatively little importance since we have a problem of two coupled states in any case. Accordingly, we define new electronic functions $\tilde{\phi}_1$ and $\tilde{\phi}_2$ as follows:

$$\begin{aligned}\tilde{\phi}_1 &= \phi_1\cos\lambda + \phi_2\sin\lambda;\\ \tilde{\phi}_2 &= -\phi_1\sin\lambda + \phi_2\cos\lambda,\end{aligned} \tag{46}$$

where λ is a function of the nuclear coordinates to be determined by convenience. Eq. (46) is the most general real orthogonal transformation of two wave functions. In terms of the new functions, we obtain $\tilde{\mathbf{v}}$, the new vector giving the derivative coupling:

$$\tilde{\mathbf{v}} = \langle\tilde{\phi}_1|\nabla\tilde{\phi}_2\rangle = \mathbf{v} - \nabla\lambda. \tag{47}$$

Ideally, one would like to choose λ so as to get rid of the derivative coupling altogether, i.e., so that $\tilde{\mathbf{v}} = 0$. According to (47), this can be achieved if we can solve the equation

$$\nabla\lambda = \mathbf{v}, \tag{48}$$

giving what is called a diabatic, or a strictly diabatic, representation [8], and several treatments in the literature are based on this idea. However, it has been pointed out [9,10] that (48) will have solutions only if curl $\mathbf{v} = 0$, a condition which is rarely, if ever, exactly satisfied.

One can work out an expression for curl $\mathbf{v}$, a component of which is given by [10]:

$$\begin{aligned}\frac{\partial}{\partial x}v_y - \frac{\partial}{\partial y}v_x &= \langle\frac{\partial}{\partial x}\phi_1|\frac{\partial}{\partial y}\phi_2\rangle - \langle\frac{\partial}{\partial y}\phi_1|\frac{\partial}{\partial x}\phi_2\rangle\\ &= \sum_{\nu=2}^{\infty}\Big\{\langle\frac{\partial}{\partial x}\phi_1|\phi_\nu\rangle\langle\phi_\nu|\frac{\partial}{\partial y}\phi_2\rangle - \langle\frac{\partial}{\partial y}\phi_1|\phi_\nu\rangle\langle\phi_\nu|\frac{\partial}{\partial x}\phi_2\rangle\Big\}.\end{aligned} \tag{49}$$

In general [10], this is of order unity, so there is no way of transforming away the derivative coupling in large regions, even approximately.

It is possible, however, to break $\mathbf{v}$ up into a sum of a removable irrotational part with zero curl, and a nonremovable solenoidal part. Under some circumstances, there can be regions in which the removable part is much larger than the nonremovable part, so that its removal takes care of the largest effect; moreover, the regions where this is so can be the important ones for determining the effect of the coupling between the states. A detailed analysis has been carried out by Thompson and Mead [11] for the triatomic X_3 system, which possesses an intersection at the equilateral triangle configuration. Defining r as a coordinate measuring the distance from the intersection, they showed that for small r the removable part is proportional to r^{-1}, while the nonremovable part is proportional to r. There is therefore a small but important region in which the removable part is by far the larger.

The questions of the correct inclusion of the nonremovable derivative coupling, and of the exact conditions under which it is permissible to neglect it, are topics of current research.

5. Adiabatic phase factors. It has already been pointed out that, when the number of electrons is even, the electronic eigenfunctions can be chosen real, so that the diagonal $\mathbf{F}$ matrix elements defined in (22) are zero. However, this ceases to hold if there is an external magnetic field present, and in a certain sense in the immediate neighborhood of the intersection of two electronic levels. Moreover, we have omitted up to this point the case of an odd number of electrons, for which each electronic level has a twofold Kramers degeneracy [12,13] for all nuclear configurations. Even when the electronic functions can be chosen real, it can be disadvantageous for some purposes to do so. These questions are related to the currently interesting topic of adiabatic phase factors [14,15,16] which we now review briefly.

Since we will only be talking here about electronic eigenfunctions and eigenvalues as functions of nuclear coordinates, and only wish to illustrate the essential points, we can simplify the notation in some respects, as in the last chapter. Again, we will just consider a three-dimensional space of nuclear coordinates $\mathbf{Q}$, and we can drop the subscript on $\hat{H}_{el}(Q)$, it being understood that the only Hamiltonian we have to talk about is the $\mathbf{Q}$-dependent electronic Hamiltonian. On the other hand, we have to consider the possibility that the electronic levels are degenerate, so the index μ must be expanded into a double index, Latin $j, k, \ldots$ for the energy level, and Greek $\alpha, \beta, \ldots$ for states within a given degenerate level. It will also be convenient to use a bracket notation $|j\alpha(\mathbf{Q})\rangle$ instead of $\phi_{j,\alpha}(\mathbf{Q}, r)$. The eigenvalue equation (13) now becomes

$$\hat{H}(\mathbf{Q})|j\alpha(\mathbf{Q})\rangle = W_J(\mathbf{Q})|j\alpha(\mathbf{Q})\rangle. \tag{50}$$

In addition, the normalization condition (14) must still be satisfied. This does not completely determine the eigenfunction, however: (14) and (50) will still be satisfied if an arbitrary $\mathbf{Q}$-dependent unitary transformation is applied to the eigenfunctions within an energy level. In the nondegenerate case, this can only be a phase factor; in the degenerate case it can entail phase factors and/or replacing the degenerate states with linear combinations of each other. The choice of this unitary transformation will affect the matrix elements of **F**, eq. (22), which are diagonal with respect to energy. The freedom of choice can be utilized in such a way as to lead to simplifications in the effective nuclear Hamiltonian, $\tilde{H}_{B0}$.

To investigate this further, consider the change in an eigenfunction when the nuclear coordinates **Q** are displaced by an infinitesimal amount:

$$|j\alpha(\mathbf{Q}+d\mathbf{Q})\rangle = \sum_{k\beta} |k\beta(\mathbf{Q})\rangle \{\delta_{k\beta,j\alpha} + i\mathcal{F}_{k\beta,j\alpha}\cdot d\mathbf{Q}\}. \tag{51}$$

Setting up a coordinate system x, y, z in the Q space, it is a straightforward matter [14,16] to integrate (51) around an infinitesimal closed rectangle in (e.g.) the xy-plane, with sides Δx, Δy. The result is that $|j\alpha(\mathbf{Q})\rangle$ is replaced by

$$|j\alpha(\mathbf{Q})\rangle' = \sum_{k\beta} |k\beta(\mathbf{Q})\rangle \{\delta_{k\beta,j\alpha} + S_{k\beta,j\alpha}\} \tag{52}$$

where

$$\mathcal{S} = \Delta x \Delta y \Big\{ i\Big(\frac{\partial}{\partial x}\mathcal{F}_y - \frac{\partial}{\partial x}\mathcal{F}_x\Big) - [\mathcal{F}_x, \mathcal{F}_y] \Big\} \tag{53}$$

We can split $\mathcal{F}$ up into parts,

$$\mathcal{F} = \mathbf{f} + \mathbf{g}, \tag{54}$$

where **f** is diagonal with respect to energy, and therefore to some extent at our disposal, while **g** is entirely off-diagonal with respect to energy, and is determined by (50) and (14):

$$\langle k\beta|\mathbf{g}|j\alpha\rangle = \frac{i}{W_k - W_j}\langle k\beta|\nabla\tilde{H}|j\alpha\rangle \tag{55}$$

If our wave functions are to be single-valued, we must have $\mathcal{S} = 0$, since $\mathcal{S}$ just denotes the change in a wave function upon traversal of a closed path returning to the starting point. It is straightforward to show that this requirement entails the following equation for the components of **f**:

$$\frac{\partial}{\partial x}f_y - \frac{\partial}{\partial y}f_x + i[f_x, f_y] = -i[g_x, g_y]_d \tag{56}$$

where the subscript d denotes the part diagonal with respect to energy. This will not always be satisfied by nonzero **f**, so the price of having $\mathbf{f} = 0$ is sometimes multiple-valued electronic eigenfunctions, which lead to most confusing complications. If **f** is not zero, we see from (20) that there will be terms even in the Born-Oppenheimer Hamiltonian of the form $\mathbf{f}\cdot\hat{\mathbf{P}}$. [In the degenerate case, the Born-Oppenheimer

Hamiltonian neglects coupling between different electronic energy levels, but cannot neglect coupling within a degenerate level].

In the nondegenerate case, the commutator on the left side of (56) vanishes between vector potential and magnetic field. The $\mathbf{f} \cdot \hat{\mathbf{P}}$ term in the effective nuclear Hamiltonian also has exactly the form of a vector potential term. In the degenerate case, $\mathbf{f}$ behaves like a generalization of the vector potential known as a "nonabelian gauge field" and used in present-day elementary particle theory [17].

If the number of electrons is even and there is no external magnetic field, so that time-reversal invariance holds, there is no degeneracy except in lower-dimensional subspaces, and the right-hand side of (56) is zero, except that the equation becomes singular at an intersection between two levels, resulting in a phase factor of π (change of sign) from a closed path around such an intersection [18]. It has been shown that this sign change can affect the energy levels of certain molecules [19,20], and that it also must be considered in the analysis of the behavior of wave functions under permutations of identical nuclei [14,21,22,23].

For the nondegenerate case without time reversal invariance, the phase factor can be worked out [14,15], and is of interest in problems with external magnetic fields and other cases where time-reversal invariance does not hold.

The phase factor in the nondegenerate case behaves, as we have said, like a nonabelian gauge field, and the resulting analogy between molecular systems and elementary particle theory has been discussed by several authors [24,25,26]. It has been pointed out [16] that a nonabelian gauge effect should be observable in an experiment involving the Stark effect in an atom in a level with $J = 3/2, 5/2$, etc. When such an atom is placed in a uniform electric field, the level is split according to the Stark formula, with two-fold degenerate levels corresponding to different values of $|M|$, the absolute value of the component of angular momentum directed positively along the field, an adiabatic rotation of the field can, through the action of the nonabelian gauge field, lead to reversal of direction of angular momentum along the field.

The study of the phase factors, both for degenerate and nondegenerate cases, is now a topic of considerable interest, not only in molecular quantum mechanics, but in other fields as well.

In these short lectures, I hope I have succeeded in showing you the importance of the Born-Oppenheimer approximation for molecular theory, and in conveying something of both its justification and its limitations. I have also tried to convince you that it still is producing fascinating problems which are topics of interesting current research.

REFERENCES

[1] M. Born and R. Oppenheimer, *Zur Quantentheorie der Molekeln*, Ann. Phys., 84 (1927), pp. 457–484.

[2] M. Born and K. Huang, *The Dynamical Theory of Crystal Lattices*, Oxford University Press, London, 1954.

[3] C.J. Ballhausen and A.E. Hansen, *Electronic Spectra*, Annu. Rev. Phys. Chem., 23 (1972), pp. 15–38.

[4] H.H. NIELSEN, *The vibration-rotation energies of molecules and their spectra in the infrared*, in *Handbuch der Physik vol. XXXVII/1*, Springer-Verlag, Berlin, 1959, pp. 173–313.

[5] S. CALIFANO, *Vibrational States*, J. Wiley and Sons, New York, 1976.

[6] C.A. MEAD, *The "noncrossing" rule for electronic potential energy surfaces: The role of time-reversal invariance*, J. Chem. Phys, 70 (1979), pp. 2276–2283.

[7] L.D. LANDAU AND E.M. LIFSHITZ, *Quantum Mechanics, Nonrelativistic Theory*, 2nd Ed., translated by J.B. Sykes and J.S. Bell, Addison-Wesley, Reading, Mass, 1965, pp. 279–282.

[8] W.D. HOBIE AND A.D. MCLACHLAN, *Dynamical Jahn-Teller effect in hydrocarbon radicals*, J. Chem. Phys, 33 (1960), pp. 1695–1703.

[9] A.D. MCLACHLAN, *The wave functions of electronically degenerate states*, Mol. Phys., 4 (1961), pp. 417–423.

[10] C.A. MEAD AND D.G. TRUHLAR, *Conditions for the definition of a strictly diabatic electronic basis for molecular systems*, J. Chem. Phys., 77 (1982), pp. 6090–6098.

[11] T.C. THOMPSON AND C.A. MEAD, *Adiabatic electronic energies and nonadiabatic couplings to all orders for system of three identical nuclei with conical intersection*, J. Chem. Phys., 82 (1985), pp. 2408–2417.

[12] E.P. WIGNER, *Group Theory*, translated by J. J. Griffin, Academic Press, New York, 1985, pp. 325–348, chapter 26.

[13] D.G. TRUHLAR, C.A. MEAD AND M.A. BRANDT, *Time-reversal invariance, representations for scattering wavefunctions, symmetry of the scattering matrix, and differential cross-sections*, Adv. Chem. Phys., 33 (1975), pp. 295–344.

[14] C.A. MEAD AND D.G. TRUHLAR, *On the determination of Born-Oppenheimer nuclear motion wave functions including complications due to conical intersections and identical nuclei*, J. Chem. Phys., 70 (1979), pp. 2284–2296.

[15] M.V. BERRY, *Quantal phase factors accompanying adiabatic changes*, Proc. Roy. Soc. London, A-392 (1984), pp. 45–57.

[16] C.A. MEAD, *Molecular Kramers degeneracy and non-Abelian adiabatic phase factors*, Phys. Rev. Lett., 59 (1987), pp. 161–164.

[17] T. CHENG AND L. LI, *Gauge Theory of Elementary Particle Physics*, Oxford University Press, Oxford, 1984.

[18] G. HERZBERG AND H.C. LONGUET-HIGGINS, *Intersection of potential energy surfaces in polyatomic molecules*, Discuss. Faraday Soc., 35 (1963), pp. 77–82.

[19] H.C. LONGUET-HIGGINS, V. ÖPIK, M.H.L. PRYCE AND R.A. SACK, *Studies of the Jahn-Teller effect II. The dynamical problem*, Proc. Roy. Soc., London, A 244 (1958), pp. 1–16.

[20] C.A. MEAD, *The molecular Aharenov-Bohm effect in bound states*, Chem. Phys., 49 (1980), pp. 23–32.

[21] C.A. MEAD, *Superposition of reactive and nonreactive scattering amplitudes in the presence of a conical intersection*, J. Chem. Phys., 72 (1980), pp. 3839–3840.

[22] S.P. KEATING AND C.A. MEAD, *Conical intersections in a system of four identical nuclei*, J. Chem. Phys, 82 (1985), pp. 5102–5117.

[23] S.P. KEATING AND C.A. MEAD, *Toward a general theory of conical intersections in systems of four identical nuclei*, J. Chem. Phys., 86 (1987), pp. 2152–2160.

[24] F. WILCZEK AND A. ZEE, *Appearance of gauge structure in simple dynamical systems*, Phys. Rev. Lett., 52 (1984), pp. 2111–2114.

[25] J. MOODY, A. SHAPERE AND F. WILCZEK, *Realizations of magnetic-monopole gauge fields: Diatoms and spin precession*, Phys. Rev. Lett., 56 (1986), pp. 893–896.

[26] H. LI, *Induced gauge fields in a nongauged quantum system*, Phys. Rev. Lett., 58 (1987), pp. 539–542.

ELECTRONIC STRUCTURE THEORY

JAN ALMLÖF †

Abstract. Fundamental aspects of computational chemistry are reviewed, with emphasis on electronic structure theory. The independent particle approach is described in some detail, along with methods for extending the treatment beyond that model. Theories and methods for treating the effects of relativity in quantum systems are also discussed. Finally, a number of different applications of these methods to problems in chemistry are presented. The purpose of this review is to introduce selected concepts of modern electronic structure theory to beginners in the field.

1. Introduction. Chemistry is often thought of as a genuinely experimental science, where knowledge is empirical and where research is conducted in stinking and hazardous laboratories. However, in chemistry as well as in many other areas of science there are alternative routes to getting new information. In general, if one knows the rules which govern a certain process, one can in principle predict the outcome of an experiment without actually performing it.

Chemical properties of molecules are known to be determined by the laws of nature known as quantum mechanics. Theoretical methods are available which, at least in principle, completely outline a computational procedure that would provide valuable input—in many cases even a complete solution—to almost any problem in chemistry where experimental techniques are traditionally used. Accordingly, quantum mechanical calculations would seem to provide a very useful source of information in almost any area of chemistry. In practice, however, most such studies have been limited by insufficient computational resources (hardware as well as software). That situation is now changing. The computing power available to the scientific community has increased by orders of magnitude over a period of relatively few years. We now face the challenge of exploiting modern computer technology as a versatile tool for chemical research on a large scale, and to make a computational approach a realistic alternative for an increasing variety of chemical research. Still, to achieve this goal a lot of algorithm development is needed.

---

† Department of Chemistry, University of Minnesota, Minneapolis, Minnesota 55455, U.S.A.

At universities and national laboratories considerable efforts are made in the development of these new approaches, which, for the first time will make it possible to simulate chemical systems of a complexity that is technologically interesting. To be successful, it is necessary that these efforts are closely coordinated with large-scale applications to realistic chemical problems, and that careful consideration is given to recent and expected progress in computer technology.

In this lecture series we will review recent progress along those lines, both in method development and in chemical applications.

2. Background. There are several reasons why one would like to compute a chemical result rather than determining it experimentally:

ACCESSIBILITY. The compound of interest may be unstable, explosive, poisonous or just very difficult to synthesize. These are all factors which could make the experiment difficult, but of course they are irrelevant to a theoretical study.

ECONOMY. While researchers in computational sciences are limited by the cost of computer time, one should realize that experimental equipment is also expensive. Normally the cost would be higher for computer equipment, but the typical time to carry out an experiment is much longer than for a computational study. In cases where a result can be obtained by either computation or experiment the former approach is often less expensive.

ACCURACY. The reliability of an experimental study is limited by experimental accuracy, which may be very difficult to improve. In a computational study, the numerical approximations used normally set the limit, and these are often easier to change.

UNDERSTANDING. The reason for observing and recording experimental results is often to understand more fundamental mechanisms. This requires an interpretation of the raw experimental data, which is often the weakest link in an experimental study. In a calculation, the physics is implicitly built into the theoretical methods, and there is usually no problem of interpretation. This is an important advantage, and the interpretation and understanding of experiments is in fact one important use of computational chemistry.

Different phenomena in physics require different mathematical models for their proper description. The mechanics that applies to a falling apple may not be appropriate to describe colliding galaxies or interactions among subatomic particles. For phenomena related to chemistry we can distinguish a few different cases, depending on the masses and velocities of the objects under consideration.

The objects with with which we deal in our daily life are considered to have high mass and low velocity on that scale. They are treated with "classical" Newtonian mechanics. If the velocity of an object is non-negligible (compared to the velocity of light) we must instead use relativistic mechanics. However, if the object is extremely small quantum mechanics must be applied (electrons are small; molecules are big; atoms are borderline cases).

The following graph illustrates schematically how different mathematical descriptions apply to systems with different characteristics. The vertical axis denotes the mass of a system under consideration, whereas the velocity of the object is shown along the horizontal axis. ('c', the velocity of light, is the upper limit for any velocity).

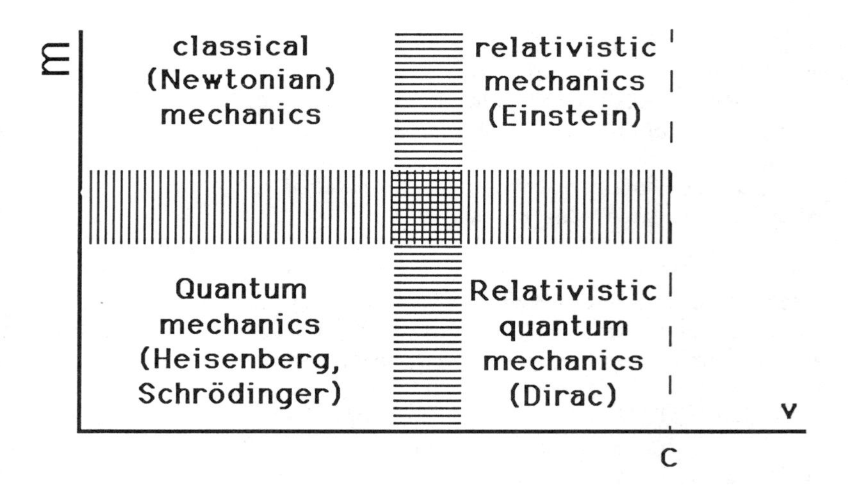

The use of large-scale simulations in chemical industry is still on a rather modest scale. A few large companies have taken such tools into use[1], but compared with the massive breakthrough (and success!) of similar techniques in the aerospace, automobile and oil-exploration industries one must conclude that development has been slow. This is certainly not due to a lack of motivation, skills or interest among chemists to use computer simulations as a source of new knowledge. On the contrary, for decades computational chemists have been at the forefront of number-crunching science. Computer hardware has traditionally been challenged to its limits by chemists struggling to apply their methods to larger and larger systems. However, the seemingly modest progress is related to the complexity of the problem.

The numerical problems encountered in computational quantum chemistry are gargantuan. Typically, one needs to solve sets of second-order differential equations with up to a thousand variables. The equations may have millions of singularities, and an accuracy of one part per billion is often required. Of course, this would be an impossible task if one could not draw massive advantage from knowledge about the physics of the problem. Fortunately that can be done, and methods are indeed being developed to cope with problems of that size.

The dramatic improvement in computer hardware performance during the last two decades is often quoted as the main reason for the rapid progress made in theoretical chemistry during the same period of time. However, it is important to realize that this improvement has by no means been uniform. Circuit technology has of course improved immensely, whereas the performance of devices for external data storage (which still must rely on mechanical parts) has shown a much slower progress. On comparing old computers to new ones it it is evident that CPU power, and recently also memory, have developed at a much faster rate than input/output and storage capacity. Consequently, new and often quite nonconventional methods and algorithms may be necessary in order to fully exploit these new technological achievements.

3. Elements of quantum mechanics. In quantum mechanical theory, a system of N particles is described by a state function (wavefunction)

$$\Psi = \Psi(r_1, r_2, r_3, \; .\; .\; .\; . \; r_N, t) \tag{1}$$

which is normally interpreted in terms of probability:

$$P(r_1, r_2, \; .\; ., t) \equiv \left| \Psi(r_1, r_2, \; .\; ., t) \right|^2 \tag{2}$$

P being the probability (density) distribution function for the system.

The wavefunction can be determined by solving the Schrödinger Equation

$$(3) \qquad H\Psi = i\hbar\partial\Psi/\partial t$$

Here, H is an operator (the Hamiltonian operator), analogous to the Hamiltonian function H = T + V in classical mechanics, T and V being the classical kinetic and potential energy, respectively. The potential V takes the same form as in classical theory, whereas T is expressed differently: Rather than the classical expression $T=mv^2/2$, we write $T=p^2/2m$, and replace the classical momentum $\mathbf{p} = (p_x,p_y,p_z)$ with its quantum mechanical analogue:

$$(4) \qquad p_x = -i\hbar\partial/\partial x$$

For a particle moving in three dimensions, the kinetic energy is thus:

$$(5) \qquad T = -\hbar^2(d^2/dx^2 + d^2/dy^2 + d^2/dz^2)/2m$$

and, consequently, for a many-particle system the Hamiltonian takes the form:

$$(6) \qquad H = -\hbar^2\sum_i(d^2/dx_i^2 + d^2/dy_i^2 + d^2/dz_i^2)/2m_i + \sum_{ij} Q_iQ_j/|r_i-r_j|$$

assuming Coulombic interaction among the particles and no external interactions.

If the system is in a stationary state, the time dependence of the wavefunction can be separated, and we can replace the original Schrödinger equation by a simpler, time-independent one:

$$(7) \qquad H\Psi = E\Psi,$$

which is the form normally used in electronic structure theory. Even after this simplification, however, Ψ is a function of 3N coordinates, N being the number of particles in our system.

(Actually, the number of coordinates is 4N. In addition to the geometrical degrees of freedom each particle has a 'spin' coordinate associated with it. In general, the space coordinates of an electron can assume all possible values; in contrast the spin is restricted to the values $\pm 1/2$. The spin arises naturally in a relativistic treatment; in non-relativistic theory the effect must be accounted for in a rather *ad hoc* fashion. The spin is sometimes visualized as an intrinsic angular momentum of the particle, but this is not a perfect analogy).

The mathematical complexity of applying this theory to chemistry may be realized by consider the benzene molecule (a small but representative system by chemical standards):

The structure formula is C_6H_6; thus the molecule has 42 electrons and 12 atomic nuclei. With these 54 particles the system has a total of 162 degrees of freedom. For many application in chemistry the required accuracy is about $10^8:1$, and apparently there is little hope that one can solve the Schrödinger equation with that accuracy. Instead, one must rely heavily on the physics of the problem to find reasonable approximations.

The Born-Oppenheimer approximation—nuclear coordinates are clamped to fixed positions $\mathbf{R}_j$ while the electronic problem is solved in the coordinates $\mathbf{R}_\alpha$—is reviewed elsewhere in this volume[2]. Even with this simplification, the electronic part of the problem

$$H_e(r;R)\Psi_e(r;R) = E_e(R)\Psi_e(r;R) \tag{8}$$

is far too complicated to be treated exactly for molecules of chemical interest. Further simplifications must therefore be sought, and additional approximations applied.

4. The independent particle model. Due to the Coulombic interaction between electrons their motions in an atom or molecule are correlated. The behavior and energy of any one electron in the system is therefore a function of the instantaneous positions of all particles in the system. From (2) we can determine the joint probability distribution - and therefore the correlations - for all the coordinates of the system. The form (6) of the molecular Hamiltonian suggests that 1) the simultaneous motion of more than two particles are rather uninteresting, since the Hamiltonian contains no three-body terms, and 2) two-particle correlations are of importance in electron structure theory.

Despite 2) above, however, one might still try the physically very simple and appealing ansatz that the positions and behavior of the electrons are completely uncorrelated, i.e. that

$$P(r_1,r_2, r_3,..) = P_1(r_1)P_2(r_2)P_3(r_3) \ . \ . \ . \tag{9}$$

which (according to (2)) would imply that the wavefunction can be written in terms of one-electron functions - *orbitals*, which are functions of the spin- as well as of the space coordinates of the electron.

$$(10) \qquad \Psi(r_1,r_2,\ldots) = \varphi_1(r_1)\varphi_2(r_2)\varphi_3(r_3)\ldots$$

This approximation does not necessarily imply that we are neglecting the electron-electron repulsion terms

$$(11) \qquad \sum_{ij} 1/|r_i - r_j|$$

in the Hamiltonian , but rather that each electron moves in the mean field — as opposed to the instantaneous field — of all the other particles. With this assumption, the Schrödinger equation can be rewritten as a set of one-electron equations defining the orbitals $\{\varphi_i\}$:

$$(12) \qquad h_i\varphi_i(r_i) = \epsilon_i\varphi_i(r_i)\,, \quad i=1,N$$

where the operators h_i are effective one-electron operators, containing only the mean-field interaction with the other electrons. The eigenvalues ϵ_i, *orbital energies*, do have a simple physical interpretation in terms of energy, although their sum is not the total energy of the system. The one-electron equations can be solved with an accuracy sufficient for practical purposes. However, using the resulting $\{\varphi_i\}$ in connection with the form (10) does not give results in satisfactory agreement with experiment. The explanation is to be sought in the probability interpretation of the wavefunction, and the fact that electrons are identical. Any permutation among identical particles must therefore leave the probability distribution unchanged. In particular, for an exchange P_{ij} of the electrons i and j we have:

$$(13) \qquad P_{ij}|\Psi|^2 = |\Psi|^2$$

and therefore (since $P_{ij}^2 = 1$)

$$(14) \qquad P_{ij}\Psi = \pm\Psi$$

Non-relativistic quantum theory provides little guidance for the choice of sign in (14). It can be experimentally verified that particles with half-integer spin obey Fermi statistics, i.e. have antisymmetric wavefunctions. Electrons have total spin=1/2, therefore the negative sign must be chosen. In other words, a many-electron wavefunction must be antisymmetric with respect to an interchange of the coordinates of any two electrons. Evidently, our product wavefunction (10) does not fulfil that requirement, which is one reason why it isn't useful for reproducing experimental results.

Antisymmetry can be imposed on the wavefunction in many different ways. A determinantal function constructed with N one-electron functions as in (15) apparently has the correct behavior upon coordinate interchanges:

$$(15) \qquad \Psi_{DET}(r_1, r_2, \ldots.) = \begin{vmatrix} \varphi_1(r_1) & \varphi_1(r_2) & \cdot\ \cdot & \varphi_1(r_N) \\ \varphi_2(r_1) & \varphi_2(r_2) & & \\ \varphi_3(r_1) & \varphi_3(r_2) & \cdot\ \cdot & \\ \cdot\ \cdot & \cdot\ \cdot & \cdot\ \cdot & \\ \varphi_N(r_1) & \cdot\ \cdot\ \cdot & & \varphi_N(r_N) \end{vmatrix}$$

It can also be shown that using a complete set of orbitals $\{\varphi_i\}$, one can construct a complete set of determinants which spans the full space of antisymmetric many-electron wavefunctions—assuming reasonable boundary conditions in the one- and many-electron spaces.

This model—using a determinant to represent the many-electron wavefunction—is referred to as the *Hartree-Fock approximation*, and the determinantal wavefunction is often called a *Slater determinant*. Even though the form was derived from the product wavefunction (10), some important differences can readily be deduced from elementary theory about determinants:

The electrons must all have different coordinates. Any two electrons having identically the same coordinates will make the corresponding two rows of the determinant identical, and the probability will therefore be zero for such an event. Note that this argument applies to the space- and spin-coordinates, the argument does not exclude two electrons with different spins from being in the same point in space. For particles with the same spin, however, this determinantal form of the wavefunction leads to a certain correlation between their positions and movements. The effect is often referred to as *Fermi correlation*, to distinguish it from that brought about by the electrostatic repulsion which is called *Coulomb correlation*.

For similar reasons, the orbitals must all be different. In fact, to give a non-zero determinant they must be linearly independent, and are usually chosen to be orthonormal.

The importance of the computational simplifications brought

about by the Hartree-Fock approximation can hardly be overemphasized. Rather than solving the differential equation in 3N dimensions, we now face the much simpler task of determining the N one-electron(3-dimensional) functions (orbitals). The equations defining these orbitals (The Hartree-Fock equations) are:

$$h_i\varphi_i = \epsilon_i\varphi_i \qquad (16)$$

which is formally no different from (12). The operators h_i (Fock operators) are given by

$$h_i\varphi_i = T_i + V_n + V_c + V_x, \qquad (17)$$

where

$$T_i = -1/2\hbar^2(d^2/dx^2 + d^2/dy^2 + d^2/dz^2)\varphi_i \qquad (17a)$$

$$V_n = -\sum_n Q_n/|r-R_n|\varphi_i \qquad (17b)$$

$$V_c = \sum_j \int dr_2\varphi_j(r_2)\varphi_j(r_2)\varphi_i(r_1)/r_{12} \qquad (17c)$$

$$V_x = -\sum_j \int dr_2\varphi_j(r_2)\varphi_j(r_1)\varphi_i(r_2)/r_{12} \qquad (17d)$$

T_i is the kinetic energy, V_n the electron-nuclear attraction (summed over all atomic nuclei n). V_c is the electron repulsion energy (summed over all orbitals j, integrated over electron coordinates r_2) for the electron under consideration. The fourth term has no classical counterpart. It is called the exchange interaction, a term which is obvious when comparing it with the Coulombic electron interaction. The exchange term arises as a direct consequence of the antisymmetry; with a wavefunction of the form (10) only the three first terms in (17) would appear.

As mentioned above, the electron has associated with it a spin coordinate in addition to its three spatial coordinates. The integrals in (17c) and (17d) are to be understood as a normal integration over space coordinates, plus an integration of the spin. Since the spin coordinate can take only two values, the procedure is rather trivial. The result of the integration is equal to if the coordinates are the same, but equal to zero if they differ. As a consequence, it is readily seen that the Coulombic repulsion between electrons will always take place, whereas exchange interaction is in effect only between electrons of the same spin.

Even though the Hartree-Fock equations (16) are formally one-electron equations it is evident from (17) that the operators h_i defining the orbitals are in themselves functions of all the other orbitals in the system. In order to solve (16) an iterative scheme must therefore be applied. After an initial guess of trial orbitals has been made, the Fock operators can be constructed and (16) can be solved. The resulting orbitals are then used to form new operators, and a new iteration can be carried out. With a skilled guess of trial orbitals and suitable damping or extrapolation schemes this iterative procedure can be brought to convergence.

As mentioned above, two orbitals can have the same spatial part only if they represent electrons with different spin. In many-electron theory it is usually convenient to associate the spin with the orbitals, rather than with the electrons. To make the distinction clear, we use the term spin-orbital for the function including the spin, and reserve the concept of orbitals for the spatial part. We can therefore rephrase the previous rule to state that at most two electrons can occupy the same orbital, and if they do they must have different(opposite) spin. The use of space-orbitals and the concept of orbital occupancy reduces the number of one-electron functions that are needed to construct a many-electron wavefunction.

5. The variational principle and the secular equation. In quantum mechanics a physical situation is defined by the Hamiltonian H, whereas the system itself is described by the wavefunction Ψ_i. The system may be in a number of different stationary states, each determined by the Schrödinger equation

$$H\Psi_i = E\Psi_i, \quad (18)$$

hence the label "i" which simply labels different states of the system. (Usually, however, we are interested in the ground state Ψ_0 ,i.e. the state with the lowest energy).

For all Hamiltonians of chemical interest, the set of eigenfunctions $\{\Psi_i\}$ forms a complete, orthonormal set. Accordingly, any trial wavefunction Θ can be expanded in $\{\Psi_i\}$:

$$\Theta = \sum_i c_i \Psi_i \quad (19)$$

We now introduce the shorthand notation: $\langle\theta H\theta\rangle = \int\theta H\theta d\tau$, and consider the energy functional

$$E_\theta = \int\theta H\theta d\tau / \int\theta\theta d\tau = \langle\theta H\theta\rangle / \langle\theta\theta\rangle \tag{20}$$

$$= \sum_{ij} c_i c_j \langle\Psi_i H \Psi_j\rangle / \sum_{ij} c_i c_j \langle\Psi_i \Psi_j\rangle = \sum_{ij} c_i^2 E_i / \sum c_i^2$$

It is readily seen that E_θ is a weighted average of the state energies E_i. If $E_0 \leq E_i$, we must also have $E_0 \leq E_\theta$. This result is completely general. Therefore, the energy calculated for a trial wavefunction in the prescribed way will always be higher than the true ground state energy of the system. Minimizing E_θ by means of adjustable parameters will always bring it closer to the correct E_0. If our trial function were an exact solution to the Schrödinger equation ($\theta = \Psi_i$) we would of course have $E_\theta = E_i$.

The variational principle suggests a technique for solving the Schrödinger equation. As a first step, we expand our (trial) wavefunction in a basis $\{\phi_i\}$:

$$\Psi \approx \theta = \sum_i c_i \phi_i \tag{21}$$

With this ansatz, we can write the energy functional as

$$E_\theta = \langle\theta H\theta\rangle / \langle\theta\theta\rangle = \sum_{ij} c_i c_j H_{ij} / \sum_{ij} c_i c_j \langle\phi_i \phi_j\rangle \tag{22}$$
$$= \sum_{ij} c_i c_j H_{ij} / \sum_i c_i^2$$

where

$$H_{ij} = \langle\phi_i H \phi_j\rangle \tag{23}$$

Since $E_0 \leq E_\theta$ for any θ we can to minimize E_θ with respect to c_k;

$$dE_\theta / dc_k = 2\sum_i c_i H_{ik} / \sum_i c_i^2 - 2E_\theta c_k \tag{24}$$

$$\sum_i H_{ki} c_i = E_\theta c_k \tag{25}$$

On matrix form, (25) becomes

(26) $$\mathbf{Hc} = \mathbf{Ec},$$

usually called *the secular equation* (The Schrödinger and Hartree-Fock equations are operator eigenvalue equations).

The computational work in solving the secular equation can be divided in two steps:

1: Construct the M^2 elements of the Hamiltonian matrix (M is the number of basis functions used to expand the wavefunction according to (21)).

(27) $$H_{ij} = \langle \phi_i H \phi_j \rangle = \int \phi_i H \phi_j d\tau$$

2: Solve **Hc** = **Ec**, i.e. diagonalize the matrix **H**, in other words find its eigenvectors and eigenvalues (or at least the lowest one).

The above discussion of the variational principle has implicitly assumed a N-electron equation. However, very similar methods can be used to determine the one-electron functions "orbitals" in (15). By applying the variational principle to an energy functional constructed with a determinantal trial wavefunction as in (15), the Hartree-Fock equation can be converted into a secular equation which can be solved with standard linear algebra techniques.

6. Basis sets. The set of molecular orbitals $\{\varphi_i\}$ (MOs) describe the probability distributions for the electron in a molecule. As expansion functions for these molecular orbitals one can use a set of atomic orbitals (AOs) $\{\chi\}$ relevant to the atoms that make up the molecule. This is the MO-LCAO approach (MOs as Linear Combinations of AOs). Its strength lies in the well-known fact that atoms retain a large part of their identity when forming a molecule. In order to determine the AOs, we note that the only atom for which the Schrödinger equation can be solved exactly is the hydrogen atom. The solutions are of the form

(28) $$\chi = Y_{l,m}(\theta,\phi)P(r)e^{-\alpha r}$$

Basis functions resembling the hydrogenic solutions—or somewhat simplified:

$$\chi' = Y_{l,m}(\theta,\phi)e^{-\alpha r} \tag{29}$$

are referred to as exponential-type or Slater-Type Orbitals (STOs).

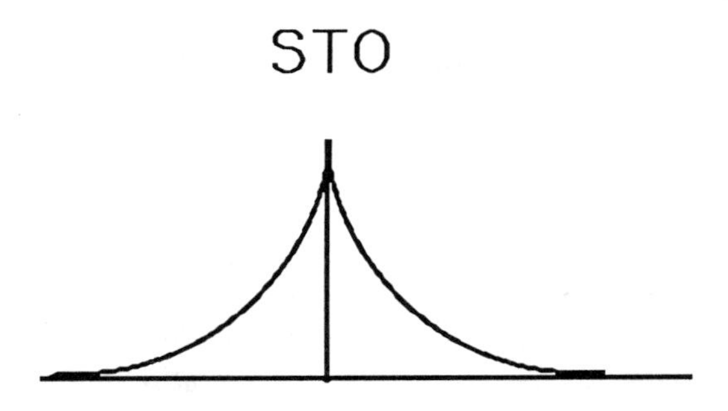

For reasons of computational efficiency, a somewhat different basis is often used; namely the Gaussian-Type Orbitals(GTOs)

$$\varphi = (x-x_0)^l(y-y_0)^m(z-z_0)^n\, e^{-\alpha(r-r_0)^2} \tag{30}$$

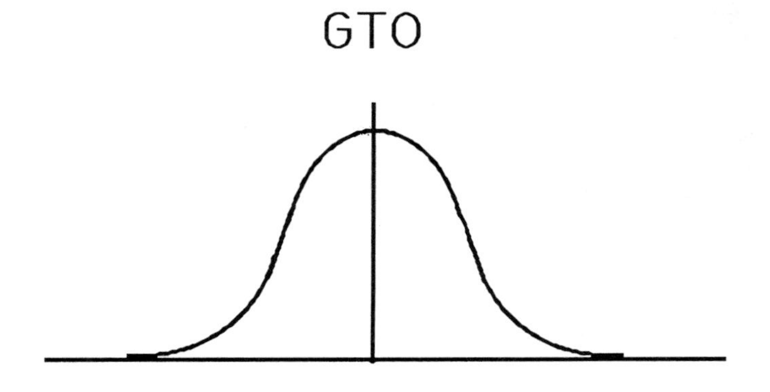

The demands on a basis set for quantum chemical calculations can be difficult to meet. The major problem in evaluating the matrix elements H_{ij} is the evaluation of the electron repulsion integrals

$$\int \varphi a(r_1)\varphi b(r_1)\varphi c(r_2)\varphi d(r_2)/r_{12}\, d\tau_1 d\tau_2 = (ab \mid cd) \quad (31)$$

We often wish to do calculations on systems with up to 50-100 atoms (or more), using 10-20 basis functions/atom. The largest calculations performed so far with these techniques involved $>10^3$ basis functions, which means that in theory there are 10^{12} such integrals to evaluate! Symmetry and pre-screening is of course extensively used to reduce the number of integrals, but still we need very fast algorithms to compute them.

The main advantage of a GTO basis is that it can be separated in its Cartesian components:

$$\varphi = \varphi x \varphi y \varphi z \quad (32)$$

$$\varphi x = (x-x_0)^l e^{-\alpha(x-x_0)^2} \quad (33)$$

The electrostatic potential can be separated as well:

$$\sqrt{\pi}/r = 2\int e^{-u^2 r^2} du \quad (34)$$

The integrals can be (partly) performed as one-dimensional integrals over the Cartesian components, and the entire integral $(ab|cd)$ reduces to a (15-fold!) sum which can be evaluated analytically.

Basis sets used in electronic structure calculations for polyatomics are almost exclusively contracted Gaussian sets, i.e. basis sets which are linear combinations of simple Gaussians. The main reason for contracting the basis sets is to reduce the number of integrals which have to be handled and stored. However, with the Direct SCF technique, and with the enormous memory anticipated to be available on future supercomputers, that reason is not as important as it used to be. There may still be valid arguments for contraction, at least as a possible option in order to maintain compatibility with large scale post-Hartree-Fock calculations, which are likely to use contracted basis sets in the foreseeable future.

7. Relativistic methods. For a particle moving at a velocity which is significant compared to that of light, the effects of relativity have to be accounted for. In order to decide whether we need to be concerned at all with these effects in chemistry, some simple qualitative comparisons are appropriate:

Consider a one-electron atom with atom number Z in its ground state. In atomic units[2] the energy is $E = -Z^2/2$. From the Virial theorem one has $T=-E$, but since $T=p^2/2m$ and $m=1$ in atomic units we conclude that $\langle v \rangle = Z$ (the brackets $\langle \ \rangle$ represent a mean value in the statistical sense). Since the velocity of light is about 137 in atomic units it is evident that relativistic effects can by no means be neglected for elements in the bottom part of the Periodic Table. To account for many effects observed for these elements and their compounds relativistic quantum mechanics must be used. Here, we will review the most basic elements of that theory.

The key equation in relativistic quantum mechanics is the Dirac Equation:

$$(35) \qquad [-i\hbar\, d/dt - V + c\, \alpha\cdot\pi + \beta\, mc^2]\Psi = 0$$

π is used for the momentum instead of **p** since, in a more general formulation the effects of external magnetic fields can also be accounted for. Here no such effects will be treated, however, and we have

$$(36) \qquad \pi = -i\hbar(d/dx\ ,\ d/dy\ ,\ d/dz)$$

$$(37a) \qquad \alpha = \begin{bmatrix} 0 & 0 & 0 & 1 \\ 0 & 0 & 1 & 0 \\ 0 & 1 & 0 & 0 \\ 1 & 0 & 0 & 0 \end{bmatrix}, \begin{bmatrix} 0 & 0 & 0 & -i \\ 0 & 0 & -i & 0 \\ 0 & -i & 0 & 0 \\ -i & 0 & 0 & 0 \end{bmatrix}, \begin{bmatrix} 0 & 0 & 1 & 0 \\ 0 & 0 & 0 & -1 \\ 1 & 0 & 0 & 0 \\ 0 & -1 & 0 & 0 \end{bmatrix}$$

$$(37b) \quad \beta = \begin{bmatrix} 1 & 0 & 0 & 0 \\ 0 & 1 & 0 & 0 \\ 0 & 0 & -1 & 0 \\ 0 & 0 & 0 & -1 \end{bmatrix}, \qquad (37c) \quad \Psi = \begin{bmatrix} \Psi_1 \\ \Psi_2 \\ \Psi_3 \\ \Psi_4 \end{bmatrix}$$

For stationary states we can derive the time-independent Dirac equation, exactly as for the non-relativistic case:

$$(38) \qquad [-V + c\, \alpha\cdot\pi + \beta\, mc^2]\Psi = E\Psi$$

This theory has proven to give results in virtually perfect agreement with experiment for one-electron systems. However, when compared to the non-relativistic theory it has a very important weakness, in that extensions to many-particle theories are by no means straightforward!

Several approaches have been tried to overcome this problem, and to develop methods that can be used to study real atoms and molecules for which relativistic effects are important. It has been demonstrated that even a simple perturbation-type account of such effects bring a significant improvement to the calculated structures. These methods are very simple to apply, and require only a fraction of the computational resources needed to compute the wavefunction and the non-relativistic energy.

8. Computational strategies. The design of efficient computational methods has always been an area of high priority within research in theoretical chemistry. Basic principles and working theories in that area were set forth in the early 50's[3]. The Self-Consistent Field (SCF) approximation, along with the Hartree-Fock method, and of course the availability of electronic computers, made it possible for the first time to predict properties for atoms and molecules of chemically interesting size. Since then, the field of computational quantum chemistry has undergone a rapid development up to its present state, where accurate predictions of properties and reactions can be made for molecules of considerable size.

Concurrently with the development of methods and algorithms, there has been impressive progress in computer hardware. Due to the organizational structure of most universities and research institutions, large-scale computer capacity was made available at a low cost and to a large group of users only with the appearance of powerful minicomputers (the VAX age). Somewhat later, the introduction of large, array-oriented mainframes—"supercomputers"—has brought even more significant changes to the field. With these remarkable tools at the hands of theoretical chemists there would be an enormous scientific potential in computational chemistry, if one could design the algorithms and computer codes to use these tools adequately.

With new generations of supercomputers appearing on the market every three-four years, it is important to look ahead to the hardware development in order to decide on a long-term strategy for method and algorithm development. At present, it is rather obvious that parallelism has to be extended to the processor level in order to continue the exponential increase in capacity that has been almost a "law of nature" for the last 20 years. Significant research efforts are now being devoted to the design of modified algorithms and computational strategies that will run efficiently in a multi-tasking environment.

Quite non-conventional methods are sometimes necessary to fully exploit new technological achievements. One example of this is the "Direct SCF" method[4]. Most quantum chemical calculations use some form of basis set; a set of auxiliary functions by means of which the electron distribution in an atom or molecule is expressed. The electron density is described by products of such functions—a total of N^2 products if N basis functions are used. Following the same reasoning the repulsion between electrons, which is the most difficult physical effect to account for in a many-particle system, would need N^4 different terms. Typically the equations occurring in a quantum chemical calculation are solved iteratively, and the N^4 electron-repulsion integrals are needed many times, 5-50 in most cases.

In traditional methods, these integrals are all calculated initially, stored on disk, and then retrieved again in each cycle of the iterative procedure. The limit set by the I/O capacity of most mainframe computers is around 150-250 basis functions, and with the N^4 dependence it is very difficult to stretch it much further, even though a lot of innovative tricks have been used. With the development of hardware for fast floating point computation and with increasingly efficient algorithms for electron-repulsion integrals, the bottleneck in computational resources for quantum chemical calculations is now set by I/O and external storage capacity rather than CPU power alone, which was previously the case. Without going into the details of the implementation, it is sufficient here to note the main philosophy of the method that are being developed:

1) In order to minimize the I/O and disk requirement, the relevant electron-repulsion integrals are recalculated in every iteration whenever they are needed, rather than retrieving them from external memory;

2) Since these data are computed "on the fly" one can utilize information available during the calculation to decide whether it is actually necessary to calculate them. Even if a certain integral is formally required by the algorithm, a simple analysis may show that for numerical reasons it will not contribute significantly to the final result.

Clearly, step 1 alone would be trivial to use, but would increase the CPU-time requirement by an order of magnitude. Step 2 is much more complicated, but its full implementation is very rewarding - in fact more and more so as the size of the problem increases! A larger and larger fraction of all the interactions can be neglected as a result of the increased distances as the size of the molecules increases. In summary, a calculation with the Direct SCF method on a large system usually takes less time than with conventional methods. In addition, of course, the I/O and disk problems are completely eliminated.

Using these techniques, one can rather routinely study systems with 300-400 basis functions or more[5], even on a minicomputer. On the fastest supercomputers available today, calculations have recently been carried out on molecules with nearly 200 atoms[6], using more than 1500 basis functions.

The above applications have all been made within the independent particle or "Hartree-Fock" approximation. In this model the wavefunction is written as a determinant of one-electron functions(orbitals), and these are then adjusted to give the best possible such one-determinant function in the sense of the variational principle. This is the mathematical definition; in physical terms the model implies that each electron moves in an effective mean electrostatic field of all the other particles, but that the motions and instantaneous positions of two or more electrons are not explicitly correlated. This assumption is of course a somewhat crude approximation, yet it is astonishing to see how well it works in many cases. For some chemical problems, however, there is a need for extending the calculations beyond the Hartree-Fock model. In particular, this is the case when the process studied involves breaking or formation of chemical bonds. It also seems to be true whenever transition-metal complexes are considered. In these cases, the correlation between the motions of electrons is too important to be approximated by a mean effective field.

The most common way to account for this correlation is to generalize the determinantal wavefunction approach. By expressing the wavefunction as a linear combination of several determinants—again using the variational principle to determine the "best" combination—one can construct trial wavefunctions which are arbitrarily close to the exact solution to the Schrödinger equation. Methods which exploit such techniques are termed Multi-Configuration SCF (or shorter MCSCF) if both the expansion coefficients for the determinants and the orbitals forming those determinants are variationally optimized. If the orbitals are not optimized the technique is called Configuration Interaction (CI). In the former case 10^4 to 10^5 determinants is a practical upper limit, in the latter case calculations with more than 10^7 determinants have been performed.

9. Applications to chemistry. Most physical properties of molecules which we want to study can be expressed as derivatives of the total energy for the molecule with respect to parameters that correspond to certain perturbations of the system. This definition often describes the experimental situation under which the property in question is observed. As an example, the dipole moment of a molecule is defined as the first derivative of the energy with respect to an external electric field, whereas a force constant for (harmonic) molecular vibration is given as a second derivative with respect to a displacement of the nuclei in the molecule. Mixed, higher derivatives are also common and can usually be given a straightforward interpretation in terms of physical properties.

The total electronic energy of a molecule is a function of the coordinates of the nuclei. This many-dimensional function is often referred to as a potential (hyper)surface. The investigation of potential energy surfaces is an important challenge both for theoretical and experimental chemistry. Systematic theoretical studies of such surfaces require efficient tools to locate and characterize stationary points. Much effort has therefore been spent on developing algorithms and computer code for first derivatives (gradients) and second derivatives (Hessians) using approximate wavefunctions.

During the last years, large efforts have been made to develop strategies for efficient calculation of energy derivatives using wavefunctions of various degree of accuracy[7]. In particular, working algorithms and computer codes have been developed which compute first and second energy derivatives at the MCSCF level[8]. These will be very useful for the mapping of potential energy surfaces of simple chemical reactions. A quite obvious extension of that work, which is now in progress in several laboratories around the world, is to generalize the treatment to third and fourth derivatives. The main purpose with that project is to study electric or mixed electric-geometrical perturbations. Interesting such properties, on which our attention will primarily focus, are e.g. β- and γ-hyperpolarizabilities.

While ordinary polarizabilities contain information about the response of molecular electric moments to an applied electric field, the hyperpolarizabilities describe how this response deviates from a simple linear dependence.

Nonlinear optics is a fairly new area of research, but it has rapidly grown to become of immense technological importance. Hyperpolarizabilities are key quantities in the search for efficient, organic nonlinear optical (ONLO) materials, where high optical nonlinearities are crucial material properties. The possibility of theoretically predicting these properties for a large variety of plausible candidate molecules would be immensely important in such work.

This goal might at first seem unrealistic. By conventional wisdom, it is extremely difficult to calculate reliable polarizabilities—and even more so hyperpolarizabilities—with any reasonable accuracy, without resorting to extremely large basis sets and highly correlated wavefunctions. However, this part of the problem is less serious for the larger molecules on which our interest is focused, which may first seem a paradox. In order to understand this fortunate condition, one must realize that the model used to study the above properties must also be able to describe electric moments and energetics of the most important, low-lying excited states. The available experience regarding calculations of polarizabilities derives almost exclusively from small molecules, for which the excited states are of an exotic character, and therefore very difficult to describe accurately with standard techniques. The situation is entirely different for the molecules considered here. We have primarily studied various linear, substituted polyacenes. For these molecules the polarizabilities in the direction along the molecule are expected to be rather extreme. A number of such molecules have already been investigated, and the calculated nonlinear electro-optical properties are being compared with experimental data.

Elemental carbon is one example of a substance that goes directly from solid to gaseous form upon heating, without passing through a liquid phase at normal pressures. In situations near the transition point, the carbon atoms form aggregates or clusters in the gaseous phase. Despite the predominant role of carbon in chemistry, little is known about such clusters, even though they have long been recognized as being of technological and scientific importance. Growing carbon clusters are responsible for soot formation during combustion, and clusters decreasing in size are equally important in coal gasification processes. Other areas where carbon clusters are of interest range from chemical catalysis to the composition of comet tails and stellar atmospheres.

It is not unexpected, therefore, that carbon clusters have been the focus of extensive experimental and theoretical research during the past 30 years. Still, the exact clustering mechanisms are not well understood. A striking observation that lacks a satisfactory explanation is the existence of 'magic numbers', i.e. the fact that in a distribution of carbon clusters, some species with a certain number of carbon atoms are much more abundant than others. Quite unusual and unexplained distributions are found for neutral clusters as well as for charged species. It now seems clear that certain of these 'magic numbers' can be strongly enhanced by changes in cluster growth conditions. In order to shed some more light into this world of phenomena which still lack a satisfactory understanding, we have been studying large clusters of carbon atoms with very accurate quantum mechanical methods.

As starting points for models of these clusters one could use e.g. the structures of solid diamond or solid graphite. We have used both[6], in addition to a few other models, and the results show some striking differences. In diamond-like clusters the properties we have considered converge very rapidly towards the bulk values, and the interior of a 148-atom cluster is virtually indistinguishable from that of pure bulk diamond for many purposes. This is encouraging, since we intend to study the effects of impurities embedded in diamond, later also in silicon, by using this cluster technique, and 148 atoms is already a considerable size system for accurate quantum mechanical calculations.

In graphite-like clusters, on the contrary, we find that many properties converge surprisingly slowly towards bulk values, and even in clusters with 180 atoms(the largest done so far) the properties are not representative for the bulk. It is clear from our study that the slow convergence is related to the conducting property of graphite. This first of all, casts new light on the chemistry of carbon clusters themselves. It also logically implies that small graphite-like clusters are not particularly useful for modeling the chemistry of intercalation compounds (stranger molecules and -atoms trapped between layers of graphite) or for processes taking place on graphite surfaces. The remarkable stability observed for the 60-atom cluster, and the proposal that it reflects an unusual stability of a specific geometrical configuration called "Buckminsterfullerene," provoked an extensive debate, and led us to investigate that hypothetical molecule in some detail[9].

The structure of these clusters is a compromise between bond "strain", which is minimized in the planar, graphite-like forms, and bond energy, which is roughly proportional to the number of bonds, and therefore favors the spheroidal structures. The conclusion from our calculations is that the bond energy considerations dominate for structures with as many as 60 atoms, and, therefore, that the "Buckminsterfullerene" structure is indeed plausible.

REFERENCES

[1] D.A. DIXON, *Molecular Modelling of Polymeric Materials Using Ab Initio Molecular Orbital Theory: The Helicity of Teflon,* Chemical Design Automation News, 2 (1987), pp.1-11.

[2] C.A. MEAD, *The Born-Oppenheimer Approximation in Molecular Quantum Mechanics,* this volume.

[3] C.C.J. ROOTHAAN, *New Development in Molecular Orbital Theory,* Revs. Mod. Phys., 23 (1951), pp.69-89; *Self-consistent Field Theory for Open Shells of Electronic Systems,* ibid. 32 (1960), pp.179-185.

[4] J. ALMLÖF , K. FAEGRI, JR. AND K. KORSELL, *Principles for a Direct SCF Approach to LCAO-MO Ab Initio Calculations,* J. Comput. Chem., 3 (1982), pp.385-399; J. ALMLÖF AND P.R. TAYLOR, *Computational Aspects of Direct SCF and MC SCF Methods,* in: "Advanced Theories and Computational Approaches to the Electronic Structure of Molecules", Reidel, Dordrecht 1984, pp.107-125.

[5] J. ALMLÖF AND K. FAEGRI, JR., *Steric Forces in Substituted Aromates: The Conformation of the Hexahalobenzenes as Determined by Ab Initio MO-LCAO Calculations,* J. Amer. Chem. Soc., 105 (1983), pp.2965-2969.

[6] J. ALMLÖF AND H.P. LÜTHI, *Theoretical Methods and Results for Electronic Structure Calculations on Very Large Systems: Carbon Clusters,* in: "A.C.S. Symposium Series 353: Supercomputer Research in Chemistry and Chemical Engineering", American Chemical Society, Washington 1987, pp.35-48.

[7] P. PULAY, *Second and Third Derivatives of Variational Energy Expressions: Applications to MC SCF Wavefunctions,* J. Chem. Phys., 78 (1983), pp.5043-5051.

[8] J. ALMLÖF AND P.R. TAYLOR, *Molecular Properties from Perturbation Theory: A Unified Treatment of Energy Derivatives,* Int. J. Quantum Chem., 27 (1985), pp.743-768; T.U. HELGAKER, J. ALMLÖF, H.J.AA. JENSEN AND P. JØRGENSEN, *Molecular Hessians for large-scale MC-SCF wavefunctions,* J. Chem. Phys., 84 (1986), pp.6266-6279.

[9] H.P. LÜTHI AND J. ALMLÖF, *Ab initio studies of the Thermodynamic Stability of the Icosahedral C_{60} molecule "Buckminsterfullerene",* Chem. Phys. Letters, 135 (1987), pp.357-360.

SPECTRAL THEORY

SPECTRA OF ATOMS AND SPECTRAL THEORY OF ATOMIC HAMILTONIANS

WILLIAM P. REINHARDT*

Abstract. An introductory overview and classification of the electronic states of simple atoms in the non-relativistic Born-Oppenheimer approximation is given, and corresponding atomic processes are illustrated by relevant experimental data. The mathematical concept of the spectrum of the atomic electronic Hamiltonian is introduced in a qualitative manner, both in terms of the Hamiltonian H itself and in terms of matrix elements of the resolvent $(z - H)^{-1}$. Properties of such matrix elements are considered as analytic functions of the complex variable z, and the pole, branch point, and cut structure of these functions are discussed. The introduction of complex scale transformations (dilatation transformations) is then introduced in terms of the spectrum of H, and subsequently in terms of the modifications of the analytic structure of the resolvent. It is seen that the latter is unchanged with respect of points of non-analyticity, but, and this is a significant result of the complex scale transformation, the branch cuts are rotated off the real axis. This alternative integral representation allows easy development of new theoretical results: the problem of the radius of convergence of the $1/Z$ perturbation expansion is discussed, as is the problem of bound states and resonances above the energy $E = 0$, defined as the threshold for complete ionization of the atom. The calculation of resonance energies is discussed, and the need for generalization of standard variational principles motivated. The Stark effect is used as an example of numerical investigation preceding the finding of rigorous results as to the effect of complex scale transformations on the spectra of atomic-like Hamiltonians. The use of dilatation analyticity in the calculation of photoabsorption cross sections and other continuum processes is indicated.

1. Introduction. The purpose of these lectures is many-fold. First to introduce non-physical scientists to an overall view of the "spectrum," in both its physical [1] and mathematical [2] senses, of a typical many–electron atom. Then to exploit the many and remarkable properties of complex scale transformations on atomic Hamiltonians, their effect on the mathematical spectrum, and their role in allowing facile computations of atomic bound, quasi-bound, and continuum processes through use of $\mathcal{L}^2$ basis functions (Lebesgue square integrable) and methods related to configuration interaction. In actual presentation of the lectures at the IMA it became evident that using a theoretical apparatus in the absence of any intuitive knowledge of what the physical states of a many–electron atom actually are did not proceed well. Thus a digression proved useful, and Section 2 of the present notes contains this material, which is very sketchy and elementary from the point of view of a physical scientist knowledgeable about the quantum theory of atomic structure, but which outlines the key aspects of these complex many–body physical systems in a way to make the motivation for later developments evident. Terms such as bound and continuum states for electronic motion are introduced, as well as the concepts of a bound state in the continuum and resonances due to both tunneling and auto-ionization processes, Several of these concepts are illustrated, and–I hope–thereby made more concrete, by actual experimental data.

*Telluride Summer Research Center, Telluride Academy, P.O. Box 2255, Telluride, Colorado 81435; and Department of Chemistry, University of Pennsylvania, Philadelphia, Pennsylvania 19104-6323.

Following this introduction (which can be skipped by those knowledgeable about atomic structure), we begin in Section 3 the meat of the lectures. The complex scale transformations of Aguilar, Combes, Balslev and Simon [3–5] are introduced, after a brief motivation based on formal variable changes, and on simple use of Cauchy's results on contour distortions for definite integrals of analytic functions. The results [see 3–5, and the reviews 2,6–8] of such scale transformations are discussed in two complementary languages: that of functional analysis of Schrödinger operators, and in terms of the analyticity of matrix elements of the resolvent, $(z-H)^{-1}$. Section 4 contains both theoretical and computational applications [see also, 6-8] of such complex dilations, and such topics as the problem of the radius of convergence of atomic $1/Z$ perturbation theory, the possibility of bound states in electronic continua of the same symmetry, and the practical computation of "resonance" eigenvalues are discussed. Section 5 discusses the use of variational approaches to calculation of resonance eigenvalues in a more general manner, indicating that new mathematical developments would be welcome. Finally, in Section 6, application of complex scale transformations to calculation of cross sections and amplitudes for atomic processes involving the continuum are introduced and shown, when used with appropriate sequence resumation methods, to give useful results in some cases and seen to require further work in others. Several of the sections are followed by "Remarks" intended to provide access to the literature, and to indicate those places where the discussion in the text needs amplification. These "remarks" are sometimes more detailed than the discussions of the body of the text, and they may be omitted on first reading if desired. Further conceptual developments in the present lectures will not depend crucially on them.

2. The atomic Hamiltonian and the mathematical and physical "spectra" of simple atoms. First we consider one-electron atoms, then the general N-electron case, and finally give a more detailed classification of states for the 2-electron case.

2.1. The hydrogen atom and hydrogen-like atoms. Much of our basic understanding of atomic structure is based on the analytically soluble two body Coulomb problem [9-13]. This is the simplest model of the hydrogen atom, consisting of two charged particles, each with no internal structure. The bound states of this simplest model atomic system are described by the $\mathcal{L}^2$ eigenfunctions of the Schrödinger equation, and the continuum or scattering states by non-normalizible formal eigenfunctions, which are bounded at infinity [10-12]. The appropriate Schrödinger equation for determination of the allowed energy levels is

$$H\Psi = E\Psi \tag{1}$$

where H is the quantum Hamiltonian, obtained from the corresponding classical Hamiltonian, which is the classical energy written in terms of position and momentum variables, by the usual [11,12] correspondence rules: $x^{op} = x$; $p^{op} = -i\hbar\frac{\partial}{\partial x}$.

In spherical polar co-ordinates the quantized Hamiltonian for motion of an electron (with charge "$-e$", and mass m_e) moving in the presence of a fixed nucleus (the Born-Oppenheimer approximation [14]) of charge Ze, and which is taken to be the co-ordinate origin is:

$$H(r,\theta,\phi) = -\frac{\hbar^2}{2m_e}\nabla^2_{r,\theta,\phi} - \frac{Ze^2}{r} \tag{2}$$

where ∇^2 is the Laplacian,

$$\nabla^2_{r,\theta,\phi} = \frac{1}{r^2}\frac{\partial}{\partial r}r^2\frac{\partial}{\partial r}\sin\theta\frac{\partial}{\partial\theta} + \frac{1}{r^2}\left(\frac{1}{\sin\theta}\frac{\partial}{\partial\theta}\sin\theta\frac{\partial}{\partial\theta} + \frac{1}{\sin^2\theta}\frac{\partial^2}{\partial\phi^2}\right). \tag{3}$$

The Schrödinger equation for the hydrogen atom is easily solved by noting that the wave function factors into a spherical harmonic $Y_{\ell,m}(\theta,\phi)$ and a radial part $R(r)$, as the Coulomb potential, $-Ze^2/r$, is spherically symmetric. The spherical harmonics are simultaneous eigenfunctions of the angular momentum operators L^2 and L_z:

$$\begin{aligned} L^2 &= -\hbar^2\left(\frac{1}{\sin\theta}\frac{\partial}{\partial\theta}\sin\theta\frac{\partial}{\partial\theta} + \frac{1}{\sin^2\theta}\frac{\partial^2}{\partial\phi^2}\right) \\ L_z &= -i\hbar\frac{\partial}{\partial\phi} \end{aligned} \tag{4}$$

which correspond to the (squared) total angular momentum, and the projection of the angular momentum on the z-axis. The radial hydrogenic wavefunction, $R_{n,\ell}(r)$ satisfies the ordinary differential equation

$$\left(-\frac{1}{2}\frac{1}{r^2}\frac{\partial}{\partial r}r^2\frac{\partial}{\partial r} + \frac{\ell(\ell+1)}{2r^2}\right)R_{n,\ell}(r) = E_n R_{n,\ell}(r) \tag{5}$$

which we abbreviate as

$$H_r R_{n,\ell}(r) = E_n R_{n,\ell}(r) \tag{6}$$

where Hartree atomic units have been adopted (*i.e.*, $\hbar = e^2 = m_e = 1$) and H_r is the radial Hamiltonian. The eigenvalues which correspond to existence of $\mathcal{L}^2$ radial wave functions for this famous problem give the energy levels [9–13] of the hydrogen atom (which are ℓ independent!)

$$E_n = -\frac{Z^2}{2n^2} \tag{7}$$

where n is the so-called principle quantum number, which is an integer running from 1 to ∞. The eigenvalues thus accumulate at $E = 0$, a symptom of the pathologies associated with the fact that the Coulomb potential is "long range," it i.e., is not relatively compact [2] with respect to the kinetic energy $\frac{-\hbar^2\nabla^2}{2m}$. These bound state eigenvalues are shown in relation to the Coulomb potential which binds them in Figure 1.

Corresponding to positive energies there are (formally at least) continuum eigenfunctions, denoted as $R_{k,\ell}(r)$ for real $k \geq 0$ such that

$$H_r R_{k,\ell}(r) = \frac{k^2}{2}R_{k,\ell}(r), \tag{8}$$

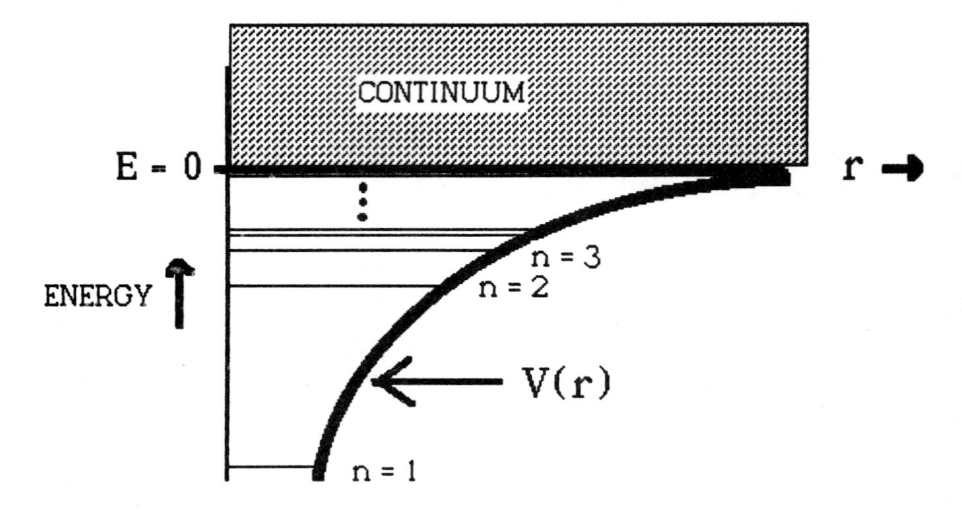

Figure 1) Bound state and continuum energy levels (parts of the point and continuous spectrum) for the hydrogen atom problem, shown in relation to the $-1/R$ Coulomb potential which "binds" them. The quantum mechanical eigenvalues accumulate at the threshold of the continuum, taken to be $E = 0$.

which, although not $\mathcal{L}^2$, do remain bounded as $r \to \infty$, and are traditionally normalized by physicists and chemists as

$$\int_0^\infty r^2\, dr R^*_{k,\ell}(r) R_{k',\ell}(r) = \delta(k - k'), \tag{9}$$

where $\delta(k - k')$ defined such that

$$\delta(k - k') = 0, \quad \forall k \neq k',$$

but

$$\int_a^b \delta(k - k')dk = 1, \text{ if } k' \in (a, b), \tag{10}$$

is the Dirac δ-function, or in more modern terms a "distribution" or generalized function. Physical scientists refer to the set of all bound and continuum eigenvalues as the "spectrum" $\sigma(H_r)$. This is illustrated in Figure 2. Mathematicians and mathematical physicists, [2] somewhat more technically, refer to the spectrum as the singularities of matrix elements of the type $\langle \phi, (z - H_r)^{-1} \phi \rangle$ on a dense set of functions, ϕ, which are assumed to be $\mathcal{L}^2$ and to belong to the appropriate Hilbert space. A more traditional energy level diagram (Grotrian Diagram) is shown in Figure 3, where it is evident that the multiplicity of the bound state eigenvalues is

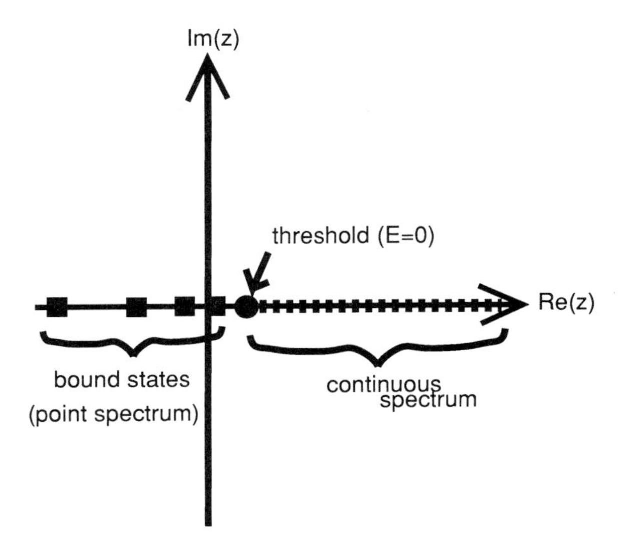

Figure 2) The mathematical "spectrum" of the hydrogen Hamiltonian shown in the complex energy plane. The squares are parts of the point spectrum, and the continuous spectrum starting at $E = 0$ is indicated as a dashed line running out to infinity along the positive energy axis. The spectrum is real, as the Hamiltonian is self adjoint. [Note that the accumulation of the point spectrum at $E = 0$ is not indicated in this Figure, see Fig. 1]

n^2, and the continuum is infinitely degenerate, in the sense that a separate branch of the continuous spectrum exists for each angular momentum, ℓ.

How does all of this relate to experiment? First it must be noted that the Hamiltonian, neglecting the finite mass of the nucleus, all relativistic effects (which include spin and other magnetic effects), and the finite spatial extent of the nucleus itself, is not exactly right. There are errors of order of a few parts in 10^4. However, for purposes of the present discussion, we ignore this and ask: are the general features of the eigenfunctions and eigenvalues of the Hamiltonian H_r correct? They surely are, and are not only of importance for atomic hydrogen, but both qualitatively and quantitatively set the stage for our understanding of more complex atoms. For example, Figure 4 shows an experimental spectrum of the Ba and Sr atoms in an energy regime where one electron is successively excited, see Figure 5, by absorption of a light quantum or photon, to higher and higher $\ell = 1$ states (called p states, for historical reasons) with ever increasing values of n. As these "Rydberg" type excited states accumulate, as predicted by eq. (7), experimental resolution eventually fails to resolve them (at about $n = 70$ for Strontium and $n = 90$ for barium in the present Figure 4) and at yet higher photon energy direct transitions transitions to the continuum are observed. Figure 6 indicates some of the observed transitions on

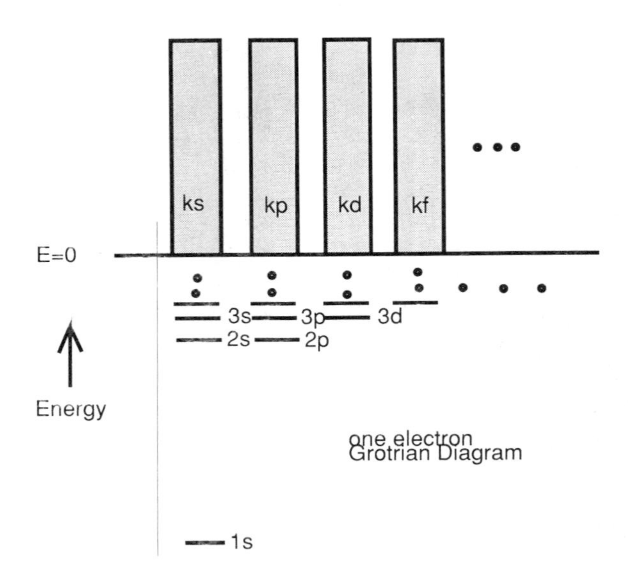

Figure 3) A Grotrian diagram for the hydrogen atom. This is yet another way of indicating the nature of the spectrum of this fundamental system. The bound and continuum levels are shown in a way which indicates that the point and continuous spectra are highly degenerate. A further degeneracy, not indicated, is the $(2\ell + 1)$ fold degeneracy of each state of angular momentum "ℓ".

the appropriate energy level diagram, in such a way as to make the correspondence with the observed spectrum as clear as possible.

We can now state several of the general objectives of the present lectures: what are the corresponding many-electron energy level schemes of atoms more complex than the hydrogen atom, or of the Rydberg excited states of Ba, above, which for all practical purposes are one-electron states? What, then, are the general features of the mathematical spectrum of N-electron atomic Hamiltonians? How does the spectrum allow understanding of the structure and dynamics of atoms? By structure we mean the bound and continuum eigenvalues and eigenfunctions, and by dynamics we imply such questions as: how does the system respond to a collision, what determines the lifetimes of quasi-bound or "resonant" states of the atomic system, or how does the system respond to being placed in an external field, such as that generated by a magnet or a laser?

Further, how does understanding of the mathematical spectrum of an atom allow us to understand various time-dependent processes which can take place as the atom is disturbed in various ways, or which follow from specific initial preparation? Dynamics is determined by the time dependent Schrödinger equation,

$$i\hbar \frac{\partial}{\partial t} \Psi(\vec{r}, t) = H \Psi(\vec{r}, t), \tag{11}$$

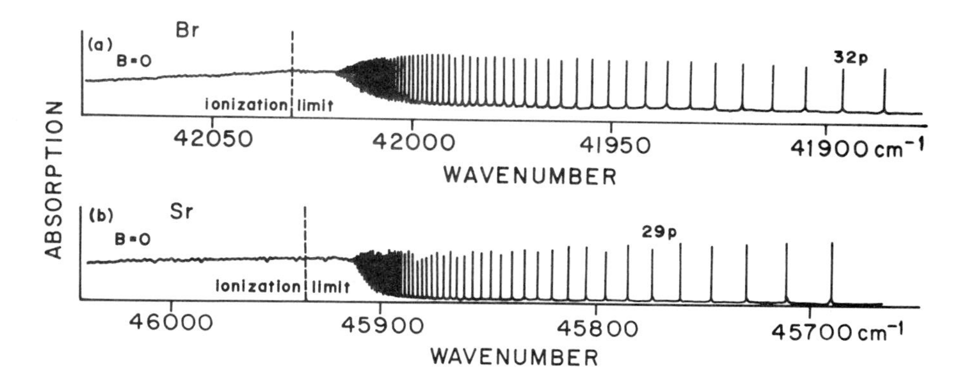

Figure 4) Experimental *absorption "spectra"* of highly excited states of the Ba and Sr (barium and strontium, respectively) atoms. The states accessed are effectively those where a single electron is excited to higher and higher "Rydberg" levels, as shown in Figure 3, which show the accumulation of eigenvalues for the hydrogen atom. This is not surprising, as the asymptotic potential is also $-1/R$ for these states, and it is this long range force which is responsible for the accumulation of the point spectrum at the threshold. These experimental "spectra" thus clearly show the nature of at least part of the mathematical "spectrum". [The experiment is carried out by changing the energy of light incident on a sample of the atoms and varying the energy of the fundamental "photons" by varying the wavelength of the light. The wave length is indicated in wave number units ("cm^{-1}") which is the number of wavelengths per centimeter, an illegal but entrenched unit of energy. As the photon energy varies, transitions occur, and thus light energy is *absorbed*, when an exact energy balance is achieved between the photon energy and the difference between initial and final energies of the atom. The spectrum is thus displayed in terms of differences between eigenvalues, see Figure 6.] These spectra are redrawn from the work of Garton et al.

and we hope to understand dynamics in terms of the static spectrum of the atomic Hamiltonian, H^{Atom}.

2.2. The general form of the N-electron atomic Hamiltonian and an indication of the resulting energy levels. The form of the electronic Hamiltonian from an N-electron atom in the non-relativistic Born-Oppenheimer approxi-

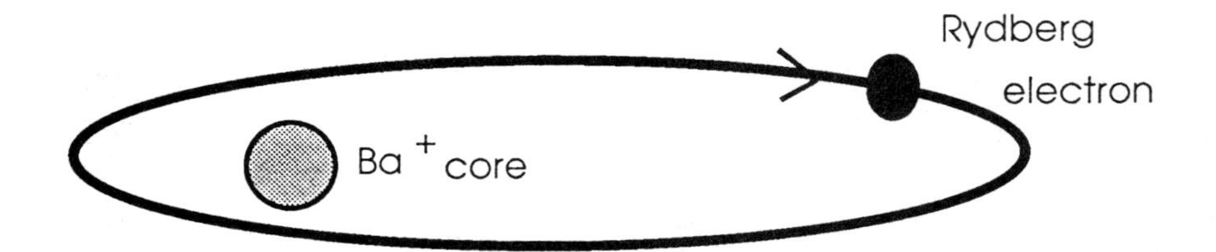

Figure 5) Rydberg states of highly excited atoms are those wherein an "outer" electron moves at large distances from the inner core of the atom. For an N electron system of overall neutrality the atomic nucleus has a charge of $+N$, and the inner $(N-1)$ electrons shield the nuclear charge at large distances. The Outer or Rydberg electron thus senses an asymptotic residual charge of $+1$, and thus a long range potential of the form $-1/R$ ensues. Thus solution of the purely Coulomb problem for atomic hydrogen has far reaching consequences, as other highly excited systems show similar features, such as accumulation of the point spectrum and the possibility of subsequent excitation to a continuum.

mation [14] follows from the physical picture of Figure 7, wherein the instantaneous positions of the electrons are shown relative to the nucleus, which is assumed to have charge Z. The potential energy of interaction of the ith electron with the nucleus and with the jth electron are both given by Coulomb's law:

$$-\frac{Z}{R_i} \equiv \frac{-Z}{|\vec{r}_i|} \tag{12}$$

and

$$\frac{1}{r_{ij}} \equiv \frac{1}{|\vec{r}_i - \vec{r}_j|} \tag{13}$$

the total electronic potential energy (PE) is thus

$$-\sum_{i=1}^{N} \frac{Z}{R_i} + \sum_{i<j} \frac{1}{r_{ij}} \tag{14}$$

where the fact that the sum over i, j is restricted to $i < j$ implies that each electron-electron interaction is counted once, and the $i = j$, or self energy of the electron, is omitted. As each electron also has kinetic energy (KE), the full Hamiltonian is

$$H^{\text{Atom}} = \sum_{i=1}^{N} -\frac{\nabla_i^2}{2} - \sum_{i=1}^{N} \frac{Z}{R_i} + \sum_{i<j} \frac{1}{r_{ij}}. \tag{15}$$

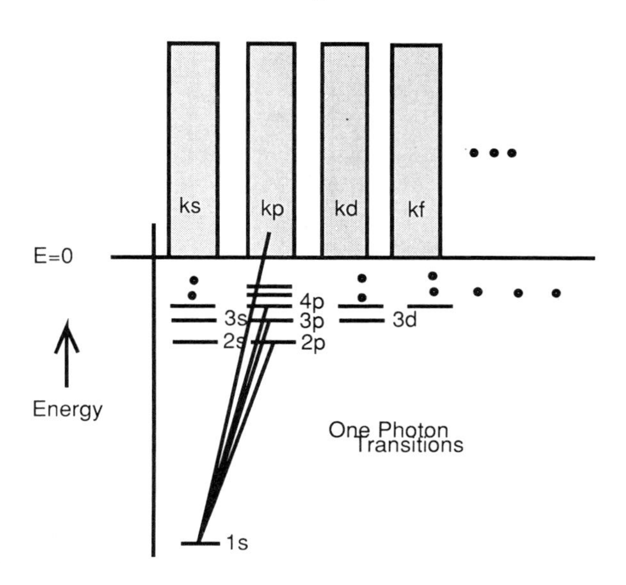

Figure 6) Grotrian diagram indicating the types of transitions observed in Figure 4). Starting from an "s" state, a single-photon transition, whose energy corresponds to the difference in energy between the initial and final states of the atom, carries the electron to successively higher energy "p" type bound states (labelled $2p, 3p, 4p, \ldots$ here). At higher energy the continuum states (labelled "kp") with asymptotic kinetic energy $k^2/2$ are accessed. These transitions correspond to those of Figure 4, except that there the transitions shown have much higher quantum numbers, $29p$ or $32p$, rather that the $2p, 3p \ldots$ shown here. The qualitative relationship is, I hope, clear.

What is the spectrum of this complicated object? Chemists and physicists have had the basic picture drilled into them since birth that most qualitative aspects of atomic structure follow from an *independent particle interpretation* of the structure of the atom, wherein each electron may be thought of as being in its own separate eigenstate of an effective "one-body" Hamiltonian. This might be the self consistent or Hartree-Fock Hamiltonian, discussed in the lectures by Almlöf in this workshop [15], or perhaps a more empirical model. In either case the independent particle model is an approximate, and usually reliable, guide to the general features to be expected in the mathematical description of the spectrum of the operator H^{Atom}. As is essential to the developments in the body of the paper to have a good qualitative "feel" for the types of features which will show up in the spectrum of H^{Atom} we give an elementary analysis of the expected energy levels of the helium (*i.e.*, two electrons interacting with a nucleus of charge $Z = 2$). To avoid inessential complexities, we first describe the expected features of what is called the s-wave radial limit model of the atom and its energy levels. This is quickly followed by a more complete

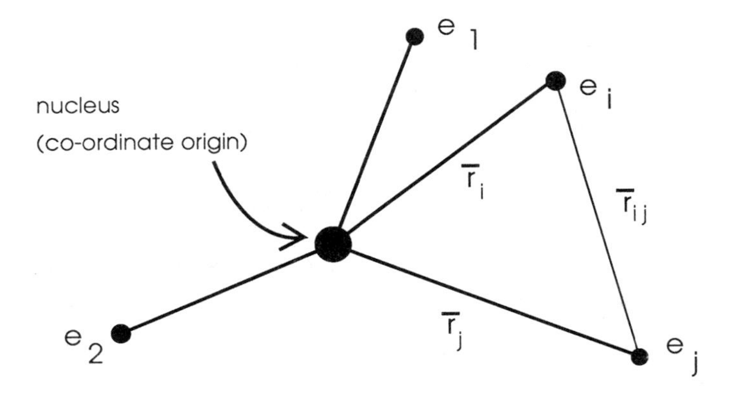

Figure 7) Co-ordinates of N-electron atom, indicating the distances of importance in the N-electron Hamiltonian.

description in Section 2.4.

2.3. Helium in the s-limit. The unperturbed spectrum. The idea of the s-limit is that each of the two electrons in the model helium atom will be assumed to have zero angular momentum. Thus each will be in a bound state described by the quantum numbers $n = 1, 2, 3 \ldots, l = 0$, $m_l = 0$. States with $l = 0$ are (again for historical reasons) called "s" states and the one-electron bound states are thus labeled $1s, 2s, 3s, \ldots, ns \ldots$. The corresponding continuum states, with momentum "k"(where usually $k = p/\hbar$ but $\hbar = 1$ in our units) and asymptotic kinetic energy $k^2/2$, are labeled ks, a sobriquet indicating both the energy of the continuum state, and the fact that the angular momentum is still $l = 0$. In this first attempt to display the spectrum of a two–electron atom, even spin will be neglected.

Taking the Hamiltonian for helium as

$$H^{\text{Helium}} = -\frac{\nabla_1^2}{2} - \frac{\nabla_2^2}{2} - \frac{2}{R_1} - \frac{2}{R_2} + \frac{1}{r_{12}} \tag{16}$$

a zero order approximation may be formed by ignoring the electron-electron interaction term

$$\frac{1}{r_{12}} \tag{17}$$

giving:

$$H^{\text{Helium}}_{\text{Approx}} = \sum_{i=1}^{2} \left(-\frac{\nabla_i^2}{2} - \frac{2}{R_i} \right). \tag{18}$$

(We will return to this type of zero order model in Section 4). This simplified Hamiltonian is separable: the wave function is an appropriately antisymmetrized (with respect to space-spin co-ordinates [16,17]) product function of eigenfunction of the individual "hydrogen-like" one-electron Hamiltonians

$$-\frac{\nabla_i^2}{2} - \frac{2}{R_i} \tag{19}$$

which are just the $R_{n,\ell}(r_i)Y_{\ell m_\ell}(\theta_i, \phi_i)$ of Section 2.1, and the bound eigenvalues are given by the sum,

$$E_{n_1\ell_1 m_{\ell_1}} + E_{n_2\ell_2 m_{\ell_2}} = -\frac{4}{2n_1^2} - \frac{4}{2n_2^2}, \qquad n_1, n_2 = 1, 2, \ldots \infty, \tag{20}$$

of hydrogenic eigenvalues. Each energy state is thus generally labeled by six quantum numbers, three for each electron. However, in our s-limit model we have only $l = 0$, $m_l = 0$, while $n = 1, 2, 3 \ldots$ Continuum states may occur with one, or both, electrons having positive energy. A Grotrian diagram for helium, within this simple model is shown in Figure 8, where a rudimentary labeling of the states is given. Figure 9 displays the mathematical spectrum corresponding to the levels shown in the Grotrian diagram. It is observed that there are bound states (point eigenvalues), continua, with energies of the form $-\frac{4}{2n_1^2} + \frac{k_2^2}{2}$, beginning at energies corresponding to the allowed energies of a single residual electron, point eigenvalues (for example, $2s^2, 2s3s, 3s^2 \ldots$) degenerate with one or more continua, and finally a "two-body" continuum (*i.e.*, both electrons are free) beginning at total energy 0. Introduction of the electron-electron interaction will of course modify the spectrum: The bound state energies must be obtained by elaborate perturbative or variational procedures (such as configuration interaction, as discussed in the Workshop), and the eigenvalues shift greatly, the doubly excited states degenerate with continua become "quasi-bound" or "resonant" states, with finite rather than infinite lifetime, yet...,yet...

Yet the general structure of the spectrum remains intact. The point spectrum still accumulates at $E = -2$, the thresholds for the various pieces of the continuous spectra still begin at $E = -2/n^2$ (as these are the exact energies of the one-electron "ion" (denoted He^+), and the two-electron continuum still begins at $E = 0$. Thus the description of the helium atom as two independent and non-interacting electrons gives a qualitative feeling for the spectrum of the full Hamiltonian. Such is also true for many electron atoms, and for the electronic structure of molecules. It must be noted that when chemists say that they *know* these things to be true, they are simply summarizing 100 years of spectroscopic and chemical experience: were a simple "independent-electron" picture not qualitatively correct life would perhaps not be so simple, and chemists and physicists would be thrown into the same flaming cauldron as mathematicians when asked to give a qualitative description of the "spectrum" of an arbitrary partial differential operator in $3N$ dimensions!

2.4 Helium P-states in the sp-limit: The unperturbed spectrum and experiment. Raising the level of complexity (and relation to reality) a bit, we now allow one of the two electrons in helium to have one unit of angular momentum (*i.e.*, $\ell = 1$, and the electron is said to be in a p-state). Assuming that the second

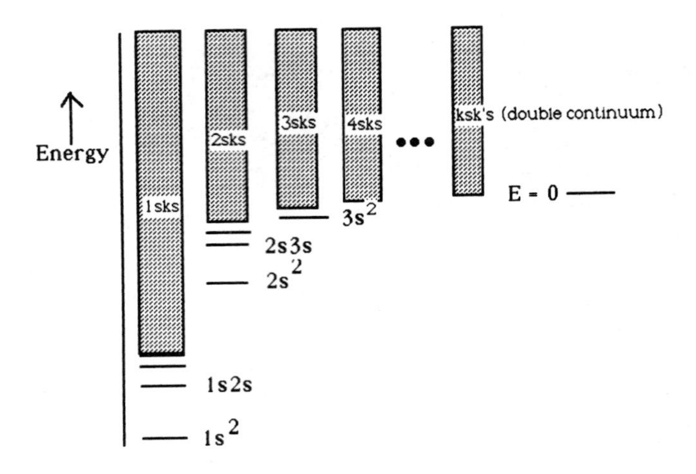

Figure 8) Grotrian diagram for the energy levels of the simple model of a two-electron atom ("helium") of eqs.(18) and (20). The spectrum is the direct product of the spectrum shown, the further simplification that both electrons have $\ell = 0$ has been made.

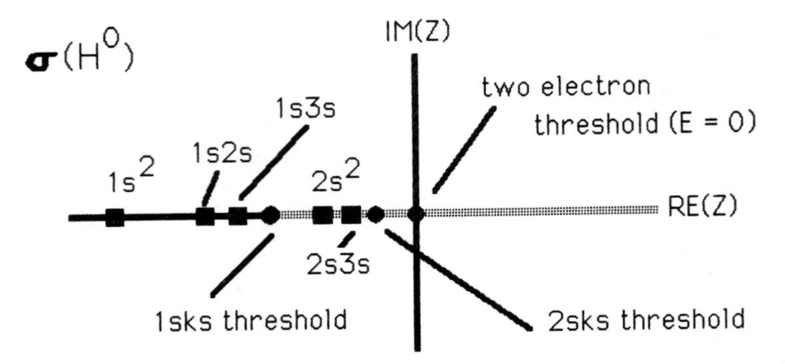

Figure 9) The mathematical spectrum for the s-limit model of helium. This spectrum corresponds to the Grotrian diagram of Figure 8. The actual (experimentally deduced) spectrum contains all of the features indicated in this simple model spectrum: bound states, continua starting at many different thresholds, and even the double excited states such as $2s^2$ as observed, these latter as resonances, of the type shown in Figure 11.

electron still, as in the s-limit example, has no angular momentum (*i.e.*, $\ell = 0$, and the electron is in an s-state), the overall angular momentum of the resulting two-electron system is one ($L = 1$), and the set of such states are referred to as P

(capital to denote total angular momentum, as opposed to the lower case letters used to denote the angular momenta of individual electrons). The bound energies are now (in the independent particle model)

$$-\frac{4}{2n_1^2} - \frac{4}{2n_2^2}, \qquad n_1 = 1, 2, 3, \ldots, \qquad n_2 = 2, 3, 4, \ldots, \tag{21}$$

and the levels are shown in the Grotrian diagram of Figure 10. Consideration of such P states of our model helium atom is relevant to experiment, as the "selection" rule governing absorption of photons (light quanta) requires that angular momenta balance. Thus the helium ground state (an S-state, with zero quanta of angular momenta) absorbing a photon of appropriate wave length, results in a P-state, (one quantum of angular momentum) as the single unit of angular momentum carried by the photon must appear in the atomic system, following destruction of the photon in the excitation process. It is these "allowed" transitions which are indicated by the dashed lines of Figure 10. The experimental [18,19] photoabsorption spectrum of helium reveals features as predicted by the Grotrian diagram of Figure 10. For example the $2s2p$, $2s3p\ldots$ series of doubly excited states give rise to resonances as shown in Figure 11.

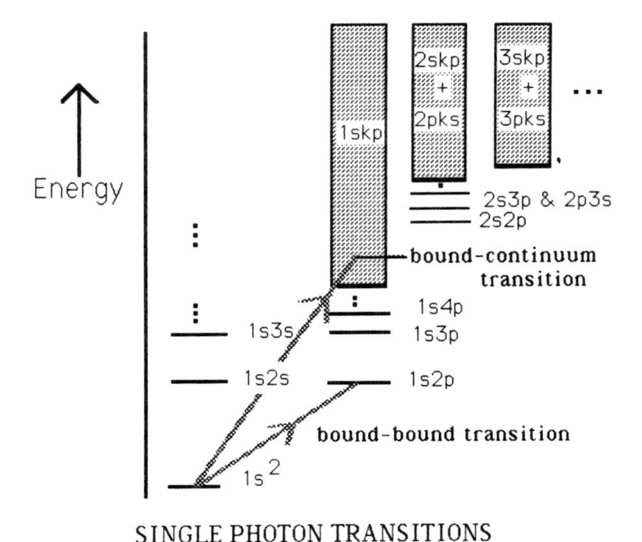

Figure 10) Grotrian showing photon induced transitions between various states of the Helium atom. The types of transitions observed correspond to those predicted, even from the simple "independent particle" picture used here to motivate the form of the actual spectrum of a two electron atom.

Remark: Helium as a fully interacting two-electron system. Figure 12A gives an instantaneous (classical) snapshot of the positions of the two electrons

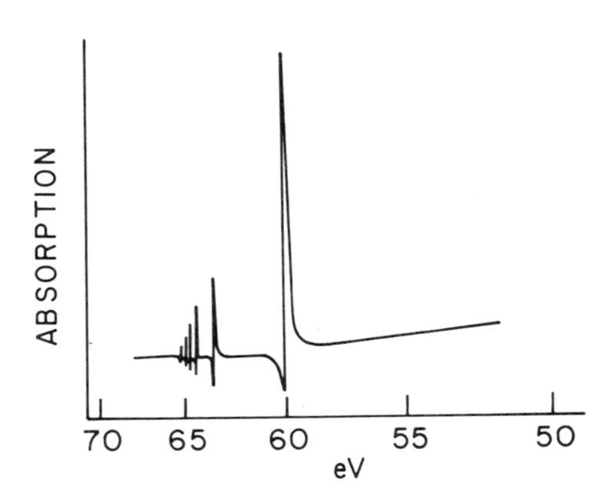

Figure 11) Photoabsorption spectrum showing the series of two-electron states $2s2p, 2s3p, 2s4p, \ldots$ embedded in the $1skp$ continuum. The intent here, as it was in Figure 4, is to indicate the correspondence between the mathematical spectrum of an atomic system, and actual experimental spectra. The figure follows the work of Madden and Coddling and is redrawn from Ref.[19].

in helium. Assuming that the nucleus is fixed (Born-Oppenheimer again) and is at the co-ordinate origin, and also assuming that their are no external fields acting on the system, an overall rotation, taking us from the spatial arrangement of Figure 12A to that of Figure 12B, will not change the energy of the system. Such overall *spherical symmetry* results in conservation of the overall angular momentum of the composite two-electron system. It is thus correct to refer to Helium as being in an S, P, or D state (corresponding to total angular momentum, L, being 0, 1 or 2 units (of $\hbar$) respectively). However, the instantanious potential energy seen by a single electron (say electron no. 1 in Figure 12A) is due to both the presence of the nucleus and the other electron, and such a potential energy, while *cylindrically symmetric* (about the line connection the two charge particles generating the potential); it is not at all spherically symmetric, and thus the angular momentum of a single electron in Helium (or in any other multi-electron atom) is not conserved. Thus while the simple ideas of the s-limit and sp-limit spectra of helium discussed above give a reasonable zero order picture (and they do!), they are not completely correct. Rather, angular momentum is constantly exchanges between the electrons in their correlated motion, and the "configuration-interaction" wave function needed to fully and accurately describe even an S-state of the atom must contain configurations where the individual electrons have arbitrarily large angular momenta, but coupled to give zero resultant for the *total* electronic angular momentum.

Remark: Does the independent-electron picture ever fail? Independent-particle models usually give a qualitative description of the states of many-particle atomic (and molecular) systems. It may be argued that this reflects the dominant

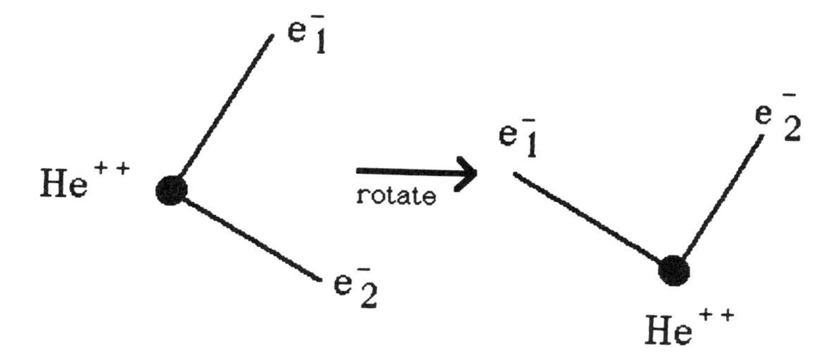

Figure 12) The energy of a two-electron system is invariant to an overall rotation about the co-ordinate center (*i.e.* the atomic nucleus). Thus overall angular momentum is conserved, giving an exact labelling of the states in terms of eigenvalues of the total angular momentum and its projection on a fixed axis. Note, however, that an individual electron does not exist in a spherically symmetric environment, and thus the angular momentum of individual electrons is not conserved in the many-body system.

role of the nuclear attraction. Thus, as the nuclear attraction decreases, and we consider the H^-, or hydrogen negative ion, which consists of two electrons and a nucleus with charge $Z = 1$, rather than $Z = 2$ as in helium, we might expect that correlated motion of the electrons might be of greater importance: Figure 13 shows the actual spectrum of H^-. Indeed things have changed! Rather than an infinite number of bound states, there is only one [20,21]! Further, the states of zero order description $1sns$, $1snp$ ... etc. do not even appear as resonances (see Ref.[22] for a discussion). Finally, the doubly excited resonances of this three–body system do not occur in the order (in energy) predicted by the hydrogenic independent electron model: for example the resonance $2s2p$ lies above $2s3p$ in energy! Rationalization requires use of alternative classifications, a quite important one being based on Lie group representations (see Refs. [23-25]).

Remark: Resonances. The concept of a resonance [26–28] has been mentioned several times. A resonant state has attributes of both bound (*i.e.*, localized in space) and free (*i.e.*, asymptotically free) states and is sometimes called quasi–bound. That is, it is a state which appears to be bound during a finite lifetime, but eventually decays away.

How do we describe such states mathematically? An ordinary bound state, corresponding to a member of the point spectrum of an atomic Hamiltonian, has an eigenvalue E_{bound} and $\mathcal{L}^2$ wave function Ψ_{bound} such that

$$H^{\text{Atom}}\Psi_{\text{Bound}} = E_{\text{Bound}}\Psi_{\text{Bound}}. \tag{22}$$

Time evolution of such a "stationary" state is given from the time–dependent

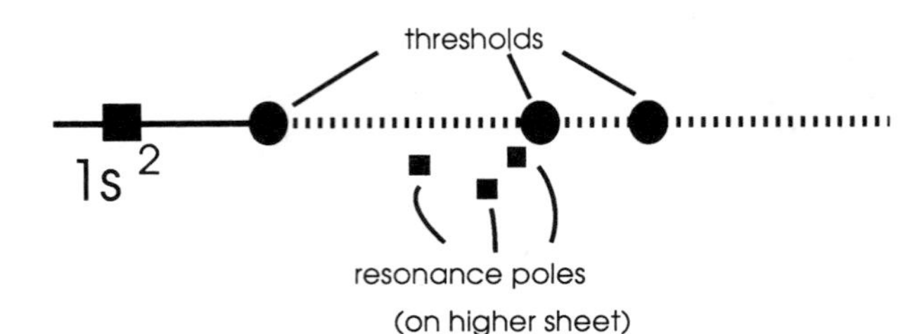

Figure 13) The spectrum of H^- (the hydrogen atom, with an extra electron!). In this case only one bound state of the composite three-body system exists, rather than the infinity of states for the Coulomb problem. There are, however, long series of resonances, giving a remnant of the simple behavior of the hydrogen atom.

Schrödinger equation, eq. (11), as

$$\Psi_{\text{Bound}}(t) = e^{-iE_{\text{Bound}}\,t/\hbar}\Psi_{\text{Bound}}(0). \tag{23}$$

As the point spectrum consists of only real numbers (H^{Atom} is self-adjoint) time evolution is only a phase, and the usual $\mathcal{L}^2$ norm is time independent:

$$\|\Psi_{\text{Bound}}(t)\| = \langle\Psi_{\text{Bound}}(t), \Psi_{\text{Bound}}(t)\rangle = \|\Psi_{\text{Bound}}(0)\|. \tag{24}$$

A simple (and almost, but not quite, correct) way to attempt to describe an initially localized, but decaying, state by assuming that $\Psi_{\text{resonance}}$ is an $\mathcal{L}^2$ eigenstate of H^{Atom} with a complex eigenvalue,

$$E = E^{\text{resonance}} - i\frac{\Gamma}{2} \tag{25}$$

with Γ real and positive.

Time evolution is then given as

$$\Psi_{\text{resonance}}(t) = e^{-i(E^{\text{resonance}} - i\frac{\Gamma}{2})t/\hbar}\Psi_{\text{Resonance}}(0) \tag{26}$$

and the norm

$$\|\Psi_{\text{resonance}}(t)\| = e^{-\frac{\Gamma t}{\hbar}}\|\Psi_{\text{resonance}}(0)\| \tag{27}$$

decays with a lifetime proportional to Γ^{-1}.

A complex eigenvalue can't be in the spectrum of the self-adjoint H^{Atom}, so this isn't a complete description, but it contains the germ of an idea to be developed in the following Sections. Is there a modified operator which does have $E^{\text{resonance}} - i\Gamma/2$ as an eigenvalue and a corresponding $\mathcal{L}^2$ eigenfunction? Yes, and it's easy to find!

3. Complex scale transformations: Spectral control.

3.1. What is the problem? In many ways the fact that the spectrum $\sigma(H^{\text{Atom}})$ of an atomic Hamiltonian is on the real axis is inconvenient. Examples:

1) In Figure 9 the unperturbed spectrum of helium, $\sigma(H^0)$, has doubly excited bound states embedded in electronic continua of the same symmetry, while, the perturbed spectrum, $\sigma(H)$, has only a continuous spectrum in the same region. What has happened to the degenerate point spectrum? The physical answer is that the embedded bound states of the unperturbed spectrum have turned into resonances, as shown in Figure 11, but then what is the signature of a resonance in the spectrum $\sigma(H)$? Titchmarsh has noted, for the case of tunneling resonances (e.g., the Stark effect), that resonances are identified as poles [26] of matrix elements of the resolvent $(z-H)^{-1}$, denoted as

$$R_\phi(z) \equiv \langle \phi, (z-H)^{-1}\phi \rangle \tag{28}$$

analytically continued (as a function of the complex variable z) onto an appropriate "non-physical" Riemann sheet.

2) Often the cross section for a photoabsorption or scattering process may be calculated from knowledge of matrix elements of the same resolvent evaluated for z in the spectrum of H. This seeming paradox, that the spectrum is defined as the singularities of $R_\phi(z)$, while it is at just those (singular) values of z that we need its value, is resolved by taking the limit from the upper half z–plane:

$$\lim_{\epsilon \to 0+} R_\phi(E + i\epsilon). \tag{29}$$

Thus for location of resonances and for calculation of continuum processes, a knowledge of the analytic properties of matrix elements of $(z-H)^{-1}$ may be very useful.

3) As yet another example, the Lippmann-Schwinger [27,28] equation of non-relativistic scattering theory, where in a free particle interacts with a center of force via a potential V, may be formally written as

$$\psi = \phi + \frac{1}{(z-H)} V\psi \tag{30}$$

where ψ is the desired scattering wave function, and ϕ its unperturbed counterpart. Thought of as an integral equation, the kernel $(z-H)^{-1}V$ is singular for physical (real) values of z. This, of course, complicates solution of this fundamental equation.

3.2. Analytic properties of $R_\phi(z)$. As it is the singularities of $R_\phi(z)$ which cause difficulty, what is the form of these singularities? A fundamental result of spectral theory (equivalent, in our notation, to Stone's Theorem [29]) is that an operator can be "resolved" in terms of its spectral projectors via a Stieltjes type integral over the spectrum, thus

$$\frac{1}{z-H} = \int_{-\infty}^{+\infty} d\sigma(E) |E\rangle \frac{1}{z-E} \langle E|. \tag{31}$$

In this notation $\langle\phi|\psi\rangle$ is a scalar product, and $|E\rangle\langle E|$ is a spectral projector. Physicists write the equivalent expansion in terms the projectors constructed from the bound and continuum eigenfunctions, and they use ordinary Riemann integral notation. Thus in physicists notation

$$\frac{1}{z-H} = \sum_{\substack{i \\ \text{Bound} \\ \text{States}}} \frac{\Psi^{+i}_{\text{Bound}}\Psi^{i}_{\text{Bound}}}{z-E_j} + \sum_{\substack{\text{channels} \\ j}} \int_{E_j^{\text{Threshold}}}^{\infty} dE_j \frac{\Psi_j^{+}(E_j)\Psi_j(E_j)}{z-E_j} \tag{32}$$

In eq. (32) the first sum is over all bound states (the point spectrum), and each integral in the second term runs over one segment of the continuum (or "channel" in the language of scattering theory), from its threshold or onset, to infinity. The sum over channels (labeled "j") acknowledges the fact that the continuous spectrum is not a monolith, but corresponds to a superposition many different continua as illustrated in the Grotrian diagrams of Figure 3.

Matrix elements of $(z-H)^{-1}$ may be similarly resolved: allowing immediate understanding of the analytic properties. Terms such as

$$\frac{\langle\phi_j|E_i\rangle\langle E_i|\phi\rangle}{z-E_i} \tag{33}$$

have simple poles for $z = E_i$, with residue $\|\langle\phi, E\rangle\|$. Contributions from each segment of the continuum,

$$\int_{E_k^{\text{Threshold}}}^{\infty} dE_k \frac{\langle\phi|E_k\rangle\langle E_k|\phi\rangle}{z-E_k}, \tag{34}$$

are analytic in the finite complex plane, with the exception of the real axis in the interval $(E_k^{\text{threshold}}, \infty)$. The point $E_k^{\text{threshold}}$ is a (usually square root) branch point, and the interval running from this branch point out to infinity along the real axis is a branch cut. This is easily visualized via an elementary result of complex analysis due to Cauchy. Given a real continuous function $f(x)$, defined for real x, an integral of the form

$$I(z) = \int_a^b \frac{f(x)}{z-x} dx \tag{35}$$

has a cut running from a to b on the real axis, and approaching this cut from the upper and lower plane gives different results: for $x_0 \in (a, b)$,

$$I(x_0 \pm i\epsilon) \equiv \lim_{z\to x_0 \pm i\epsilon} \int_z^b \frac{f(x)}{z-x} dx = \mathcal{P}\int_a^b \frac{x(x)}{x_0 - x} dx \mp i\pi f(x_0), \tag{36}$$

$\mathcal{P}$ denoting the Cauchy principal value of the integral, with the limit $\epsilon \to 0$ understood in the notation of [30]. The fact that the real axis in the interval (a, b) is a branch cut is clearly evidenced by the multivaluedness of the integral. We also note that the discontinuity across the cut is pure imaginary and has the value

$$2\pi i f(x_0). \tag{37}$$

Based on this preliminary discussion, the analytic properties of $R_\phi(z)$ for a typical atomic Hamiltonian may be simply described. $R_\phi(z)$ is analytic except for poles corresponding to the point spectrum and branch points corresponding to each continuum channel threshold. Branch cuts, which begin at each threshold, run out along the real axis to $+\infty$. That is, as per the definition of the spectrum, $R_\phi(z)$ is non-analytic for z in $\sigma(H^{\text{Atom}})$. No surprise.

But wait: while poles and branch points are actual points of non-analyticity, and thus properties of the function, aren't branch cuts arbitrary, in the sense that they may be chosen for our convenience as long as they perform the job of forcing single valuedness in the cut plane? Why then are the cuts on the real axis? The answer is that the cuts are segments of the continuous spectrum of H^{Atom}, and thus on the real axis, as H^{Atom} is self-adjoint (or we at least assume it has a self adjoint extension, see Reed and Simon [2]). Are we forced to live with this? The cut in the integral of eq. (35) may be shifted at will as long as $f(x)$ is analytic in x (this is, to be sure, a rather stronger assumption than continuity of $f(x)$ on the real axis). Cauchy's theorem then allows distortion of the contour, without changing the value of the function $I(z)$, unless the new contour is pulled across our observation point: see Figure 14. Elementary arguments of analytic continuation give the immediate result that as the contour distorts, parts of the second sheet are exposed, and the corresponding parts of the first sheet are now hidden. Thus the analytic function $I(z)$ has an infinity of single valued integral representations, each differing in choice of contour, $\mathcal{C}$, and in which part of the two sheeted complex plane we choose to display. The conclusion: Branch cut placement is a property of the integral representation, not of the analytic function itself.

Why don't we have this freedom to move branch cuts when working with matrix elements of the resolvent $(z-H)^{-1}$? Of course, we do: suppose that we make the distortion

$$\int_{E_j^{\text{Threshold}}}^{\infty} dE \frac{\langle \phi, E_j\rangle\langle E, \phi_j\rangle}{Z - E_j} \to \int_{\mathcal{C}_j} d\tilde{z} \frac{\langle \phi, \tilde{z}\rangle\langle \tilde{z}, \phi\rangle}{z - \tilde{z}}. \tag{38}$$

The cut is now along the contour $\mathcal{C}_j$, see Figure 15 for several choices of contour distortion. This may be done as long as $\langle \phi, E\rangle$ has an analytic continuation to complex E. We can thus envisage a spectral resolution of the form

$$\frac{1}{z-H} = \sum_{\substack{i \\ \text{Bound} \\ \text{States}}} \frac{\Psi_i^+ \Psi_i}{z - E_i} + \sum_{\substack{\text{channels} \\ j}} \int_{\mathcal{C}_j} d\tilde{z}_j \frac{\langle \phi, \tilde{z}_j\rangle\langle \tilde{z}_j, \phi\rangle}{z - \tilde{z}_j} \tag{39}$$

which has the same poles and branch points as the representation of eq. (32) but has the cuts off at least part of the real axis.

An immediate question comes to mind: is there an operator H^{**} whose spectrum generates the representation of eq. (39) directly? Clearly H^{**} cannot be self-adjoint, as its spectrum will be complex. A formal answer is that H^{**} is defined by its resolution

$$H^{**} = \sum_{\substack{i \\ \text{Bound} \\ \text{States}}} \Psi_{\text{Bound}}^{+i} \Psi_{\text{Bound}}^{i} + \sum_{\substack{\text{channels} \\ j}} \int_{\mathcal{C}_j} dz_j |\tilde{z}_j\rangle \langle \tilde{z}_j| \tag{40}$$

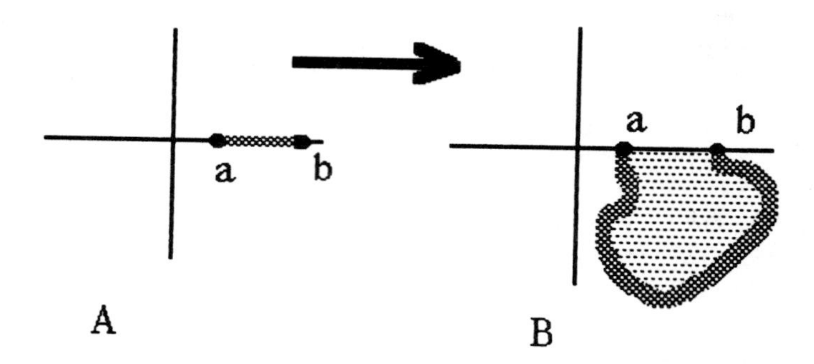

Figure 14) Cauchy contour distortion. The integration contour of the integral in Eqn 35, may be distorted via Cauchy's theorem, provided that $f(z)$ is analytic. If this is the case, the new integral representation "exposes" part of a higher sheet, as shown in the shaded area of part B of the figure. This amounts to choice of a different branch cut location.

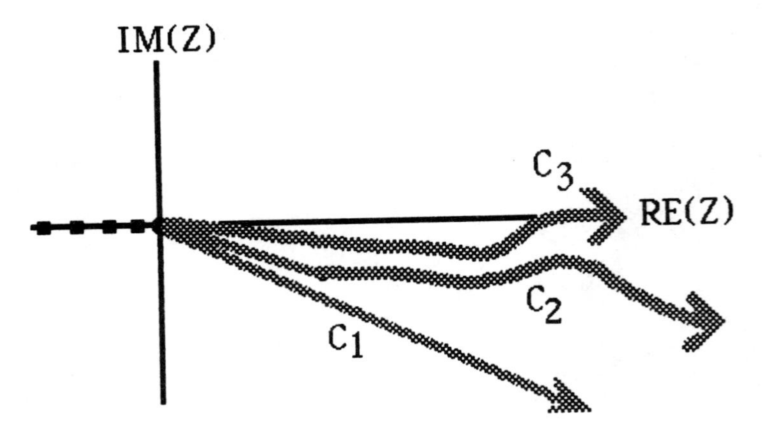

Figure 15) Various choices if integration contour, all running from a threshold (branch point) on the real axis, out towards plus infinity in the right half complex plane. Under proper assumptions regarding analyticity of the integrands, any of these contours could be used to produce an integral representation of a matrix element of $(z - H)^{-1}$, with each representation having a correspondingly different cut location

which, provided that the necessary analytically continued function exist, and are known, might even provide a useful working *ansatz*. However, if all matrix elements of the form $\langle \phi, \tilde{z}_j \rangle$ are known the Schrödinger problem is solved in any case, and we need not proceed further. As they are not known, determining the analyticity properties of the discontinuities $\langle\phi, \tilde{z}_j\rangle \langle\tilde{z}_j, \phi\rangle$ along the distorted contours, $\mathcal{C}_j$ is tricky business. Eqs. (39, 40) are thus apparently of only formal utility.

However, with some sacrifice of flexibility, an operator which immediately generates a more general spectral resolution is easily found. Such an operator may be generated directly via the method of complex scaling of interparticle co-ordinates.

3.3. The dilatation analytic many–body coulomb Hamiltonian. Without further ado we state the results of the rather amazing properties of the effect of complex scaling of the interparticle distances on the non-relativistic Coulomb Hamiltonian. These results have been proved, and discussed by Combes, Balslev, and Simon, and reviewed by Reinhardt, Junker, and Ho [3-8].

We begin using the language of spectral theory. If the interparticle distances in the atomic Hamiltonian are scaled as $r \to re^{i\phi}$, the new Hamiltonian, $H^{\text{Atom}}(\phi)$ becomes

$$(41) \qquad H^{\text{Atom}}(\phi) = e^{-2i\phi}(\text{Kinetic Energy}) + e^{-i\phi}(\text{Potential Energy})$$

due to the particular form of the Coulomb interaction. The effect of such a transformation on the spectrum of H^{Atom} is shown in Figure 16. In summary:

1) the point spectrum on the real axis is invariant to the transformation
2) thresholds (points at which continua begin) are invariant to the transformation
3) segments of the continuous spectrum associated with each threshold are rotated by and angle 2θ into the lower half plane (assuming positive θ)
4) as segments of continua rotate (as a function of increasing θ) complex members of the point spectrum may appear and disappear as the moving continua expose them, and others sweep over them
5) the point spectrum can only have accumulation points at thresholds.

It is to be noted at once that once the continuous spectrum is off the real axis, the $E + i\epsilon$ limits which complicate quantum scattering theory may be taken with impunity, as the resolvent is no longer singular on the real axis.

We also note that the new complex eigenvalues of $H^{\text{Atom}}(\theta)$ are to be associated with resonances. The normal oscillatory time evolution of a usual bound state eigenfunction of the time independent Schrödinger equation is given by $e^{-Et/\hbar}$ where E is real as H is self-adjoint. The analogous time evolution given a complex eigenvalue (of $H^{\text{Atom}}(\theta)$, not H^{Atom}!) is still $e^{-i\tilde{E}\,t/\hbar}$, where now

$$(42) \qquad H^{\text{Atom}}(\theta)\Psi(\theta) = \tilde{E}\Psi(\theta).$$

Separating E into its real and imaginary parts, $\tilde{E}^{\text{res}} = E \quad - i\,\Gamma/2$ gives a quantum probability which decays exponentially as $e^{-\Gamma t/2}$ as discussed at the end of Section 2. We thus wish to associate the a point eigenvalue of $H^{\text{Atom}}(\theta)$ with a negative imaginary part with a decaying or resonant state.

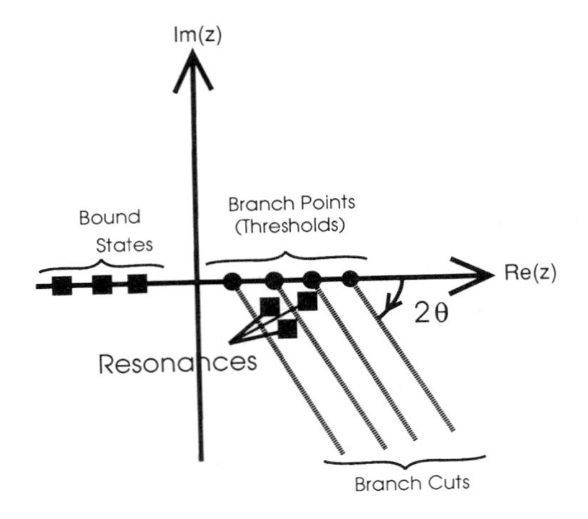

Figure 16) The spectrum of the "rotated" atomic Hamiltonian $H(\theta)$. The point spectrum and threshold are invariant to the complex rotation of co-ordinates. The continua rotate about their respective thresholds, exposing higher sheets where "resonance" eigenvalues of $H(\theta)$ may lie.

Let us now rephrase these results in language appropriate to consideration of a Hilbert space matrix element of the resolvent. What happens to $R_\phi(z)$ as $H^{\mathrm{Atom}} \to H^{\mathrm{Atom}}(\theta)$? The results are easily stated for ϕ belonging to a dense set in the Hilbert space:

1) Poles and branch points of $R_\phi(z)$, as befits singularities of an analytic function, are invariant.
2) Branch cuts of $R_\phi(z)$ rotate about their thresholds, to an angle 2θ in the lower half plane (again θ is positive).
3) Complex poles of $R_\phi(z)$ may be exposed as a cut passes over, and subsequently hidden by passage of another cut.
4) Poles can only accumulate at thresholds.
5) For values of $\theta \leq \pi$, $R_\phi(z)$ represents the same analytic function, different values of θ simply giving views of different portions of the function on its many Riemann sheets.

Thus, for example, the complex poles of $R_\phi(z)$ exposed for nonzero θ are the same as those which could be obtained by appropriate analytic continuation of $R_\phi(z)$ from the upper half plane. However, the latter poles are not obtained by direct solution of an eigenvalue problem. In a dilatation transformation the analytic continuation has been performed on the operator, rather than on the spectral resolution of its resolvent.

As will be seen in subsequent sections, this alteration of the spectrum of H^{Atom} allows both theoretical and computational advances.

Remark. Why is the point spectrum invariant? Point eigenvalues correspond to $\mathcal{L}^2$ eigenfunctions of H. Invariance of the point spectrum is easy to grasp for the two body problem: Consider the hydrogen atom radial Hamiltonian:

$$H_r = -\frac{1}{2}\frac{1}{r^2}\frac{d}{dr}r^2\frac{d}{dr} - \frac{1}{r}. \tag{43}$$

The normalized ground, or $1s$, eigenfunction of this operator is $2e^{-r}$ with eigenvalue $-\frac{1}{2}$ (in atomic units). Suppose we scale the Hamiltonian. As $r \to re^{i\theta}$, $H_r \to H_r(\theta)$:

$$H_r(\theta) = -e^{-2i\theta}\frac{1}{r^2}\frac{d}{dr}r^2\frac{d}{dr} - \frac{e^{-i\theta}}{r}. \tag{44}$$

However, $2e^{-r}$ remains a formal eigenfunction of the complex scaled $H_r(\theta)$ if we make the same variable change:

$$e^{-r} \to e^{-re^{i\theta}} \tag{45}$$

and thus *the eigenvalue is unchanged*. Note that $e^{-re^{i\theta}}$ is $\mathcal{L}^2$ as long as $\theta < \pi$. and thus is a perfectly acceptable bound state eigenfunction. One can wonder why we ever thought that the co-ordinates in a Schrödinger operator were real! Extending this result to the case of an arbitrary number of particles takes some effort [2-5], although the idea is the same.

Remark. Why does the continuum rotate, and why by 2θ? The boundary conditions on the "eigenfunctions" corresponding to the continuous spectrum are different from those for bound states, and it is this difference which prevails. Continuum eigenfunctions are not $\mathcal{L}^2$, but rather only bounded at infinity. For example, the regular eigenfunctions of the radial kinetic energy $-\frac{1}{2}\frac{1}{r^2}\frac{d}{dr}r^2\frac{d}{dr}$ behave asymptotically as $\sin(kr)$ with eigenvalue $k^2/2$. If we scale the kinetic energy operator

$$KE \to e^{-2i\theta}KE \tag{46}$$

this formal solution is still asymptotically $\sin(kre^{i\theta})$ but, ... and its a crucial BUT, ... these are only bounded at infinity *iff* $k \to ke^{-i\theta}$. The continuum eigenvalue allowing such a bounded solution thus transforms as $k^2/2 \to e^{-2i\theta}k^2/2$. As all continua essentially come from the fact that at large separation the Hamiltonian becomes simply a relative kinetic energy (the potentials die off), this treatment of the kinetic energy eigenvalues suggests the general result, that the continua are rotated by 2θ into the lower half plane. Proving all of this is rather more complex. The Coulomb potential is not relatively compact with respect to the kinetic energy, so the simple argument here is suggestive but not conclusive; further extension to the N-body Coulomb case requires delicate discussion or range/domain issues, so that for example, not all N-body Schrödinger operators with analytic interparticle potentials fall in the class of dilatation analytic operators. We are fortunate that the Coulomb Hamiltonian does fit into this forgiving class [3-5].

4. Examples: Exploitation of dilatation analyticity. We will give three examples.

4.1 Radius of convergence of the $1/Z$ perturbation expansion. The sequence of two-electron systems, with nuclear charges $Z = 1$, $Z = 2$, $Z = 3$ ... etc. up to 10, and even higher (or in chemical parlance, H^-, He, Li^+, ... Ne^{+8} ...) is of great importance in plasma physics and astrophysics, as the two-electron $1s^2$ configuration is of special "closed shell" stability, and thus a common end state when conditions lead to highly stripped ions. This iso-electronic series is also of theoretical interest, as the energy level structure and dynamical properties of two-electron systems can be studied experimentally as a function of the integer parameter Z. Empirical and computational correlations can be studied, the latter not restricted to integral Z.

A particularly attractive computational approach which attempts to unify the treatment of the whole isoelectronic series (at least in the non-relativistic approximation) is the $1/Z$ perturbation expansion [31,32]. Not only is the whole isoelectronic sequence treated at once, but the unperturbed energies and wave functions are those of the completely uncoupled Hamiltonian of eq. (18), as used to motivate the structure of the spectra of two-electron atoms in Section II. The usual Hamiltonian for the two-electron atom of nuclear charge Z is, of course,

$$H(\vec{r}_1, \vec{r}_2) = -\frac{1}{2}\nabla^2_{r_1} - \frac{1}{2}\nabla^2_{r_2} - \frac{Z}{|\vec{r}_1|} - \frac{Z}{|\vec{z}_2|} + \frac{1}{|\vec{r}_1 - \vec{r}_2|}. \tag{47}$$

Making the (this time real) dilation, let $r_i \to Zr_i \equiv x_i$., where Z is the nuclear charge, rather than an arbitrary scaling parameter. In terms of the scaled, x, variables

$$\begin{aligned} H(\vec{r}_1, \vec{r}_2) &= Z^2\left(-\frac{1}{2}\nabla^2_{x_1} - \frac{1}{2}\nabla^2_{x_2} - \frac{1}{|\vec{x}_1|} - \frac{1}{|\vec{x}_2|} + \frac{1}{Z}\frac{1}{|\vec{x}_1 - \vec{x}_1|}\right) \\ &\equiv Z^2 H_Z(\vec{x}_1, \vec{x}_2) \end{aligned} \tag{48}$$

where H_Z is the x-space operator of eq. (48). The so called "1/Z" expansion results if we now write

$$H_Z^0 = -\frac{1}{2}\nabla^2_{x_1} - \frac{1}{2}\nabla^2_{x_2} - \frac{1}{|\vec{x}_1|} - \frac{1}{|\vec{x}_2|} \tag{49}$$

and treat $|\vec{x}_1 - \vec{x}_2|^{-1}Z^{-1}$ as a perturbation. The usual Rayleigh-Schrödinger series for the ground state energy is thus written

$$E_Z = E^0 + \frac{1}{Z}\epsilon^{(1)} + \frac{1}{Z^2}\epsilon^{(2)} + \frac{1}{Z^3}\epsilon^{(3)} + \ldots. \tag{50}$$

Once the perturbation coefficients $\epsilon^{(i)}$ are known, estimates of the energy for any value of Z, and thus for the whole isoelectronic sequence, may be made via the rescaling of the energy $E = Z^2 E_Z$. As the perturbation coefficients $\epsilon^{(i)}$ are Z-independent, they need only be calculated once and, as $(1/Z)^n$ decreases rapidly for large Z, only a few coefficients should be needed for accurate treatment of large Z ions. One can also ask, how small can Z get before the series diverges? We first note that as Z decreases from $1+\alpha$ to 1 the number of bound states goes from infinity to one (again, see Hill [20,21] for a proof that H^- has only one bound state of ^1S symmetry) which is a result of the fact that for $Z > 1$ there will always be a residual attractive Coulomb potential of the form

$$-\frac{\alpha}{r_2}, \tag{51}$$

that is, a "far away" electron "2" sees a nuclear charge of $1+\alpha$, which is asymptotically screened by the unit charge of the "inner" electron. An attractive Coulomb potential with charge of any finite magnitude always has an infinite number of bound eigenvalues, which accumulate at the threshold.

The spectrum of the atomic Hamiltonian is thus severely disrupted in the limit $Z = 1$. The $1/Z$ expansion for excited states may well cease to contain useful information for $Z < 1$. However, for $Z = 1$ the H^- ground state is still a point eigenvalue (see Figure 13), well below the onset of the continuum, and the usual ideas of Kato-Rellich [32–34] perturbation theory suggest that E_Z should be analytic in Z near $Z = 1$. It is then at least possible that the $1/Z$ expansion for the ground state converges for $Z = 1$. Whether or not this is so will depend on the rate of decrease of the $\epsilon^{(i)}$ themselves.

Stillinger and coworkers [35–37], in a series of interesting and provocative papers, carried out an analysis of the convergence properties of the $1/Z$ expansion using the approximately twenty $\epsilon^{(i)}$ which had been numerically determined by Midtdal [38] in 1969. The series is, in fact, convergent for $Z = 1$, and thus gives useful results for the whole isoelectronic series, beginning at $Z = 1$. However, Stillinger asked a more interesting question: as Z decreases to values below 1, the coupling parameter $(= 1/Z)$ is larger than one, and at some point the series cannot converge. What determines the actual radius of convergence? What is wanted is the nearest singularity to $\lambda = 0$ in the complex $\lambda \equiv \frac{1}{Z}$ plane. Using the available $\epsilon^{(i)}$, Stillinger proposed the ansatz

$$E(\lambda) = (\lambda - \lambda^{**})^{\eta} \tag{52}$$

for the form of this singularity, and found a singularity of $E(\lambda)$ at a real value of $\lambda = \lambda^{**}$. The startling observation made by Stillinger was that this value of λ is larger than the value of $\lambda \equiv \lambda^{\text{crit}}$ leading to $E(\lambda)$ becoming degenerate with the threshold for the continuous spectrum. Stillinger suggested that in the interval $(\lambda^{\text{crit}}, \lambda^{**})$ that there might well be a "bound state in the continuum," that is a member of the point spectrum embedded in the continuous spectrum of the same symmetry, as illustrated in Figure 17. As effective non-integral values of Z sometimes arise in transport of electrons in condensed media, non-integral Z is possibly relevant to more than just mathematical speculation, and as the proposed bound state exists for a continuous range of coupling constant it does not result from special cancellation of effects, and thus might have previously unanticipated experimental consequence. Note at once that we cannot apply the usual Kato-Rellich ideas of analyticity of $E(\lambda)$ to investigate whether a singularity of the form of eq. (52) might or might not occur for real λ near λ^{**}. This is simply because Kato-Rellich Theory does not apply to an embedded eigenvalue. However, as indicated in Figure 18 the situation changes entirely [39] if we employ the method of complex scaling to the Hamiltonian H_Z. An isolated member of the point spectrum between the thresholds at E_1 and E_2 must now be an analytic function of the perturbation $\lambda = (1/Z)$, as the usual ideas of Kato-Rellich perturbation theory now do apply (Simon [5]). Thus $E(\lambda)$ cannot become non-analytic on the real axis, except at values of λ at which the point eigenvalue is degenerate with a threshold. Reinhardt [39] suggested that *if* the radius of convergence were determined by a singularity in the complex λ plane on the real axis, and near λ^{crit} that it must be at λ^{crit}, itself. (Note that dilation analyticity does not require that the radius of convergence be determined by a singularity on the real axis, it only places restrictions on the analytic properties of $E(\lambda)$!) This being the case, one would concluded that the $\epsilon^{(i)}$ available in 1969 were either not accurate enough, or not known to high enough order to allow a reliable estimate of the radius of convergence. One would also not search for bound states in the continua of many electron systems.

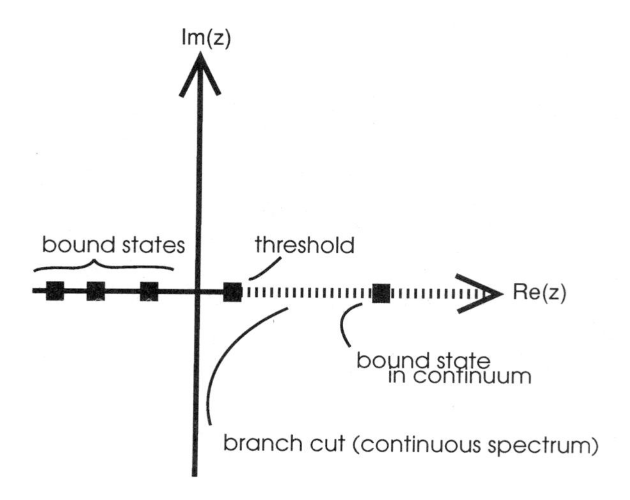

Figure 17) A bound state embedded in the continuum. It is not evident that such a non-isolated part of the point spectrum is analytic in a perturbation.

Stillinger, quite appropriately, commented that the matter was not quite settled. Examples are known (Stillinger [40], Simon [41]) where the radius of convergence of the Rayleigh-Schrödinger series is much larger than the range of existence of a member of the point spectrum for the corresponding operator. He thus suggested

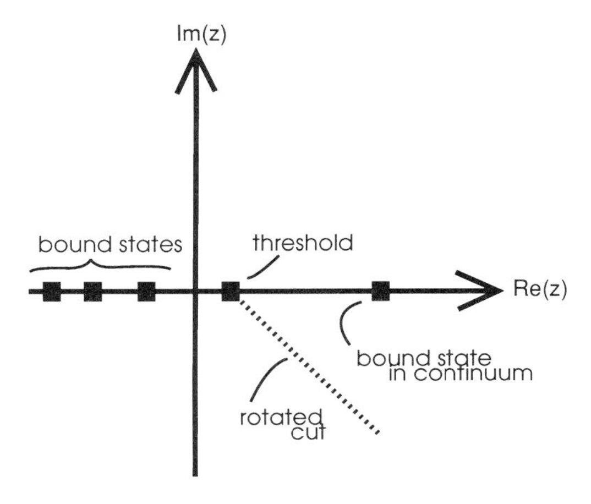

Figure 18) A bound state in the continuum for a dilation analytic system. In this case the complex scaling of the Hamiltonian moves the continuous spectrum away from the real axis, and the formerly embedded bound state in the continuum is now simply part of the point spectrum of $H(\theta)$, and if isolated must be analytic in small perturbations.

the necessity of a simultaneous re-analysis of the convergence properties of the $1/Z$ series for the ground state energy and a full analysis of the $1/Z$ expansion of the norm, $||\Psi_{1s^2}||$, of the corresponding wave function. Such an exhaustive study has now been completed by Baker, Freund, Hill and Morgan [42], who have examined terms through order 401 in the $1/Z$ expansion for both $E(\frac{1}{Z})$ and $||\Psi_{1s^2}||\,(\frac{1}{Z})$, and additionally carried out a detailed asymptotic analysis of the convergence properties of the series. Both of the expansions have the same radius of convergence, seemingly determined by a singularity on the real axis, and the radius of convergence is, to within numerical precision, that determined by the value of where E_{1s^2} becomes degenerate with the threshold. This is a pleasing result: the radius of convergence is determined by the physics.

Remark: Bound states in continua. It is important to note that isolated bound states in continua can occur. In two body potential scattering, potentials such as $\sin(r)/r$, which are not analytic at infinity, support an isolated member of the point spectrum embedded in a continuum. For analytic potentials, this is not possible (see Simon [43]). In many particle systems, accidental cancellation of matrix elements can give rise to resonances of arbitrary narrow width. It should also be noted that in discussing the possibility of atomic bound states in continua, we are referring to states of a single electronic symmetry, where the electron-electron interaction, $1/r_{ij}$, can induce transitions, not to the case of bound states, such as the infamous $2p^2\,{}^3P^e$ state of H^-, which is degenerate with continua of different parity, and thus not coupled to them within the framework of the type of atomic

Hamiltonian envisaged here. Inclusion of "weak" interactions in H^{Atom}, could give such states a finite lifetime.

4.2. Use of the numerical range and dilatation analyticity to bound atomic spectra. It was mentioned in the previous sub-section that in two body potential scattering with analytic potential, that bound states in the continuum do not exist (Simon [43]). A simple demonstration of this result follows for the two body Coulomb problem using the relation between the spectrum, and numerical range of the appropriate Schrödinger operator. The demonstration is of interest as it extends, almost immediately, to the N-electron case, and provides a useful first approach to the spectra of more complex operators (for example, Cerjan *et al.* [44]), Herbst [45]).

The numerical range, N, of a operator A (which is not necessarily self-adjoint) is defined as

$$N(A) = \{z : z = \langle \phi, A\phi \rangle \text{, for all } \phi \text{ in the appropriate Hilbert space}\} \tag{53}$$

The numerical range is of interest is, modulo technical details,

$$\sigma(A) \subseteq \overline{N(A)}. \tag{54}$$

That is, the spectrum is a subset of the closure of the numerical range. We now show that for the Coulomb Schrödinger operator

$$H = -\frac{1}{2}\nabla^2 - \frac{1}{|\vec{r}|} \tag{55}$$

the point spectrum is empty in the right half plane $\Re e\{z\} \geq 0$. That is, there are no bound states (which would be in the continuum) at positive energy. There are also no resonances in pure Coulomb scattering, as these would correspond to complex eigenvalues in the right half plane. The proof of these assertions follows from the fact that the Coulomb is dilatation analytic: consider

$$H(\theta) = -\frac{e^{-2i\theta}}{2}\nabla^2 - \frac{e^{-i\theta}}{|\vec{r}|} \tag{56}$$

and now choose $\theta = \pi/2$, whereupon the operator becomes

$$H\left(\frac{\pi}{2}\right) = -\left(-\frac{\nabla^2}{2} - \frac{i}{|\vec{r}|}\right). \tag{57}$$

The numerical range is constructed by considering expectation values of $H\left(\frac{\pi}{2}\right)$ with respect to a complete set of $\mathcal{L}^2$ functions:

$$-\langle \phi, -\frac{\nabla^2}{2}\phi \rangle + i\langle \phi, -\frac{1}{|\vec{r}|}\phi \rangle. \tag{58}$$

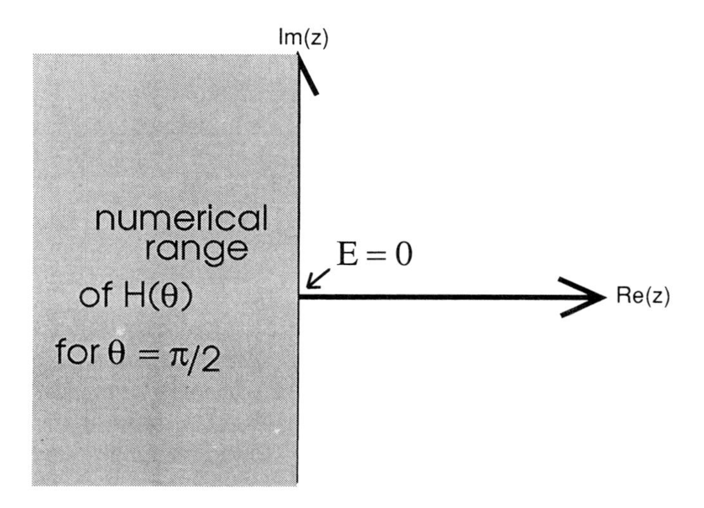

Figure 19) The numerical range for a dilatation analytic system with all continua rotated to run out the negative real axis. This simple estimate shows that atoms have no bound states, or resonances, above $E = 0$, this being the energy for complete breakup of the system.

However, as the kinetic energy $-\frac{\nabla^2}{2}$ is positive definite, $-\langle \phi, -\frac{\nabla^2}{2}\phi \rangle$ is always real and negative. Clearly $i\langle \phi, -\frac{1}{|\vec{r}_1|}\phi \rangle$ is pure imaginary. $N(H(\pi/2))$ is thus the left half plane $\mathcal{R}e\{z\} \leq 0$, as shown in Figure 19. The spectrum of $H(\pi/2)$ is thus empty in the corresponding right half plane. However, dilatation analyticity indicates that if H itself, *i.e.* $H(\theta = 0)$, had bound states or resonances for $E \geq 0$ they would be part of the point spectrum, of $H(\theta) = \pi/2$. Thus they cannot occur.

The same theorem holds [46] for the full N-electron atomic case, where $E = 0$ is defined as the energy of the fully stripped system, with all particles at rest at infinite separation. Atoms or ions thus cannot have bound states in the continuum, or even resonances (if these correspond to poles of the analytically continued resolvent) above $E = 0$. Reinhardt[6] gives a more complete discussion of this point.

4.3. Numerical calculation of positions and widths of atomic resonances. Determination of the locations in energy, E^{res}, and widths, $\Gamma/2$ (and corresponding lifetimes), of resonances is often of great interest in atomic physics, and it is of practical importance in plasma diagnostics and astrophysics. The fact that $H(\theta)$ has point eigenvalues corresponding to the resonances, with imaginary parts giving the width, or lifetime information, and with $\mathcal{L}^2$-eigenfunctions (as a function of the rotated co-ordinates) suggests direct variational calculation. Practicalities of this operation have received much discussion (see the reviews of Reinhardt, Junker, and Ho [6–8]), and the recent work of Moiseyev, Davidson, and Engdahl [47]. We thus only indicate the major issues, and then attempt to indicate that not as much as one might expect is known about variational principles.

Approximations to resonance eigenvalues maybe carried out by forming the matrix representation of $H(\theta)$ in an $\mathcal{L}^2$-basis of functions (just as in the method of configuration interaction) followed by diagonalization of the resulting complex symmetric matrix. More explicitly, for Coulomb Hamiltonians

$$H(\theta) = e^{-2i\theta}(\text{Kinetic Energy}) + e^{-i\theta}(\text{Potential Energy}). \tag{59}$$

Thus given an appropriate basis $\{\phi_i\}$ of $\mathcal{L}^2$ functions of the desired spatial and spin symmetry, and fully antisymmetrized, $H(\theta)$ has the matrix representation

$$\overline{H}(\theta) = e^{-2i\theta}\overline{KE} + e^{-i\theta}\overline{PE} \tag{60}$$

where $\overline{KE}$ and $\overline{PE}$ are the usual matrix representations of the kinetic and potential energy in the basis. Thus, for example

$$(\overline{KE})_{ij} = \langle \phi_i, \left(\sum_i -\frac{\nabla_i^2}{2}\right) \phi_j \rangle \tag{61}$$

with the usual Hermitian scalar product. The matrices $\overline{KE}$ and $\overline{PE}$ will (in practical situations) be real symmetric, and thus $\overline{H}(\theta)$ is complex symmetric. The left and right matrix eigenfunctions of share common complex eigenvalues, and are related by simple transposition, rather than by the usual Hermitian conjugation, which requires transpositions and taking the complex conjugate. The matrix eigenvectors are thus normed as

$$\chi_i^T \chi_j = \delta_{ij} \tag{62}$$

rather than the usual

$$\chi_i^\dagger \chi_j = \delta_{ij} \tag{63}$$

which would be appropriate for a complex Hermitian matrix.

Brave souls (the present author included) have thus just rewritten real symmetric matrix eigenvalue routines (such as Givens, Householder, and LR decompositions) to accept complex input, with nary a complex conjugate. While this may well fail in pathological cases, we have observed no such case in a problem arising in atomic physics. This direct method can often give good zero order results, and allow an assessment of the utility of the method, as the real symmetric matrices $\overline{KE}$ and $\overline{PE}$ are often available from standard codes written for usual configuration interaction calculations. The only caveat is that as $\Gamma/2$ is often several orders of magnitude smaller than E^{res}, a rather high level of convergence must be attained if $\Gamma/2$ is needed with high precision, as the error in the real and imaginary parts of the complex eigenvalue are often comparable. Examples are in the referenced reviews.

With this direct method as a starting point, let us begin to generalize. A first remark is to note that the same complex symmetric matrices which arise by taking matrix elements of $H(\theta)$ in a basis of $\mathcal{L}^2$ functions with real radial dependences, may be obtained by taking matrix elements of H itself in a basis of complex radial functions, provided that a special scalar product is introduced. As an illustration, consider the one-body Coulomb Hamiltonian:

$$H(r,\theta,\phi) = -\frac{1}{2r^2}\frac{d}{dr}r^2\frac{d}{dr} + \frac{L^2}{2r^2} - \frac{1}{r}. \tag{64}$$

As a basis take $\Psi(r,\theta,\phi) = Y_{l,m}(\theta,\phi)R_{n,l}(r)$. A reasonable choice for the radial functions is

$$e^{-\alpha r}L_n^{2\ell+1}(\alpha r), \qquad n = 0,1,2,\ldots, \tag{65}$$

where the $L_n^{2\ell+1}(\alpha r)$ are generalized Laguerre polynomials which are a complete, $\mathcal{L}^2$, discrete set. These radial functions, $R_{n\ell}(r)$ are also real-analytic in r, a point which will be important in what follows. The $Y_{l,m}$ are the usual spherical harmonics, with the usual Hermitian norm

$$\int_{4\pi} d\Omega Y^*_{\ell,m}(\Omega)Y_{\ell',m'}(\Omega) = \delta_{\ell\ell'}\delta_{mm'}. \tag{66}$$

A matrix element of $H(\theta)$ is thus calculated as (the $*$ on $R_{n_i\ell_i}$ is omitted as this is real)

$$H_{ij}(\theta) = \int_0^\infty r^2 dr \int_{4\pi} d\Omega R_{n_i,\ell_i}(r)Y^*_{\ell_i,m_i}(\Omega)H(re^{i\theta},\Omega)R_{n_j,\ell_j}(r)Y_{\ell_j,m_j}(\Omega). \tag{67}$$

The angular integrations may be done in standard fashion; giving

$$H_{ij}(\theta) = \int_0^\infty r^2 dr\, R_{n_i\ell_i}(r)H_r(re^{i\theta})R_{n_j\ell_j}(r)\delta_{\ell\ell'}\delta_{mm'}. \tag{68}$$

We now note that

$$H_{ij}(\theta) = \int_0^\infty e^{-3i\theta}r^2 dr R_{n_i\ell_i}(re^{-i\theta})H_r(r)R_{n_j\ell_j}(re^{-i\theta})\delta_{\ell\ell'}\delta_{mm'}. \tag{69}$$

That is, we can use the real basis functions $R_{n,\ell}(r)$ and the dilated $H(\theta)$, or H itself and the dilated basis functions $e^{-\frac{3i\theta}{2}}R_{n,\ell}(re^{-i\theta})$ ...PROVIDED..., and it is a big provided..., in using the change of dummy variables needed to derive the equivalence of eqs. (68) and (69), *we don't take the complex conjugate of the complex radial dependence of the analytically continued basis functions!* This is an odd situation (and very confusing at first) in that if basis functions are complex for real interparticle distances, the usual definition of the Hermitian scalar product applies. However, complex conjugation is not applied to those parts of the basis where complex variables appear as interparticle distances become intrinsically complex. This is just an application of Cauchy contour distortion freedom, and stated another way, if complex numbers arise in a contour distortion complex conjugates are never taken. We can thus choose to work with a complex scaled basis (and a special scalar product, where not all "i"'s are equivalent) or with a real basis and the dilated $H(\theta)$ to the same effect. One can choose to view this choice as analogous to the difference between the active and passive views of co-ordinate changes: Do we move the object, or the co-ordinates? However, once the basis is complex how much freedom do we have? It is at this point that full development is not yet in hand, but we can give several illustrations of the perhaps surprising results which can arise.

The first surprising result is that Hartree-Fock calculations with real H and real basis functions can yield complex eigenvalues! The Hartree-Fock model, discussed earlier in the workshop [15] by Almlöf, finds single-particle wave functions, not as eigenfunctions of the simple Coulomb problem, but as self-consistently determined eigenfunctions of the Fock operator F, defined as

$$F(\vec{r}) = -\frac{1}{2}\nabla_r^2 - \frac{Z}{|\vec{r}|} + \sum_{i=1}^{N}[2J(\vec{r}) - K(\vec{r})]. \tag{70}$$

F itself requires knowledge of its own eigenfunctions:

$$F(\vec{r})\chi_i(\vec{r})_N = \epsilon_i\chi_i(\vec{r}) \tag{71}$$

since, for example,

$$J(\vec{r}) = \sum_{i=1} \int d^3\tilde{r}\frac{\chi_i^*(\vec{r})\chi_i(\vec{r})}{|\vec{r} - \vec{\tilde{r}}|}. \tag{72}$$

Thus implementation of the Hartree-Fock picture requires solution of non-linear equations, which chemists refer to as self-consistency: each electron moves in the average potential generated by the motion of the $N-1$ others. All are then adjusted (variationally) to give the best single configurational energy and the resulting Euler problem is eq. (71).

In actual atomic and molecular applications a matrix version of the Hartree-Fock equations, introduced by Roothaan [48], is often used (actually almost always for molecular applications!). Suppose the unknown $\chi_i(\vec{r})$ are are expanded in a fixed basis, $\phi_\alpha(\vec{r})$

$$\chi_i(\vec{r}) = \sum_\alpha c_i^\alpha \phi_\alpha(\vec{r}). \tag{73}$$

Then the self consistent solution of the non-linear equations involves determination of the coefficient vectors $\{\overline{c}_i \equiv (c_i^1, c_i^2, c_i^3 \ldots)\}$; this is Roothaan's [48] matrix Hartree-Fock.

Usually the Hartree-Fock method is used to find an approximate ground-state wave function for the N-electron system. However, there are cases where scattering states in the Hartree-Fock picture have resonances (in electron scattering applications, the Hartree-Fock approximation is equivalent to what is called the "static-exchange" approximation) and we would like to find them using complex scale transformations. We could start with $H(\theta)$, and rederive the complex scaled Hartree-Fock equations (see [49]). Or, following the earlier discussion, we could take the complex scaling into the radial basis, as

$$\phi_\alpha(\vec{r}) = R_\alpha(re^{i\theta})Y_{\ell,m}(\theta,\phi) \tag{74}$$

and calculate matrix elements without complex conjugation of this radial analytic continuation but taking the usual conjugates of the spherical harmonics.

McCurdy, Lauderdale, and Mowrey [49] have done something rather different and, at first sight, quite surprising. Keeping H real, and using only the real basis functions, they implemented an analytic continuation by allowing the expansion coefficients $\{\overline{c}_i\}$ to be complex, keeping the usual Hermitian scalar product for the basis functions $\phi_\alpha(\vec{r})$, but then ignoring complex conjugation of the (complex) $\overline{c}_i$'s.

This is in the spirit of using the biorthogonal vectors of the complex symmetric matrix $\overline{H}(\theta)$ discussed above. Excellent approximate resonance eigenvalues were obtained, by this method, where neither the basis functions for Hamiltonian were scaled. Rather a complex variation of the (radial) expansion coefficients was carried out, with a redefinition of the scalar product!

5. The complex variational principle of Stillinger and Herrick. Suppose we write a variational functional of the form

$$I(\overline{c},\overline{\alpha}) = \langle\chi^{\text{Trial}}_{\overline{c},\overline{\alpha}}, H\chi^{\text{Trial}}_{\overline{c},\overline{\alpha}}\rangle \Big/ \langle\chi^{\text{Trial}}_{\overline{c},\overline{\alpha}}, \chi^{\text{Trial}}_{\overline{c},\overline{\alpha}}\rangle \tag{75}$$

where the trial function has both linear, $\overline{c}$, and non-linear, $\overline{\alpha}$, parameters. For example, in a one-dimensional radial problem take

$$\chi(r) \equiv \sum_i c_i\phi_i(\alpha_i r) \tag{76}$$

and construct the expectation value of H. Suppose now we allow free variation of the linear (*i.e.*, the c_i) and non-linear (*i.e.*, α_i) parameters, including allowing these parameters to become complex, but *ignoring the taking of complex conjugates.* That is we again use a real scalar product with complex functions! Requirement of stationarity with respect to all parameters has, in the hands of Stillinger [36,37] and Herrick [50], led to prediction of complex resonance eigenvalues. This positive result is clearly related to McCurdy's observation, above, that neither the basis nor Hamiltonian need be scaled, as long as a modified (real) scalar product is used with complex variations of coefficients! Now, however, we consider a seeming paradox: scaling the Hamiltonian as $H(r \to re^{i\theta})$ rigorously gives the results of the theory of dilatation analyticity: for an appropriate range of θ complex resonance eigenvalues are exposed and may be found in the point spectrum of $H(\theta)$. Using complex basis functions with $r \to re^{-i\theta}$ with an appropriately modified scalar product gives the same result. Just using a modified symmetric "real analytic" scalar product also seems to contain the same results. But, what if we take $H(\theta)$ and also scale the basis $r \to re^{-i\theta}$? Clearly the effects cancel, and only the real spectrum of H is obtained. Thus a general variational functional of the type $I(\overline{c}, \overline{\alpha})$ of eq. (75) and used without complex conjugation but with free complex variation of the $\overline{c}$ and $\overline{\alpha}$ contains the spectrum of both H and $H(\theta)$!!! Can we then get any result at all? Junker[7], McCurdy[31], and Reinhardt[6] have discussed some of these sometimes confusing issues. Löwdin[52] has begun a more serious mathematical investigation analyzing the range and domain of H and $H(\theta)$ on the same footing, to begin to provide appropriate constraints on general variational principles of the type of eq. (75). Moiseyev and Hirschfelder[53] have recently introduced a more general scale transformation, $r \to re^{i\theta f(r)}$, to achieve an extended variational principle. There is clearly room for development here, and contributions from functional analysis may set the stage for a revolution in our understanding of variational principles.

Remark. It is useful to stress again a crucial point: the new scalar product is part Hermitian and part complex symmetric. Those parts of the wave function which are not associated with interparticle scaling still must be complex conjugated! Reinhardt [6] has suggested the name Generalized Dilatation Transformation (GDT) to describe this odd situation, where only complex dilations are not conjugated, and intrinsically complex parts of the wave function such as those related to angular parts of the wave function are used with the normal Hermitian scalar product.

6. Dilatation analyticity in calculation of continuum processes. We will consider both the photeffect and the scattering of a particle by a spherical potential.

6.1. Discretization, complex scaling, contour distortion, and the "E+iε" limit: Calculation of photoeffect cross sections. So far, we have considered the use of complex scale transformations to calculate complex eigenvalues, which give the energy and width (or lifetime) of resonances, and as a useful tool in the functional analysis of Coulomb Schrödinger operators. However, a motivation of the use of contour distortion and complex scaling techniques has always been to allow calculation of scattering amplitudes and amplitudes and cross sections in those cases where direct co-ordinates space enforcement of boundary conditions is not convenient. Such is the case when strong non-spherically symmetric potentials are involved, as in the interaction of an electron with a molecular frame. It is also the case when the number of channels grows rapidly with increasing energy, becoming infinite as the target system dissociates or undergoes "break-up". Thus for example, we would like to be able to calculate the cross section for a photoabsorption process such as

$$\text{photon} + H^- \to H^+ + e^- + e^- \tag{77}$$

or the amplitude for the specific state-to-state scattering process

$$e^- + H(n, \ell, m) \to H(n', \ell', m') + e^- \tag{78}$$

but in an energy range where an infinity of other scattering processes are energetically allowed. For the case of molecular photoabsorption processes direct computation of the continuum wavefunction satisfying both the Born-Oppenheimer

Schrödinger equation in the body-fixed frame, and simultaneously the usual asymptotic forms of many channel scattering theory is possible in theory, but computationally extraordinarily difficult. This has led to moment theory methods, discussed in the lectures by Langhoff[54] at the workshop, which allow calculation of molecular photoeffect cross sections without explicit enforcement of spatial multichannel boundary conditions.

Taking the photo-effect as an example, suppose we are interested in a system whose energy levels are displayed in Figure 20. Shown is a typical situation wherein a selection rule determines that of all the continua energetically accessible for the indicated range of photon energies, only one, the kth, is accessible via a single photon absorption. If "T" is the transition operator describing the coupling of the photon field to the atomic system, the probability per unit time per unit flux of photons per atom, or more simply the "cross section," $\sigma(\omega)$, for absorption of a photon with subsequent ionization of the atom is proportional to the absolute square of the amplitude $\langle \Phi_{\text{Bound}}, T\Psi_k(E_k)\rangle$ where energy balance is given by the requirement that $E_k = E_{\text{Bound}} + \hbar\omega$, $\Psi_j(E_j)$ is the appropriate continuum function at energy E_j and $\hbar w$ is the photon energy.

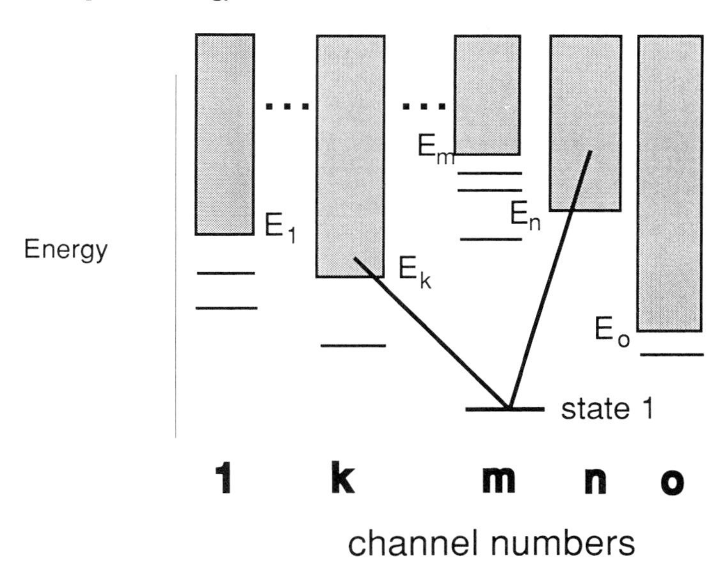

Figure 20) Grotrian diagram showing selection rules, indicating that photoabsorption only accesses certain of the channels open at a fixed energy. At low energy only channel "k" is accessed, at higher photon energy channel "n" opens.

We now note that this cross section is given by the imaginary part of a matrix element of the resolvent $(z - H)^{-1}$, taken with respect to the $\mathcal{L}^2$ function $T\Phi_{\text{Bound}}$. This is an important result, is it is just such matrix elements which have special properties under dilatation transformations. To show the result, use may be made of the formal identity of eq. (32):

$$\frac{1}{z-H}=\sum_{\substack{i\\ \text{Bound}\\ \text{States}}}\frac{\Psi^{+i}_{\text{Bound}}\Psi^{i}_{\text{Bound}}}{z-E_i}+\sum_{\substack{\text{channels}\\ j}}\int_{E_j^{\text{Threshold } j}}^{\infty}dEj\,\frac{\Psi_j^{+}(E_j)\Psi_j(E_j)}{z-E_j}. \tag{79}$$

Taking the expectation value of both sides with respect to the $\mathcal{L}^2$ function $T\Phi_{\text{Bound}}$, where $\Phi_B \equiv \Phi_{\text{Bound}}$ is the initial state wave function, and again, T is the transition operator, we have

(80)

$$\langle T\Phi_B,\frac{1}{z-H}T\Phi_B\rangle=\sum_{\substack{i\\ \text{Bound}\\ \text{States}}}\frac{\langle\Phi_B,T\Phi_i\rangle\langle\Phi_i,T\Phi_B\rangle}{z-E_i}+\int_{E_k^{\text{Threshold}}}^{\infty}dE_k\frac{\langle\Phi_B T E_k\rangle\langle E_k,T\Phi_B\rangle}{z-E_k}$$

this latter simplification occurring because of the selection rule. Taking the limit as $z\rightarrow E+i\epsilon$, we have (see eq. (36) and the discussion of that equation)

$$\lim_{z\rightarrow E+i\epsilon}\langle T\Phi_B,\frac{1}{z-H}T\Phi_B\rangle=-\pi|\langle\Phi_{\text{Bound}},T\Psi_k(E_k)\rangle|^2 \tag{81}$$

which is proportional to $\sigma(\omega)$. Were other channels energetically allowed, at say a higher photon energy, as illustrated in Figure 20, the equivalent result would be

$$\lim_{z\rightarrow E+i\epsilon}\langle T\Phi_B,\frac{1}{z-H}T\Phi_B\rangle=\sum_{\substack{\text{allowed channels}\\ \text{open at}\\ \text{energy}\hbar w}}-\pi|\langle\Phi_B,T\Psi_k(E_m)\rangle|^2. \tag{82}$$

That is $Im\langle T\Phi_B,(z-H)^{-1}T\Phi_B\rangle$ is proportional to the total photoabsorption cross section, which is the sum of all the cross sections of the individual allowed photoabsorption processes. There is thus the possibility of calculating total cross sections in the presence of infinities of channels without explicit recognition of each, provided, that is, that we can calculate $(z-H)^{-1}$ in the $z\rightarrow E+i\epsilon$ limit, without using a channel–by–channel resolution of the operator, such as that of eq.(80).

A first attempt: Suppose we simply diagonalize H, the atomic Hamiltonian, in a finite subset of a complete $\mathcal{L}^2$ basis. As the basis is $\mathcal{L}^2$ it does not extend to infinity, and thus scattering boundary conditions are not satisfied. Can we calculate matrix dimensions of the type $\langle\Phi_B T,(z-H)^{-1}T\Phi_B\rangle$ and take the $z\rightarrow E+i\epsilon$ limit? The spectrum of the matrix representation of H in a finite basis is shown in Figure 21. If $\overline{H}$ is the matrix representation in a subspace, and $\overline{\chi}_i$ and $\overline{E}_i$ are the matrix eigenvectors eigenvalues, then

$$\langle T\Phi_B,(z-\overline{H})^{-1}T\Phi_B\rangle=\sum\frac{|\langle\Phi_B,T\overline{\chi}_i\rangle|^2}{z-\overline{E}_i}. \tag{83}$$

Now, taking the limit $z\rightarrow E+i\epsilon$ gives either a real result (and thus no contribution to the cross section) or if $E=\overline{E}_i$, one of the matrix eigenvalues, the limit is singular. [As indicated in the "Remark" following this section, all is by no means

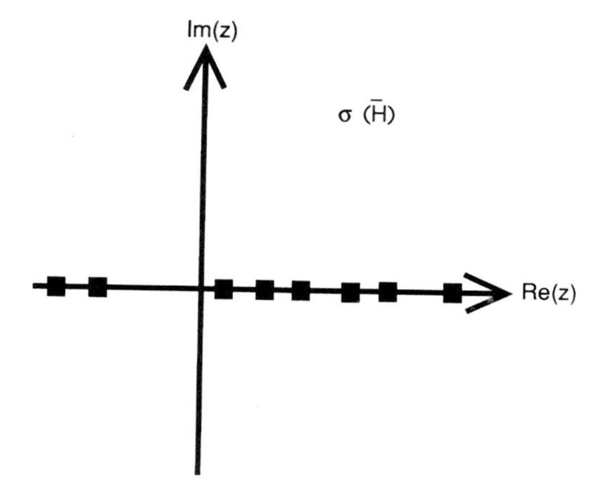

Figure 21) Discretization (*i.e.* diagonalization in an appropriate $\mathcal{L}^2$ basis) of the spectrum of an atomic Hamiltonian. The continuum is gone, as finite matrices have a purely discrete spectrum. Discretization thus complicates extraction of continuum information.

lost in the representation of eq. (83), but we do not digress at this point.] The spectral resolution of $(z-H)^{-1}$ does not seem to be of help in obtaining a value of $Im(\Phi_B, T(z-H)^{-1}T\Phi_B)$. But, dilatation analyticity to the rescue!

Suppose, introducing a unitary transformation to carry out the complex scaling, we take

$$
\begin{aligned}
\langle T\Phi_B,(z-H)^{-1}T\Phi_B\rangle &= \langle T\Phi_B, U^{\dagger}(\theta)U(\theta)(z-H)^{-1}U^{\dagger}(\theta)U(\theta)T\Phi_B\rangle \\
&= \langle U(\theta)T\Phi_B,(z-H(\theta))^{-1}UT\Phi_B\rangle
\end{aligned}
\tag{84}
$$

as an alternative integral representation. This representation is the same analytic function (if H is dilatation analytic) and if the functions $U(\theta)(T\Phi_B)$, which are the analytic continuation of $T\Phi_B$ to complex co-ordinates, are bounded at infinity. This new representation has the advantage that putting z directly on the real axis is no longer a problem: the new representation has no branch cut there! Matrix discretization of $H(\theta)$ now proceeds as shown in Figure 22. It is immediately seen that in the new representation, discretization of the continua, as occurs in taking a finite matrix representation of $\overline{H}(\theta)$, now occurs of the real axis, and thus sensible results may be obtained by taking $z \to E + i\epsilon$ as illustrated in Figure 23 from Ref. [55].

Remark. Direct calculation of the cross section by diagonalization of $\overline{H}$ in a finte $\mathcal{L}^2$ basis. The Stieltjes techniques of Langhoff, discussed at the workshop, use the residues $|\langle \Phi_B, T\overline{\chi_i}\rangle|^2$ and matrix eigenvalues $\overline{E}_i$ as input to construct moments of the photoeffect cross section, which are subsequently inverted to give

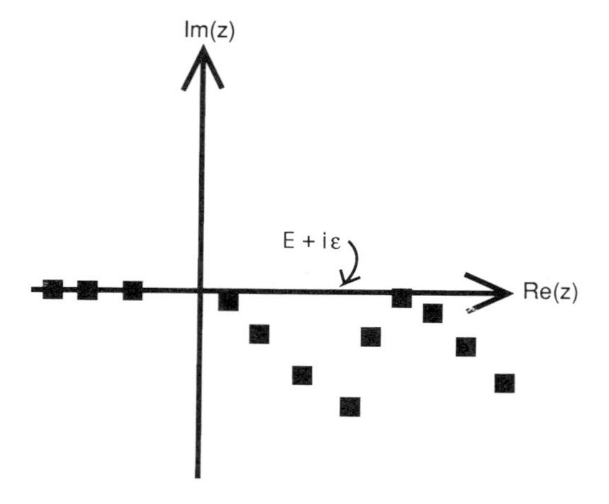

Figure 22) Discretization of $H(\theta)$. The matrix eigenvalues which, somehow, correspond to the continua as opposed to resonances or bound states now clearly show themselves, as the matrix spectrum suggests the spectrum of the actual operator, as shown in Figure 16. Scattering and photoabsorption cross section calculation is thus facilitated, as the $E + i\varepsilon$ limit is no longer singular

the cross section. (See Langhoff's contribution to these proceedings.) Alternatively, discretization of H may be interpreted as generating a generalized Gaussian integration, and analysis of the scheme yields the cross section (see Yamani and Reinhardt [56], Reinhardt [57]). Still another approach, but one of less intrinsic stability, is to use the representation of eq. (83) at complex z, and then carry out an approximate analytic continuation back to the real axis obtaining the $E + i\epsilon$ limit, as discussed by Schlessinger [58], Reinhardt, Oxtoby, and Rescigno [59], Broad and Reinhardt [60], and Reinhardt(61). Detailed discussion of these alternative methods would take us too far afield at this point.

6.2. Calculation of scattering amplitudes: Short- and long-range potentials. The calculation of scattering amplitudes using the tricks of Section 5 above is a much more delicate matter. Without wishing to derive the equations of formal non-relativistic scattering theory (for which see Newton [27] or Taylor [28]) we simply note that if matrix elements of the form

$$\langle \Phi_j(E_j), V(z-H)^{-1} V \Phi_k(E_k) \rangle, \tag{85}$$

where $\Phi_j(E_j)$ and $\Phi_k(E_k)$ are continuum functions of the unperturbed system (*i.e.*, plane waves or their spherical counterparts) and V is the interaction potential, could be evaluated as

$$\langle \Phi_j(E_j), VU^\dagger(\theta)U(\theta)(z-H)^{-1}U^\dagger(\theta)V(\theta)U\Phi_k(E_k) \rangle, \tag{86}$$

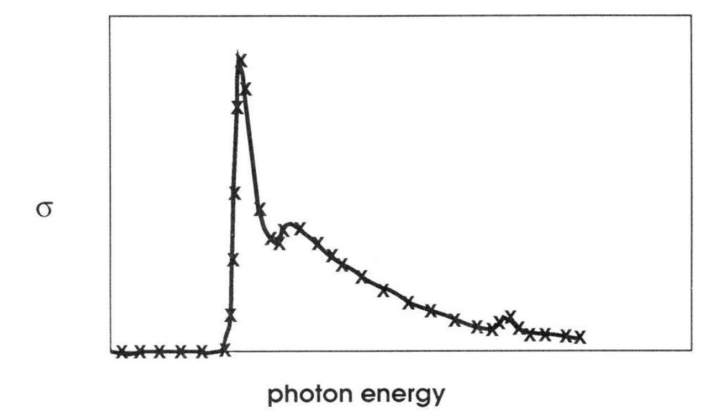

Figure 23) An example of a photoabsorption spectrum of a two-electron system as calculated from a discretized Hamiltonian via a complex scaling technique. In this case an "energy averaged" low resolution spectrum was desired, and the resulting smooth curve (Kerner, Ref. [55]) is compared with experiment (with the same energy resolution) of van der Wiel *et al.* (Ref. [18]). It is evident that discretization of an atomic Hamiltonian does not prevent extraction of continuum of "scattering" information. See also the lectures of Langhoff, this volume.

the $z \to E + i\epsilon$ limit would cease to be problematic, and scattering amplitudes and cross sections could be easily calculated. This is not at all straightforward in atomic physics. To see why, it is useful to introduce a more explicit co-ordinate space notation, which we do for a radial potential scattering problem. Consider the matrix element

$$(87) \qquad \int_0^\infty r^2\, dr \int (r')^2 dr' j_\ell(k_j r) V(r) G(z,r,r') V(r') j_\ell(k_k r')$$

where $G(z,r,r')$ is the integral kernel or Green's function representation of $(z - H)^{-1}$. We can now distort the contours so:

$$(88) \quad e^{i6\theta} \int_0^\infty r^2 dr \int_0^\infty (r')^2 dr' j_\ell(k_j re^{i\theta}) V(re^{i\theta}) G(z, re^{i\theta}, r'e^{i\theta}) V(r'e^{i\theta}) j_\ell(k_k r'e^{i\theta})$$

where now $G(z, re^{i\theta}, r'e^{i\theta})$ is the co-ordinate representation of $(z - H(\theta))^{-1}$. Implicit in this result is that there are no points of non-analyticity in the sector defined by the lines $(0 \to \infty)$ and $(0 \to e^{+i\theta}\infty)$, *and* that the contours may be closed at infinity along the arc $e^{i\theta}(\infty)$ as runs from 0 to its final value. In nuclear problems, the nucleon-nucleon forces are short ranged, and the potentials used are exponentially bounded so that $V(r) < \exp(-\alpha r)$ for large r. Thus factors in the integrand such as $\sin(kre^{i\theta})V(re^{i\theta})$ still allow convergence of the integral, as long as θ is not too large, in spite of the exponential growth of the spherical Bessel functions $j_\ell(kre^{i\theta})$

as $r \to \infty$. Typical applications to scattering (including break-up processes in which two particles collide and three emerge) in the presence of such short-range potentials are given in by Hendry [62]. However, the situation is quite different in atomic and molecular physics and chemistry: even the potential between neutral atoms only falls off as r^{-6}! The potential between an electron and a neutral atom goes as r^{-4} at large distances, and of course the potential between charged particles is the Coulomb Law, r^{-1}. A product of the form $\sin(kre^{i\theta})(re^{i\theta})^{-n}$ (for any $n > 0$) is not sufficiently well behaved as $r \to \infty$ to allow closing of the contour in r-space in eq. (88), and thus dilatation analyticity of H is not enough to allow calculation of scattering amplitudes for long ranged potentials, which for our purposes are those which are not exponentially bounded. Or so it seemed through the early 1980's.

In 1983–84 Johnson and Reinhardt [63] attempted to circumvent the problem of long-range potentials by using a partial resolution of the identity in an N-term expansion is terms of a subset of a complete discrete set of $\mathcal{L}^2$ functions:

$$\tilde{1}_M \equiv \sum_{i=1}^{M} \chi_i(r)\chi_i(\vec{r}\,') \tag{89}$$

where $\langle \chi_i, \chi_j \rangle = \delta_{ij}$. Then, inserting this approximate resolution into the desired amplitude, gives

$$\langle \Phi_j(E_j), VU^{\dagger}(\theta)U(\theta)\tilde{1}_M(z - H)^{-1}\tilde{1}_m U^{\dagger}(\theta)U(\theta)V\Phi_k(E_k)\rangle \tag{90}$$

which may now be evaluated term by term using contour distortion, as the functions $\chi_i(\vec{r})$ are exponentially bounded. The question is, to what limit does this procedure converge as M increases, and the resolution of eq. (89) becomes complete? Figure 24 shows convergence of the amplitude for a model problem with an attractive r^{-4} potential as a function of the number of $\mathcal{L}^2$ functions in the resolution of the identity (and in the diagonalization of $H(\theta)$). The exact amplitude (determined by a phase shift analysis) is indicated, and it is evident that there is a regime of pseudo convergence, followed by divergence. However, employing a Shanks sequence to sequence transformation [64] on the divergent sequence of approximants gave the results of Table I. This type of results opens the possibility of calculation of atomic scattering amplitudes (Johnson and Reinhardt [65]), and photoabsorption amplitudes (Johnson and Reinhardt [63]) exploiting the dilatation analyticity of the Coulomb Hamiltonian. The divergent expansions thus contain the requisite information for the construction of scattering amplitudes in the presence of long-range forces. One can only ask, isn't there a better way?

7. Discussion. The interplay of functional analysts, atomic physicists, and computational chemists has led to more than a decade of quite fruitful advances in the theory of the electronic structure of atoms and in development of computational techniques. Is there anything left to do? Are there new advances? Of course, of course!

To cite a few examples: initial application of complex co-ordinate techniques to the direct calculation of Coulomb amplitudes has thus far only been partially

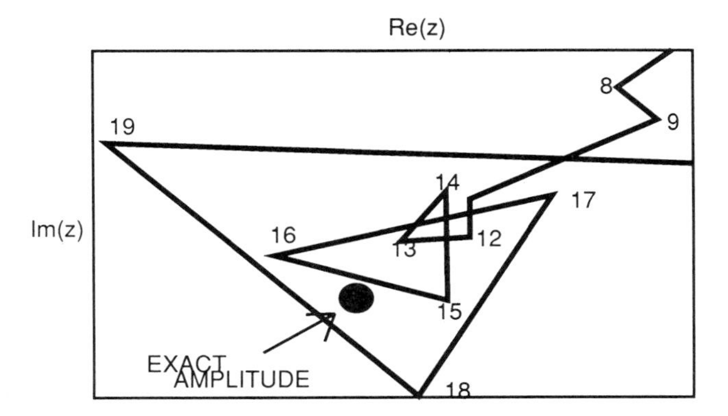

Figure 24) Pseudo-convergence and ultimate divergence of a computed approximate scattering amplitude for long range interaction of two structureless particles. The divergence is due to an illegal use of a complex scaling, where it was known before hand that the required analyticity did not exist. The exact (complex) amplitude is shown as a black dot, and the "trajectory" has corners with numbers indicating the size, M, of the basis used in calculation of the approximate amplitude. For values of M between 12 and 16 an intriguing stationarity is seen, followed by strong divergence as M continues to grow. As seen in the Table, however, this sequence of approximate amplitudes may be summed, via a Shanks' sequence-to-sequence transformation, and the essentially exact result obtained.

successful (Johnson *et al.* [66]). If practical methods for strongly convergent computation of Coulomb amplitudes could be found, Coulomb wave operator methods (Dollard [67], Chandler and Gibson [68,69], and Zorbas [70]) could be used to calculate amplitudes for processes such as

$$\hbar\omega + H^- \to H^+ + 2e^- \tag{91}$$

which are not currently amenable to calculation.

Rescigno and McCurdy [71] have quite recently found methods for carrying out local distortion of the energy spectrum, rather that the global rotations of the continua obtained from $H(\theta)$ itself. These methods still only involve matrix diagonalization, although with specifically chosen complex basis functions, and not surprisingly, see Section 4, involve use of a non-standard scalar product with respect to the complex basis function! Initial results (McCurdy and Rescigno [71]) indicate better numerical stability than with earlier implementations. Jansen op de Haar and Miller [72] have recently seen how to rationalize and incorporate the McCurdy-Rescigno ansatz into more conventional variational scattering theory. The problem of long range forces also goes away.

$[M, M]$	$\mathrm{Re}(\langle\phi, V(E+i\varepsilon-H)^{-1}V\phi\rangle)$	$\mathrm{Im}(\langle\phi, V(E+i\varepsilon-H)^{-1}V\phi\rangle)$
$[1,1]$	-0.005705158	-0.010671417
$[2,2]$	-0.006252684	-0.010955857
$[3,3]$	-0.006332018	-0.010954464
$[5,5]$	-0.006371141	-0.010962266
$[10,10]$	-0.006378950	-0.010970626
$[15,15]$	-0.006379279	-0.010971503
$[19,19]$	-0.006379305	-0.010971533

Table I.) Diagonal $[M, M]$ Padé approximants (a.k.a. Shanks' extrapolants) formed from the divergent expansion of eg. (89), as illustrated in Figure 24. As the L^2 resolution of the identity becomes complete (as $M \to \infty$) the amplitude of eg. (90) diverges, but is summable, as illustrated here. The $M = 40$ Padé gives an amplitude which gives, in tern, a scattering phase shift accurate to 5 significant figures. Numerical results from Ref. [63].

One expects even more to happen in the future.

Acknowledgments: The author is most grateful for the input of his many co-workers and colleagues from the mathematical, physical, and chemical sides. Thanks are due to D. Truhlar for organization of a stimulating Workshop at the IMA, and to the scientific and administrative staff of the IMA for their superb efforts on behalf of the speakers and attendees. The hospitality of the Telluride Summer Research Center is gratefully acknowledged.

The work reported here has been supported by the National Science Foundation.

REFERENCES

[1] E.U. Condon and G.H. Shortley, *The Theory of Atomic Spectra*, Cambridge Univ. Press, London, 1953.
[2] M. Reed and B. Simon, *Methods of Modern Mathematical Physics*, Vols. I–IV, see especially Vol. IV, pages 51–66, Academic Press, New York, 1972–1978.
[3] J. Aguilar and J. M Combes, Commun. Math. Phys., 22 (1971), pp. 269–279.
[4] E. Balslev and J. M. Combes, Commun. Math. Phys., 22 (1971), pp. 280–294.
[5] B. Simon, Ann. Math., 97 (1973), pp. 247–274.
[6] W. P. Reinhardt, Annu. Rev. Phys. Chem., 33 (1982), pp. 223–255.
[7] B. R. Junker, Adv. Atomic Molec. Phys., 18 (1982), pp. 207–263.

[8] Y. K. Ho, Phys. Reports, 99 (1983), pp. 1–68.
[9] E. Schrödinger, Ann. Phys., 79 (1926), pp. 1–12. Printed in an English translation in E. Schrödinger, *Collected Papers on Wave Mechanics*, Chelsea, New York, 1982, pp. 1ff.
[10] E.C. Titchmarsh, *Eigenfunction Expansions Associated with Second Order Differential Equations*, Part I, Oxford, 1962, pp. 98–100.
[11] L. Schiff, *Quantum Mechanics*, McGraw-Hill, New York, 3rd ed., 1968.
[12] K. Gottfried, *Quantum Mechanics, Vol. I*, Benjamin, New York, 1966.
[13] W. Pauli, Zeit. Phys., 36 (1926), pp. 336–363.
[14] C. A. Mead, *this volume, chapter 1, and references therein.*
[15] J. Almlöf, *this volume, chapter 2, and references therein.*
[16] I. Shavitt, *this volume, chapter 10, and references therein.*
[17] J. Paldus, *this volume, chapter 11, and references therein.*
[18] C. Backx, R.R. Tol, G.R. Wight, and M.J. van der Wiel, J. Phys. B., 8 (1975), p. 2050-2057.
[19] R.P. Madden and K. Coddling, Astrophys J., 141 (1965), pp. 364–375.
[20] R. N. Hill, Phys. Rev. Letts., 38 (1977), pp. 643–646.
[21] R. N. Hill, J. Math. Phys., 18 (1977), pp. 2316–2330.
[22] E. Herbert, T.A. Patterson, D.W. Norcross and W.C. Lineberger, Astrophysical J., 191 (1974), pp. L143–L144.
[23] D. R. Herrick and O. Sinanoglu, Phys Rev., A 11 (1975), pp. 97–110.
[24] C. E. Wulfman, Chem. Phys. Lett., 40 (1976), pp. 139–141.
[25] M. Kellman and D. R. Herrick, Phys Rev, A 22 (1980), pp. 1536–1551.
[26] E.C. Titchmarsh, *Eigenfunction Expansions Associated with Second Order Differential Equations*, Part II. Oxford, (1958), chapter XX.
[27] R. G. Netwon, *Scattering Theory of Waves and Particles*, 2nd ed., Springer, New York, 1982.
[28] J. R. Taylor, *Scattering Theory: The Quantum Theory of Non-relativistic Collisions*, Wiley, New York, 1972.
[29] M.H. Stone, *Linear Transformations in Hilbert Space*, American Mathematical Society, Providence R.I., (1932), chapters II, IV.
[30] G. F. Carrier, M. Krook, and C.E. Pearson, *Functions of a Complex Variable*, McGraw Hill, New York, (1966), Chapter 3.
[31] E. A. Hylleraas, Zeit. F. Physik, 65 (1930), pp. 209–225.
[32] T. Kato, J. Fac. Sci. Univ. Tokyo Sect. I, 6 (1951), pp. 145ff.
[33] T. Kato, *Perturbation Theory for Linear Operators*, 2nd Ed., Springer, New York, (1976) Chapter 7.
[34] F. Rellich, *Perturbation Theory of Eigenvalue Problems*, Gordon and Breach, New York, 1969.
[35] F. H. Stillinger, J. Chem. Phys., 45 (1966), pp. 3623–3631.
[36] F. H. Stillinger D. K. Stillinger, Phys. Rev. A, 10 (1974), pp. 1109–1121.
[37] F. H. Stillinger and T. A. Weber, Phys. Rev. A, 10 (1974), pp. 1122–1130.
[38] J. M. Midtdal, Phys. Rev., 138 (1969), pp. A1010–A1014.
[39] W. P. Reinhardt, Phys. Rev. A, 15 (1977), pp. 802–805.
[40] F. H. Stillinger, Phys. Rev. A, 15 (1977), p. 806.
[41] M. Klaus and B. Simon, Ann. Phys., 130 (1980), pp. 251–281.
[42] J. Baker D. E. Freund, R. N. Hill and J. D. Morgan, III, Phys. Rev. A (in press).
[43] B. Simon, Math. Ann., 207 (1974), pp. 133-138.
[44] C. Cerjan, W. P. Reinhardt, and J. E. Avron, J. Phys. B, 11 (1978), pp. L201-L205.
[45] I. Herbst, Commun. Math. Phys., 64 (1979), pp. 279–298.
[46] W. Hunziker, in *The Schrödinger Equation*, Ed. W. Thirring and P. Urban, Springer, New York, 1977, pp. 43–72.
[47] E.R. Davidson, E. Engdahl, and N. Moiseyev, Phys. Rev. A, 33 (1986), pp. 2436–2439.
[48] C. C. J. Roothaan, Rev. Mod. Phys., 23 (1951), pp. 69–89.
[49] C. W. McCurdy, J. G. Lauderdale, and R. C. Mowrey, J. Chem. Phys., 75 (1981), pp. 1835–1842.
[50] D. R. Herrick, unpublished work and private communications.
[51] C.W. McCurdy, in *Resonances in Electron-Molecule Scattering, van der Waals Complexes and Reactive Chemical Dynamics*, ACS Symposium Series 263, D. Truhlar, Ed., American Chemical Society, Washington, 1984, pp. 17–34.

[52] P.-O. LOWDIN, *Quantum Theory Project Note 503*, University of Florida, Gainesville, unpublished.
[53] N. MOISEYEV AND J.O. HIRSCHFELDER, J. Chem. Phys., 88 (1988), pp. 1063–1065.
[54] P. W. LANGHOFF, *this volume, chapter 4, and references therein.*
[55] D. KERNER, Ph.D. Thesis, University of Colorado, Boulder, 1985, unpublished.
[56] H. A. YAMANI AND W. P. REINHARDT, Phys. Rev. A, 11 (1975), pp. 1144–56.
[57] W. P. REINHARDT, Computer Phys. Commun., 17 (1979), pp. 1–21.
[58] L. SCHLESSINGER, Phys. Rev., 167 (1968), pp. 1411–1423.
[59] W. P. REINHARDT, D. W. OXTOBY, AND T. N. RESCIGNO, Phys. Rev. Lett., 28 (1972), pp. 401–403.
[60] J. T. BROAD, AND W. P. REINHARDT, J. Chem. Phys., 60 (1974), pp. 2182–83.
[61] W.P. REINHARDT, Computer Phys. Commun., 6 (1973), pp. 303–15.
[62] J. A. HENDRY, Nucl. Phys. A, 198 (1972), pp. 391–396.
[63] B. R. JOHNSON AND W. P. REINHARDT, Phys. Rev. A, 29 (1984), pp. 2933–2935.
[64] D. SHANKS, J. Math. Phys., 34 (1955), pp. 1–42.
[65] B. R. JOHNSON AND W. P. REINHARDT, Phys. Rev. A, 28 (1983), pp. 1930–1944.
[66] B. R. JOHNSON, W. P. REINHARDT, C. W. MCCURDY, AND T. N. RESCIGNO, Phys. Rev. A, 32 (1985), pp. 1998–2009.
[67] J. D. DOLLARD, J. Math. Phys., 5 (1964), pp. 729–738.
[68] A. G. GIBSON AND C. CHANDLER, J. Math. Phys., 15 (1974), pp. 1366–1377.
[69] C. CHANDLER AND A. G. GIBSON, J. Math. Phys., 15 (1974), pp. 291–294.
[70] J. ZORBAS, J. Math. Phys., 17 (1976), pp. 498–502.
[71] C. W. MCCURDY AND T.N. RESCIGNO, Phys. Rev. A, 31 (1985), pp. 624–633.
[72] W.H. MILLER AND B.M.D.D. JANSEN OP DE HAAR, J. Chem. Phys., 86 (1987), pp. 6213-6220.

CLASSICAL MANY-BODY SYSTEMS

STIELTJES METHODS FOR SCHRODINGER SPECTRA: HILBERT-SPACE APPROXIMATIONS TO THE DISCRETE AND CONTINUUM EIGENSTATES OF SPATIALLY ANISOTROPIC HAMILTONIAN OPERATORS[a]

PETER W. LANGHOFF[b]

"In his work on quantum theory, Schrödinger was led to a type of eigenvalue problem with a spectrum of an entirely different structure. This spectrum consists of a continuous and a discrete part; the discrete part does not extent to infinity but has a finite point of accumulation."

Courant and Hilbert, 1931

Abstract. An account is provided of recent developments in explicit Hilbert-space construction of the discrete and continuum electronic eigenstates of polyatomic molecules. Particular attention is focused on reduction of the (infinite) directional degeneracy of the molecular electronic continuum, accomplished employing so-called interaction-prepared states, on enforcement of asymptotic boundary conditions in the absence of an explicit asymptotic solution of the Schrödinger equation, on construction of convergent Hilbert-space approximations to the relevant discrete and continuum states, and on evaluation of appropriate cross sectional expressions for photoabsorption and ionization processes. The interaction-prepared continuum states adopted are seen to transform irreducibly under the point-group symmetry of the molecular Hamiltonian, and to be otherwise closely related to the more familiar bound electronic states of molecules. Lanczos and moment methods are described for construction of L^2 approximations to these states, as well as to the more conventional bound electronic states, and for determining the appropriate discrete or continuum wave function normalization. Explicit expressions are provided for molecular photoionization cross sections differential in the direction of ejected electron, expressed entirely in terms of interaction-prepared states. Recent computational applications of the formalism to selected diatomic and polyatomic molecules are also provided.

---

a Work supported in part by grants from the National Science Foundation (CHE83-04021 and CHE86-14344) and the U.S. Department of Energy (DE-FC05-85ER250000).

b Department of Chemistry and Supercomputer Computations Research Institute Florida State University, Tallahassee, FL 32306 USA. Present address: Department of Chemistry, Indiana University, Bloomington, IN 47405 USA

1.1 Pedagogical orientation. These lecture notes describe Ritz approximations to the discrete and continuum eigenstates of self-adjoint (Hamiltonian) operators [1], with particular reference to spatially anisotropic systems, such as electrons in polyatomic molecules [2,3], in which cases there are no apparent coordinates for separation of variables. Although variational calculations in Hilbert space are commonly employed for the bound states of such quantum systems [4,5], dating from the early work of Hylleraas [6], applications of these methods to scattering states with the attendant enforcement of appropriate asymptotic boundary conditions and continuum or Dirac delta-function normalization [7,8] are perhaps less familiar. Indeed, the Schrödinger wave mechanics [9] is apparently more commonly employed for scattering problems than is the matrix mechanics of Heisenberg [10], following a historical precedent set by Born [11], Sommerfeld [12], and others [13]. Anisotropic potentials are treated in this standard Schrödinger development much as are central potentials by adopting partial-wave expansions and radial integrations of the resulting coupled equations [14]. By contrast, the methods described here are based on matrix representations of Hamiltonian or other operators in basis sets of functions that transform irreducibly under the point-group symmetry of the potential. The central issues and extensions addressed in the text include computationally convenient resolution of the infinite degeneracy of the continuum, specification and enforcement of asymptotic boundary conditions, formulation of explicit expressions for differential cross sections, and construction of convergent approximations to the required discrete and continuum states of the Hamiltonian - the Schrödinger spectrum - in the appropriate Hilbert space.

Although much of the present development is based on standard material, particularly the Stieltjes-Tchebycheff moment theory as described and refined, for example, in the works of Shohat and Tamarkin [15], Wall [16], Akhiezer [17], Vorobyev [18], and Baker [19], applications to scattering states and related physical matters require additional considerations. Moreover, some of the material provided here - resolution of the infinite degeneracy of the continuum, construction of directed-wave solutions, and the incorporation of Feshbach methods for resonances into the Stieltjes development - has not appeared previously in the literature. These issues and extensions have been studied largely in collaboration with capable students (M.R. Hermann, C.L. Winstead, and B.W. Fatyga), who, however, bear no responsibility for what barbarisms may appear in the text.

1.2 Introductory remarks. Photoabsorption and ionization processes are intimately connected with the origins of the quantum theory [20], and find application in a surprising variety of scientific contexts [21]. Recent refinements of sychrotron-radiation light sources [22], and in electron-impact techniques [23], have led to quantitatively reliable measurements of the absolute cross sections describing these processes for a considerable number of molecules [24]. Accordingly, theoretical and computational approaches have been devised to account for the spectral variations of the measured cross sections, in which connection the increasing availability of supercomputers has played an important role [25-28]. An account is given here of aspects of these theoretical and computational developments, with particular reference to explicit Hilbert-space methods [29-32].

In the first (Born-Oppenheimer) approximation [33], molecular photoabsorption and ionization processes refer to electrons moving in an anisotropic potential perturbed by an oscillating dipolar electric field, leading to the familiar cross-sectional expression in terms of electric dipole matrix elements between fixed-nuclei electronic eigenstates [13,14,20]. Ritz methods have long been employed in construction of approximations to the required discrete molecular eigenstates, which transform irreducibly under the fixed-nuclei point-group symmetry of the molecule [2,3]. By contrast, the method of partial waves has generally been adopted in constructing the electronic continuum states required in scattering or ionization processes [14]. Although these methods can be highly satisfactory [25,26], there is evidently a theoretical and computational discontinuity in standard treatments of the two distinct spectral regions of molecular photoabsorption/ionization cross sections. Conceptual and computational advantages are afforded by unified treatments of both spectral regions on a common basis [29-32]. In particular, the slowly convergent partial-wave expansions of continuum states, particularly for large anisotropic molecular targets, can be avoided by adopting the basis sets familiar from bound-state studies [34], which are intrinsically of many-center form.

In order to devise a satisfactory Hilbert-space approach to construction of both discrete and continuum Schrödinger states, a number of troublesome issues must be resolved. First, the differing normalization conventions - Kronecker delta for discrete states and Dirac delta function for continuum states - must be incorporated in the development [7]. Second, the enforcement of directed-wave boundary conditions required for calculating differential cross sections [13], and the more general resolution of the infinite degeneracy of molecular continuum states, must be accomplished. Third, methods for resolving the degeneracy of continuum states in the presence of two or more open energetic channels must be devised. Fourth, computational expressions and methods suitable for the differential photoionization cross sections of polyatomic molecules in sensible dynamical approximations, and at a useful degree of spectral resolution, must be formulated. Finally, the nature of convergence of a Hilbert-space development for Schrödinger spectra must be established. The extent to which these issues have been satisfactorily resolved to date is reported in the following sections.

Section 2 provides a brief review of the absorption and ionization cross sections of interest, expressed in terms of the standard discrete and continuum Schrödinger eigenstates commonly employed for their evaluation [20]. Particular attention is addressed to the boundary conditions satisfied by the continuum states, and to the corresponding resolutions of degeneracy these choices imply, with particular reference to directed-wave, l-wave K-matrix, l-wave S-matrix, and eigenchannel boundary conditions [8]. The so-called fixed-nuclei approximation [14] is employed throughout in order to focus the development on issues of central interest here. Interaction-prepared states are defined in Section 3. These are seen to provide a reduction of the infinitely degenerate continuum of an anisotropic Hamiltonian to a single nondegenerate continuum, and to also provide the necessary and sufficient information for construction of differential cross sections. The interaction-prepared states correspond closely to the customary bound states of molecular point-group symmetry, providing an opportunity to deal with both spectral regions on a common basis. Hilbert-space methods for constructing L^2 approximations to interaction-prepared states are described, and the strong convergence of the approach for self-adjoint operators is indicated [35]. Normalization procedures for the nondegenerate continuum states of Section 3 are reported in Section 4 employing methods adopted from the classical theory of moments [15-19]. The relevant cross sections appear in the development as density functions, or spectral measures, and also provide the appropriate factors for normalization of the interaction-prepared states. Extensions of the development to Feshbach-Fano situations, in which cases the lineshapes of highly structured resonance cross sections can be obtained from the moment-theory development, are also reported [36]. Illustrative applications to diatomic and polyatomic molecules in single-channel approximation are reported in Section 5 to give some of the flavor of recent calculations, and concluding remarks are made in Section 6.

2. Molecular photoexcitation and ionization processes - aspects of the standard development employing Schrodinger eigenstates. Cross sectional expressions are given in this Section, the required Schrödinger eigenfunctions and their boundary conditions are indicated, and standard calculational techniques are briefly described employing the fixed-nuclei approximation.

2.1 Cross sections for absorption and ionization. Quantum mechanics provides a complete description of the photoabsorption process

$$M(i) + h\nu(\hat{\boldsymbol{\varepsilon}}) \rightarrow M(f), \tag{2.1}$$

where i and f refer to initial and final states of the molecule M, with $h\nu(\hat{\boldsymbol{\varepsilon}})$ representing the quantum energy of incident monochromatic light of frequency ν having polarization direction given by the unit vector $\hat{\boldsymbol{\varepsilon}}$ [13,20]. The cross section describing the efficiency of absorption of $h\nu(\hat{\boldsymbol{\varepsilon}})$ by the fixed-nuclei molecule M(i) is [20,37]

$$\sigma_i(h\nu,\hat{\boldsymbol{\varepsilon}}) = \sum_f |\langle\Phi_f|\boldsymbol{\mu}\cdot\hat{\boldsymbol{\varepsilon}}|\Phi_i\rangle|^2\delta(E_f - E_i - h\nu), \tag{2.2}$$

where $\boldsymbol{\mu}$ is the dipole interaction operator, given by the vector sum of electron position coordinates $\mathbf{r}_j$ multiplied by the electronic charge e [13]

$$\boldsymbol{\mu} = \sum_{j=1}^{N} e\,\mathbf{r}_j\ . \tag{2.3}$$

The E_f and Φ_f, f = 1,2,..., are Schrödinger eigenvalues and eigenfunctions, respectively, satisfying the Sturm-Liouville problem [1]

$$(H - E_f)\Phi_f = 0, \tag{2.4}$$

having boundary conditions discussed below, with H the appropriate (self-adjoint) Hamiltonian operator. In the present development, H describes electrons moving in the anisotropic potential provided by the laboratory - frame-fixed molecular nuclei [14,38,39]. Unless otherwise indicated, the fixed nuclei have relative positions given by their ground-state equilibrium values, regardless of the angular orientation of the rigid molecular frame. The relevant electronic wave functions and corresponding cross sections depend parametrically on this orientation unless an angle average is performed.

Averaging Eq. (2.2) over all directions of polarization of incident light of frequency ν for a fixed-nuclei molecule gives the cross section for unpolarized light in the form

$$\sigma_i(h\nu) = (1/3)\sum_{k=1}^{3} \sigma_i(h\nu, \hat{\varepsilon}_k), \tag{2.5}$$

where $\hat{\varepsilon}_1$, $\hat{\varepsilon}_2$, and $\hat{\varepsilon}_3$ are the three body-frame principal-axis polarization directions [33], and the $\sigma_i(h\nu, \hat{\varepsilon}_k)$ are the corresponding body-frame cross cross sections for these polarization directions, calculated employing Eq. (2.2) and appropriate body-frame wave functions. Equation (2.5) is also appropriate for an ensemble of randomly oriented fixed-nuclei molecules in the presence of linearly polarized light. In Figure 1 is shown a portion of the absorption cross section of an ensemble of randomly oriented water molecules (H_2O) in the presence of linearly polarized light as obtained from both measured and calculated values. The indicated state assignments are based on Eq. (2.2) and electronic calculations in fixed-nuclei approximation [40].

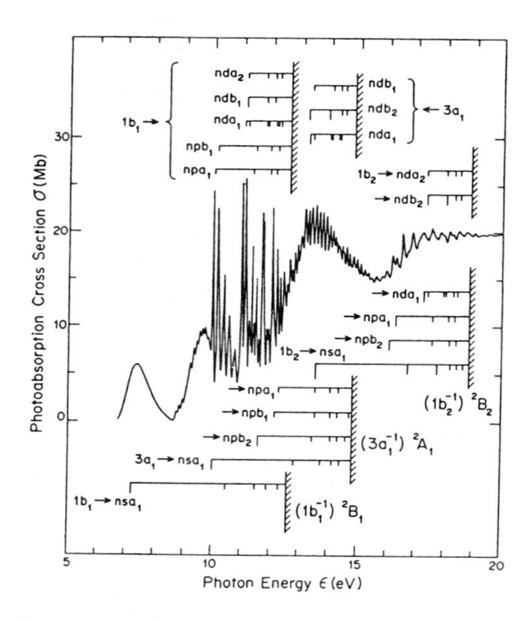

Figure 1. Total photoabsorption cross section of a randomly oriented ensemble of ground-state H_2O molecules obtained from synchrotron-radiation measurements and appropriate calculations discussed below [40]. ($1\text{Mb} = 10^{-18}\text{cm}^2$).

At sufficiently high photon energy, ionization processes of the type

$$(2.6) \qquad M(i) + h\nu(\hat{\varepsilon}) \rightarrow M^{+}(t) + e^{-}(\mathbf{k}_t)$$

occur [20], where $M^{+}(t)$ is a molecular ion in the final electronic state t and $e^{-}(\mathbf{k}_t)$ is an ejected electron with asymptotic ($r \rightarrow \infty$) linear momentum $\mathbf{k}_t$. The differential cross section describing the efficiency of the fixed-nuclei molecule M(i) in this process is

$$(2.7) \quad d\sigma_{i\rightarrow t}(h\nu, \hat{\varepsilon}, \mathbf{k}_t) = d\Omega_{\mathbf{k}_t} \int_{E_t}^{\infty} dE|\langle\Phi_{\mathbf{k}_t}^{(-)}|\boldsymbol{\mu} \cdot \hat{\varepsilon}|\Phi_i\rangle|^2 \delta(E - E_i - h\nu),$$

where E_t is the threshold energy for production of the ion $M^{+}(t)$, $E = E_t + (hk_t/8\pi)^2$ is the total final-state energy of the system $[M^{+}(t) + e^{-}(\mathbf{k}_t)]$, and $d\Omega_{\mathbf{k}_t}$ is a differential solid angle element in the direction $\mathbf{k}_t$ [13]. The Schrödinger eigenfunction $\Phi_{\mathbf{k}_t}^{(-)}$ satisfies

$$(2.8) \qquad (H - E)\, \Phi_{\mathbf{k}_t}^{(-)} = 0,$$

subject to special boundary conditions specified by the (-) superscript discussed below [41]. This function describes the ejected electron moving asymptotically with momentum $\mathbf{k}_t$, as well as the state of the fixed-nuclei molecular ion $M^{+}(t)$. Integration of Eq. (2.7) over all angles of the ejected electron gives [42-49],

$$(2.9) \qquad \sigma_{i\rightarrow t}(h\nu, \hat{\varepsilon}) = [\sigma_{i\rightarrow t}(h\nu)/4\pi]\ [1 + \beta_{i\rightarrow t}(h\nu)P_2(\cos\theta)],$$

the partial-channel photoionization cross section for production of fixed-nuclei molecular ions $M^{+}(t)$ from the fixed-nuclei molecule M(i). Here,

$$(2.10) \qquad \sigma_{i\rightarrow t}(h\nu) = (1/3)\sum_{k=1}^{3} \sigma_{i\rightarrow t}(h\nu, \hat{\varepsilon}_k)$$

is the molecular-orientation-averaged partial-channel photoionization cross section for production of $M^{+}(t)$, expressed in terms of the principal-axis, body-frame photoionization cross sections $\sigma_{i\rightarrow t}(h\nu, \hat{\varepsilon}_k)$, and

$$(2.11) \qquad \beta_{i\rightarrow t}(h\nu) = 2\{\sigma_{i\rightarrow t}(h\nu, \hat{\varepsilon}_3) - [\sigma_{i\rightarrow t}(h\nu, \hat{\varepsilon}_1) + \sigma_{i\rightarrow t}(h\nu, \hat{\varepsilon}_2)]/2\}/\sigma_{i\rightarrow t}(h\nu)$$

is the so-called integrated-detector angular anisotropy factor [47]. In Eq. (2.9) θ is the angle between the light polarization direction ($\hat{\xi}$) and the direction of the major symmetry axis ($\hat{\xi}_3$) of the oriented molecule, assuming a random orientation of the other two axes in the plane perpendicular to $\hat{\xi}_3$. In Figure 2 are shown the photoionization cross sections for production of the three lowest electronic ionic states of a randomly oriented ensemble of ground-state nitrogen (N_2) molecules in the presence of linearly-polarized light of fixed direction [50]. In this case, the principal-axis cross sections in the directions $\hat{\xi}_1$ and $\hat{\xi}_2$ are identical, with $\beta_{i\to t}(h\nu)$ of Eq. (2.11) referring to the cross sectional anisotropy for polarizations along and perpendicular to the molecule axis. It should be noted that the anisotropy factor of Eq. (2.11) is distinct from that which refers to the angular distribution of electrons ejected from an ensemble of randomly oriented molecules in the presence of ionizing radiation having a fixed linear polarization vector [20], which is discussed in Section 3 below.

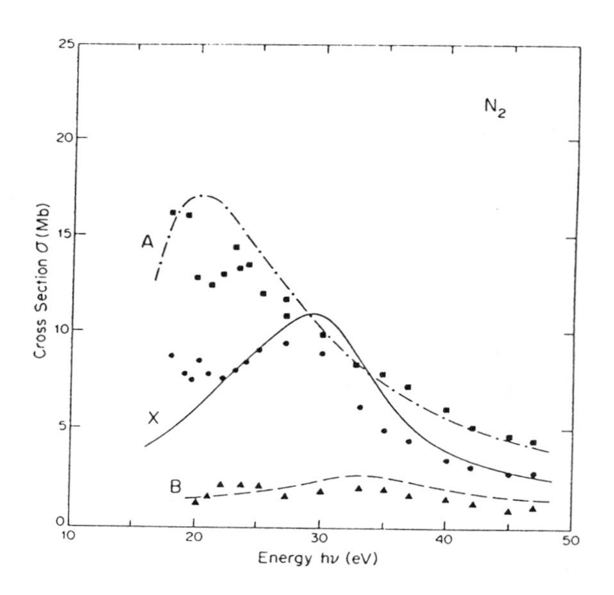

Figure 2. Partial-channel photoionization cross sections of Eq. (2.10) for production of the three lowest-lying ionic states $(3\sigma_g^{-1})X$, $(1\pi_u^{-1})A$, $(2\sigma_u^{-1})B$ of a randomly oriented ensemble of ground-state N_2 molecules, obtained from measured data (points) and calculated values (lines) [50].

Finally, the total photoionization cross section in fixed-nuclei approximation is given by the expression

$$(2.12) \qquad \sigma_i(h\nu) = \sum_t \sigma_{i\to t}(h\nu) .$$

where $\sigma_{i\to t}(h\nu)$ are the molecular-orientation-averaged partial cross sections indicated above. This quantity is obtained for the randomly oriented ensemble of ground-state N_2 molecules of Figure 2 by adding the three partial cross sections shown there.

The results depicted in Figures 1 and 2 for H_2O and N_2, respectively, are typical of most molecules in that there is a discrete portion of the spectrum (Figure 1), wherein the final states are labeled according to energy and molecular point-group symmetry, and there are continua beginning at accumulation points, above which the final states are labeled by threshold energies E_t and the directions of electron ejection $\underset{\sim}{k}_t$. These latter directions are integrated over in obtaining the partial cross sections shown in Figure 2. The Schrödinger forms of the states required in Eqs. (2.2) and (2.7) are described further in the following Section.

2.2 Schrodinger eigenstates for anisotropic potentials. The bound-state functions Φ_f and continuum states $\Phi^{(-)}_{\underset{\sim}{k}_t}$ are evidently required in order to evaluate the matrix elements of Eqs. (2.2) and (2.7). To focus attention on those aspects of the situation of interest here, the following development treats the electronic degrees of freedom of a one-electron Hamiltonian $H = T + V$, where T is the kinetic-energy operator and V is the anisotropic (non-central) potential-energy operator appropriate for a fixed-nuclei molecule. Since V can be nonlocal and possibly energy dependent in the present development, no real limitation is made by restriction to a one-electron Hamiltonian, although it is perhaps helpful to have the simplest case of a local potential in mind during the course of the development.

The bound electronic states Φ_f of Eq. (2.4) transform irreducibly under the point group of the potential V, and are generally so classified [2,3]. In the case of Figure 1 (H_2O), for example, the point group is C_{2v}, and there are four (nondegenerate) irreducible representations a_1, a_2, b_1, b_2 labeling the states [33]. The additional discrete-state labels (s,p,d,...) of Figure 1 refer to the angular momenta of zeroth-order atomic states which are split (eg., $p \to pa_1$, pb_1, pb_2) by the anisotropic molecular field - these provide additional quantum labels for the higher Rydberg states. In the so-called united-atom limit, in which case the two hydrogen nuclei coalesce with the oxygen nucleus forming a neon atom, these labels refer to the orbital angular momenta of the electrons, and in this limit are good quantum numbers [33]. The allowable discrete eigenstates Φ_f of the self-adjoint Hamiltonian H are furthermore Lebesgue square integrable (L^2) and orthogonal [1]

$$\langle \Phi_j | \Phi_k \rangle = \delta_{jk}, \tag{2.13}$$

having norms set to unity by convention [7]. As a consequence of the asymptotic boundary conditions implicit in Eq. (2.13), only a subset of solutions of the second-order Sturm-Liouville problem of Eq. (2.4) are acceptable; these must also be regular at the Coulomb singularities of the molecular potential corresponding to the fixed positions of the atomic nuclei [33].

Rayleigh-Ritz methods are widely employed in constructing approximations to the Hilbert space of bound states $\{E_f, \Phi_f ; f = 1,2,\ldots\infty\}$ [1,4,5]. In these methods, an orthonormal, complete, denumerably infinite set of functions $\{\chi_j ; j = 1,2,\ldots\infty\}$ satisfying the appropriate boundary conditions is chosen as a basis for the calculation. A common choice for molecules can be formed from the set of Cartesian Gaussian functions in three-dimensions (R^3) [34],

$$g_j = x^l y^m z^n e^{-\alpha_j |\underset{\sim}{r} - \underset{\sim}{r}_c|^2}, \tag{2.14}$$

with $j \rightarrow (\alpha_j, l, m, n, \underset{\sim}{r}_c)$, where $\underset{\sim}{r}_c$ specifies an atomic or other center, l,m,n are integers, and $\alpha_j = \alpha\beta^j$ is a real exponential parameter. For fixed l,m,n, and $\underset{\sim}{r}_c$ values, the g_j with α and β real form a denumerably infinite complete set in the appropriate limit [51]. More generally, a number of different centers and α,β values are employed in calculations, giving rise to questions of "overcompleteness" or near linear dependence. These difficulties are avoided by forming the overlap or metric matrix $\underset{\sim}{S}$

$$(\underset{\sim}{S})_{jk} = \langle g_j | g_k \rangle, \tag{2.15}$$

for a chosen basis set and determining its eigenvalues (s_j) and eigenvectors (χ_j). Those eigenvectors having very small eigenvalues are discarded [32,52], and the procedure repeated subsequent to the addition of more Gaussians to the basis set. In this way, an arbitrary large set of orthonormal functions can be devised which is linearly independent in a large but finite portion of R^3 and complete in the appropriate limit. By judicious choice of the l, m, n values at the various centers, the resulting "canonical" basis $\{\chi_j ; j =$

1,2,...} can be partitioned into subsets of good symmetry type transforming irreducibly under the point group of interest, although this is not strictly necessary. Finally, the Hamiltonian matrix $\underset{\sim}{H}$

(2.16) $$(\underset{\sim}{H})_{jk} = \langle \chi_j | H | \chi_k \rangle$$

is constructed and its eigenvalues and vectors are obtained from the linear equations

(2.17) $$(\underset{\sim}{H} - \underset{\sim}{I}\tilde{E}_f) \cdot \underset{\sim}{\Phi}_f = 0,$$

where $\underset{\sim}{I}$ is the unit matrix and $\underset{\sim}{\Phi}_f$ is a column eigenvector. The "pseudospectrum" of energies and functions $\{\tilde{E}_f, \tilde{\Phi}_f \, ; \, f = 1,2,..\}$, where

(2.18) $$\tilde{\Phi}_f = \sum_j (\underset{\sim}{\Phi}_f)_j \chi_j \; ,$$

converges monotonically from above to the correct discrete eigenvalues and functions with increase in basis, in accordance with the Ritz-Hylleraas-Unheim principle [8]. A typical example is shown in Table 1 (following page), which depicts a Gaussian basis set for the carbonyl sulfide (OCS) molecule [53], providing excitation and ionization cross sections described further below (Section 5). The computational results shown in Figure 1 are also obtained following the foregoing development in an appropriately chosen Gaussian basis set [40]. The single-channel potentials commonly employed in such molecular excitation and ionization calculations are obtained from the noncentral and nonlocal (Coulomb and exchange) fields acting on a single active electron removed from a close-shell ground-state Hartree-Fock determinental function, an approach designated in the following as the static-exchange approximation [14,32].

In contrast to the case of bound electronic states described above, the Schrödinger states of the continuous or essential spectrum of H are obtained from a standard development in which partial-wave expansions and boundary conditions play a major role [8,13,14]. This choice of method also relates closely to the required resolution of the infinite (or directional) degeneracy of the continuum at any energy E. That is, the specific states required for use in Eq. (2.7) are those which behave asymptotically as directed Coulomb waves with incoming radial wave boundary conditions [41,54], conditions that can be conveniently met employing partial-wave expansions [13,14]. The required states, moreover, must be Dirac delta-function normalized on the energy scale

Table 1
Exponents for supplemental Cartesian Gaussian basis functions employed in OCS static-exchange calculations [a]

σ functions (87)3[b]			π functions (59) [b]		δ functions (35) [b]
s	p	$d_{x^2}, d_{y^2}, d_{z^2}$	p_x	d_{xz}	d_{xy}
centered at oxygen atom					
0.125000	0.125000	0.425000	2.00000	2.00000	1.00000
0.062500	0.062500	0.212500	0.982034	0.982034	0.500000
		0.106250	0.482195	0.482195	0.250000
			0.236766	0.236766	0.125000
centered at carbon atom					
0.062500	0.125000	0.375000	1.50000	1.50000	0.800000
	0.062500	0.187500	0.750000	0.750000	0.400000
		0.093750	0.375000	0.375000	0.200000
			0.187500	0.187500	0.100000
centered at sulfur atom					
0.062500	0.062500	0.920000	1.50000	0.800000	1.50000
0.031250	0.031250	0.230000	0.756042	0.400000	0.756042
		0.115000	0.381066	0.200000	0.381066
			0.192068	0.100000	0.192068
			0.096807	0.050000	0.096807
			0.048794	0.025000	0.048794
centered at center-of-mass of molecule					
0.023438	0.046875	0.046785	1.50000	1.50000	200000
0.011719	0.023438	0.023438	0.756042	0.756042	1.17662
0.005859	0.011719	0.011719	0.381066	0.381066	0.692212
0.002930	0.005859	0.005859	0.192068	0.192068	0.407234
0.001465	0.002930	0.002930	0.096807	0.096807	0.239579
0.000732	0.001465	0.001465	0.048794	0.048794	0.140946
0.000366	0.000732	0.000732	0.024593	0.024593	0.082920
	0.000366	0.000366	0.012396	0.012396	0.048782
			0.006248	0.006248	0.028699
			0.003149	0.003149	0.016884
			0.001587	0.001587	0.009933
			0.000800	0.000800	0.005844
			0.000403	0.000403	0.003438
			0.000203	0.000203	0.002023
					0.001190
					0.000700
					0.000412
					0.000242

[a] Functions employed in addition to the 73 contracted Gaussians that comprise the more compact atomic-orbital basis.

[b] Value in parentheses gives the total number of static-exchange pseudostates obtained in the indicated symmetry.

$$\langle \Phi_{k}^{(-)} | \Phi_{k'}^{(-)} \rangle = \delta(E - E') \, \delta(k - k') \quad , \tag{2.19}$$

as well as in the asymptotic direction of the ejected electron momentum [7], in order to be employed in the matrix element of Eq. (2.7) [41]. For notational convenience the channel label t is dropped from the vector k in the following unless required for clarity.

It is convenient to describe the standard development for the construction of scattering states in the context of a Green's function treatment, which incorporates the boundary conditions in a transparent fashion [8]. The required solutions are written in this development in the form

$$\Phi_{k}^{(-)} = \phi_{k}^{(-)} + G(E-i\delta)V\phi_{k}^{(-)} \tag{2.20}$$

where $\phi_{k}^{(-)}$ is the reference directed Coulomb wave [54], continuum normalized on the energy scale, and $G(E-i\delta)$ is the lower-rim limit $(\delta \to 0)$ of the full Hamiltonian Green's function [8]

$$G(z) = (z-H)^{-1}. \tag{2.21}$$

It is perhaps helpful to recall that [8]

$$G(E-i\delta) = P_c(E-H)^{-1} + i\pi\delta(E-H), \tag{2.22}$$

where P_c is the Cauchy principal value and the second term on the right-hand side provides the incoming (-) scattered flux in the context of Eq. (2.20) [8]. The asymptotic $(r \to \infty)$ behavior of $\Phi_{k}^{(-)}$ is [13,54]

$$\Phi_{k}^{(-)} \to [(2\pi)^{-3}k/2]^{\frac{1}{2}}[e^{i[k\cdot r + \theta(k,r)]} + [f^{(-)}(\hat{r},k)/r]e^{-i[kr-\theta(k,r)]}], \tag{2.23}$$

where the second term, with $f^{(-)}(\hat{r}, k)$ the so-called scattering factor, provides the appropriate angular distribution of incoming flux [41]. The additional phase term $\theta(k,r) = (1/k)\ln(kr - k \cdot r) - \sigma(k)$, where $\sigma(k) = \arg \Gamma(1 + i/k)$ is the complex phase of the Γ function, arises due to the Coulombic nature of the potential in the large r limit [13,54].

Alternative sets of states to those of Eq. (2.20), based on l-wave expansion and traveling- or standing-wave boundary conditions, are generally constructed first in computational developments, providing a basis from which the former are obtained by unitary transformation [8,13,14,25,26,28,43-49]. The standing-wave states satisfy the equation

$$(2.24) \qquad \Phi_E^{(l,m)} = \phi_E^{(l,m)} + G(E)V\phi_E^{(l,m)}$$

where $\phi_E^{(l,m)}$ is obtained from the l, m partial wave of $\phi_{\mathbf{k}}^{(-)}$ and G(E) is the principal value Green's function [8]. Solutions of this type are of so-called K-matrix normalized partial-wave form, since they behave asymptotically ($r \to \infty$) as [8]

$$(2.25) \qquad \Phi_E^{(l,m)} \to (\pi k)^{\frac{1}{2}} (1/r) \sum_{l',m'} Y_{l'}^{m'}(\hat{\mathbf{r}})\{\delta_{ll'}\delta_{mm'} \sin[kr + \theta_l(k,r)]$$

$$+ K(l,m\,;l',m') \cos[kr + \theta_l(k,r)]\} .$$

Here, $\theta_l(k,r) = - l\pi/2 - (1/k)\ln(2kr) + \arg\Gamma(l + 1 - i/k)$, and $K(l,m;\, l',\, m')$ is the K-matrix element which provides a measure of the strength of the anisotropic coupling between waves in the l,m and l', m' channels. The solutions of Eqs. (2.24) and (2.25) are related to those of Eqs. (2.20) and (2.23), and to l-wave solutions satisfying traveling radial wave boundary conditions, by unitary transformations which are of a form that is largely independent of the particular molecular potential under study [13,43-49]. That is, particular linear combinations of the partial-wave solutions of Eqs. (2.24) and (2.25) are required in order to form the asymptotic directed waves of Eqs. (2.20) and (2.23) [41], which requirements relate only to the asymptotic forms of the wave functions involved.

As a consequence of the asymptotic boundary conditions employed, the solutions $\Phi_{\mathbf{k}}^{(-)}$ and $\Phi_E^{(l,m)}$ do not appear to necessarily transform irreducibly under the point group of V. Both sets of solutions, however, can be written as function of good symmetry type in the context of Eq. (2.7) by restricting their partial-wave expansions to particular values [38,39]. Such solutions correspond to different combinations of degenerate states at a given energy within each irreducible symmetry type, and refer asymptotically to the two different quantization schemes of linear or angular momentum. It is useful to also consider linear combinations of degenerate scattering states at a given energy of a particular canonical form, constructed as linear combinations of the $\Phi_E^{(l,m)}$ which bring the K matrix to diagonal form at each energy [55,56]. These so-called eigenchannel states satisfy the equation [8]

$$\Phi_E^{(\alpha)} = \phi_E^{(\alpha)} + G(E)V\,\phi_E^{(\alpha)} \quad , \tag{2.26}$$

where $\phi_E{}^{(\alpha)}$ is obtained from the $\phi_E^{(l,m)}$ and the unitary transform ($\underset{\sim}{U}$) that diagonalizes the K matrix [55,56]. Consequently, the solutions of Eq. (2.26) are generally constructed after the $\Phi_E^{(l,m)}$ are obtained. The asymptotic ($r \to \infty$) form of the $\Phi_E^{(\alpha)}$ is [8,55,56]

$$\Phi_E^{(\alpha)} \to (\pi k)^{-\frac{1}{2}}\,(1/r)\sum_{l,m} Y_l^m(\hat{\underset{\sim}{r}})\,U_{l,m;\alpha}\,\{\sin\,[kr + \theta_l(k,r)] + \tan\,(\delta_\alpha)\cos\,[kr + \theta_l(k,r)]\}, \tag{2.27}$$

where the tangent of the eigenchannel phase δ_α is an element of K in diagonal form, exhibiting the standing-wave behavior of these solutions. It should be noted that the eigenchannel states are of point-group symmetry type as a consequence of the Lippmann-Schwinger equation relating the K matrix and the potential [8,55]. They also constitute an enumeration of the degenerate above-threshold solutions at a given energy which is similar to the l-wave labeling (sa_1, pa_1, da_1, ...) employed for the below-threshold states depicted for H_2O in Figure 1. That is, within each symmetry type (a_1, a_2, b_1, b_2) the label α specifies a particular eigenchannel state of an infinite number of such states at a given total energy.

The cross sectional expressions of Section 2.1 take particularly simple forms when expressed in terms of the eigenchannel solutions of Eqs. (2.26) and (2.27). Specifically, the body-frame, principal-axis, partial-channel cross sections appearing in Eqs. (2.9) to (2.12) take the form

$$\sigma_{i\to t}\,(h\nu,\,\hat{\underset{\sim}{\varepsilon}}_k) = \int_{E_t}^{\infty} dE \sum_{\alpha_t} |\langle\Phi_E^{(\alpha_t)}|\underset{\sim}{\mu}\cdot\hat{\underset{\sim}{\varepsilon}}_k|\Phi_i\rangle|^2\,\delta(E - E_i - h\nu) \quad , \tag{2.28}$$

from which the laboratory-frame cross sections and anisotropy factors are conveniently obtained. The directed-wave solutions of Eqs. (2.20) and (2.23) must be formed as linear combinations of the $\Phi_E^{(\alpha)}$ in order to evaluate the differential cross section of Eq. (2.7) [57]. It is helpful to recall in this connection that the $\Phi_E^{(l,m)}$ and $\Phi_E^{(\alpha)}$ are generally not delta-funtion normalized on the energy scale [8], although this convention is assumed for purposes of analysis in the present development.

The standard Schrödinger development indicated above has been implemented in a number of contexts and approximations [25,26,28,43-49,56-60], in some cases employing basis sets of the type of Eq. (2.14) to represent potential or Green's-function matrix elements [25,26]. These developments have in common an attempt to construct the solutions $\Phi_E^{(l,m)}$ and $\Phi_{\mathbf{k}}^{(-)}$, and in some cases the $\Phi_E^{(\alpha)}$, although the details of the procedures differ in each case. By contrast, in the present development the explicit construction of these states is avoided, although the same physically relevant information is obtained from the calculations.

3. Resolution of continuum degeneracy - definition and construction of interaction-prepared states. The computational development for bound states indicated in Section 2.2 adopts at the outset the notion of point-group symmetry reduction by employing basis functions of good symmetry type. Diagonalization of the Hamiltonian in a basis of mixed symmetry will, in any event, provide good-symmetry solutions [33]. In the case of nondegenerate representations, these bound states require no further enumeration, since this is provided by their distinct energies. As indicated above, however, angular momentum labels arising from united-atom considerations are commonly employed to distinguish among the diffuse Rydberg states of a particular symmetry type [2,3]. In the standard construction of scattering functions, specific steps are taken to completely enumerate the degenerate states at each energy by adoption of particular boundary conditions. The eigenchannel solutions, in particular, provide a computationally convenient infinite set of such solutions at each energy which are of good-symmetry type. Since only very specific linear combinations of the eigenchannel solutions are required to evaluate the matrix elements of Section 2.1, as indicated in Eq. (2.28), it would be desirable to construct these combinations directly, circumventing the standard development of Section 2.2. Moreover, it would be useful to devise an explicit Hilbert-space procedure for this purpose which would extend the bound-state methods employed below threshold to the scattering states above threshold, providing continuity of method and calculated results across the ionization thresholds, with both spectral intervals treated on a common basis.

3.1 Definition of interaction-prepared states. Some insight into achieving an approriate degeneracy reduction of the continuum Schrödinger states is obtained by examining the matrix element of the Hamiltonian resolvent with respect to a test function for a particular principal-axis polarization direction ($\hat{\varepsilon}_k$). The test function employed is

$$\Phi \equiv (\mu \cdot \hat{\varepsilon}_k)\, \Phi_i \tag{3.1}$$

and the resolvent expectation value is

$$\begin{aligned} R_\Phi(z) &= \langle \Phi | (z-H)^{-1} | \Phi \rangle \\ &= \sum_f^\infty \frac{|\langle \Phi_f | \Phi \rangle|^2}{z - E_f} + \sum_t \int_{E_t}^\infty dE \sum_{\alpha_t}^\infty \frac{|\langle \Phi_E^{(\alpha_t)} | \Phi \rangle|^2}{z - E} \; . \end{aligned} \tag{3.2}$$

Here, a sum over energetic channels t, as well as over the eigenchannel functions $\Phi_E^{(\alpha_t)}$ associated with each energetic channel, has been employed in forming the continuum part of the spectral representation of the Green's function. Introducing delta-function normalized, dipole-prepared continuum states as the specific linear combinations of eigenchannel functions [61,62]

$$\Phi_E^{(d_t)} \equiv \sum_{\alpha_t} \Phi_E^{(\alpha_t)} \langle \Phi_E^{(\alpha_t)} | \Phi \rangle / \left(\sum_{\alpha_t} |\langle \Phi_E^{(\alpha_t)} | \Phi \rangle|^2 \right)^{\frac{1}{2}}, \tag{3.3}$$

Eq. (3.2) can be written in the form

$$R_\Phi(z) = \sum_f^\infty \frac{|\langle \Phi_f | \Phi \rangle|^2}{z - E_f} + \sum_t \int_{E_t}^\infty dE \, \frac{|\langle \Phi_E^{(d_t)} | \Phi \rangle|^2}{z - E} \; . \tag{3.4}$$

It is clear from the residue of Eq. (3.4), or from inspection of Eqs. (2.28) and (3.3), that the principal-axis, partial-channel cross sections are obtained in the form

$$\sigma_{i \to t}(h\nu, \hat{\varepsilon}_k) = \int_{E_t}^\infty dE |\langle \Phi_E^{(d_t)} | \Phi \rangle|^2 \, \delta(E - E_i - h\nu). \tag{3.5}$$

Moreover, the spectral representation of G(z), when employed in any matrix element involving Φ on either the left or right side, takes the form

$$G(z) = \sum_f^\infty \frac{|\Phi_f\rangle\langle\Phi_f|}{z - E_f} + \sum_t \int_{E_t}^\infty dE \, \frac{|\Phi_E^{(d_t)}\rangle\langle\Phi_E^{(d_t)}|}{z - E} \; . \tag{3.6}$$

As a consequence, the differential cross section of Eq. (2.7) is obtained from the matrix element

$$(3.7) \qquad \langle\Phi|\Phi^{(-)}_{\mathbf{k}_t}\rangle = \int_{E_t}^{\infty} dE' \, \langle\Phi|\Phi^{(d_t)}_{E',t}\rangle\langle\Phi^{(d_t)}_{E',t}|1 + V(E - E' - i\delta)^{-1}|\phi^{(-)}_{\mathbf{k}_t}\rangle,$$

where the (nonsingular) discrete sum has been cancelled against an identical contribution from the reference-wave term employing the closure relation.

The foregoing development indicates the dipole interaction prepared states of Eq. (3.3) provide the necessary and sufficient information for evaluation of the relevant cross sections of Section 2. It would be useful to devise practical means for construction of these states at each energy without obtaining first all of the degenerate eigenchannel solutions required in the defining relation.

3.2 Hilbert-space construction of interaction-prepared states. The dipole interaction-prepared states $\Phi^{(d)}_E$ of Eq. (3.3) will be seen in the following development to constitute the continuum portion of an invariant subspace $H_s(\Phi)$ of the Hilbert space H_s in which H is a closed linear (self-adjoint) operator [35,63]. It is convenient to construct the entire subspace $H_s(\Phi)$, as it will be seen to include the discrete dipole allowed states Φ_f, as well as the continuum dipole-interaction prepared $\Phi^{(d)}_E$ states. Additionally, it is helpful to regard the canonical Hilbert space $\{\chi_j \; ; \; j = 1,2,...\}$ or the pseudostate Hilbert space $\{\tilde{\Phi}_j \; ; \; j = 1,2,...\}$ described above as suitable basis sets for the construction of $H_s(\Phi)$.

The development begins with construction of a unity normalized representation (v_1) of the test function Φ of Eq. (3.1) in the available basis of L^2 functions. This starting vector v_1 is then employed to determine a set of n orthonormal vectors $\{v_j \; ; \; j = 1,2,...n\}$ which bring the matrix representation of an operator A(H), specified below, to tridiagonal form [17]. This is accomplished recursively employing Lanczos methods [64,65] starting from diagonal or nondiagonal forms of H in either of the two basis sets of functions [29-32]. The resulting nxn tridiagonal matrix representation of A(H) is called $\mathbb{A}^{(n)}$, having diagonal and off-diagonal elements of the forms

$$(3.8) \qquad \alpha_j = \langle v_j|A(H)|v_j\rangle, \; j = 1,2,...n,$$

$$(3.9) \qquad \beta_j = \langle v_{j+1}|A(H)|v_j\rangle, \; j = 1,2,...n-1.$$

These values are obtained from the recursive procedure for determining the v_j, which employs the equation [32],

(3.10) $\beta_j v_{j+1} = (A(H)-\alpha_j)v_j - \beta_{j-1}v_{j-1}$, $j = 1,2,\ldots n-1$,

with $v_o=0$ and v_1 given by the test function Φ. The nature of the orthonormal vectors v_j is made clear by writing them in the form [17]

(3.11) $v_j = q_j[A(H)]\ \Phi/\langle\Phi|\Phi\rangle^{1/2}$,

where $q_{j+1}(E)$ is a characteristic polynomial of the matrix $\mathbb{A}^{(j)}$ [32]

(3.12) $q_{j+1}(E) = (-1)^j \det |\mathbb{A}^{(j)} - E\mathbb{I}^{(j)}| / (\beta_1\beta_2\ldots\beta_j)$,

satisfying the recurrence relation [15]

(3.13) $\beta_j q_{j+1}(E) = (A(E) - \alpha_j)q_j(E) - \beta_{j-1}q_{j-1}(E)$, $j= 1,2,\ldots n-1$,

with $q_o = 0$, $q_1 = 1$ as starting conditions [32]. The v_j correspond to an orthogonal rearrangement of the power basis $A(H)^j\Phi$ for $j = 1,2,\ldots$, the recursive Lanczos procedure providing a computationally convenient and stable method for constructing the elements of this space [63-65]. To insure that this basis forms an infinite-dimensional Hilbert space ($j\to\infty$), the operator $A(H)$ can be chosen as [32]

(3.14) $A(H) = (H - E_i)^{-1}$,

giving an infinite dimensional power or Cauchy basis of finite norm for all j [29].

Taking all linear combinations of the finite number of functions v_j, $j = 1,\ldots n$, provides a linear manifold L_Φ. An arbitrary element in L_Φ can be written [17]

(3.15) $$\Phi_E^{(n)} = \sum_{j=1}^{n} q_j(E)\ v_j$$

$$= \sum_{j=1}^{n} q_j(E)\ q_j[A(H)]\ \Phi/\langle\Phi|\Phi\rangle^{\frac{1}{2}} ,$$

where $q_j(E)$ and $q_j[A(H)]$ are the above indicated polynomials. The metric of these elements is

(3.16) $$\langle \Phi_E^{(n)} | \Phi_{E'}^{(n)} \rangle = \sum_{j=1}^{n} q_j(E)\, q_j(E')$$

which is identified as the polynomial kernel $K^{(n)}(E,E')$ of the $q_j(E)$ [17]. In the limit $n \to \infty$ the roots of $q_n(E)$ converge to eigenvalues E_f below threshold ($E<E_t$), whereas the roots become dense above threshold ($E \geq E_t$) [15]. As a consequence of this behavior, it can be shown that for eigenvalues E_i and E_j of H and A(H)

(3.17) $$K^{(n\to\infty)}(E_i,E_j) = \delta_{ij}/[\Gamma^{(n\to\infty)}(E_i)/\langle \Phi | \Phi \rangle],$$

where [15]

(3.18) $$\Gamma^{(n)}(E_i) = [\sum_{j=1}^{n} q_j(E_i)^2]^{-1},$$

is the quadrature weight associated with the point E_i. Moreover, the kernel $K^{(n\to\infty)}(E,E')$ vanishes for all other E and E' below threshold. Therefore, the states [32]

(3.19) $$\Phi_{E_f}^{(n)} = \Gamma^{(n)}(E_f)^{\frac{1}{2}} \sum_{j=1}^{n} q_j(E_f) q_j[A(H)]\Phi / \langle \Phi | \Phi \rangle^{\frac{1}{2}}$$

converge to the Φ_f when limit points of the manifold of L_Φ are included. This is demonstrated by inserting the set of discrete eigenstates Φ_k, $k = 1,2,\ldots\infty$, in Eq. (3.19), giving the expression

(3.20) $$\Phi_{E_f}^{(n)} = \Gamma^{(n)}(E_f)^{\frac{1}{2}} \sum_{k=1}^{\infty} [\sum_{j=1}^{n} q_j(E_f) q_j(E_k)]\Phi_k [\langle \Phi_k | \Phi \rangle / \langle \Phi | \Phi \rangle^{\frac{1}{2}}] .$$

The term in brackets on the right-hand side becomes the Kronecker delta δ_{fk} [Eq. (3.17)] in the limit $n\to\infty$, so that

(3.21) $$\Phi_{E_f}^{(n\to\infty)} \to \Gamma^{(\infty)}(E_f)^{-\frac{1}{2}}\, \Phi_f \langle \Phi_f | \Phi \rangle \langle \Phi | \Phi \rangle^{\frac{1}{2}}$$

$$= \Phi_f,$$

where the quadrature weight of Eq. (3.18) satisfies

(3.22) $$\Gamma^{(\infty)}(E_f) = |\langle \Phi_f | \Phi \rangle|^2 \langle \Phi | \Phi \rangle,$$

providing the appropriate factor to insure unity normalization [32].

The above-threshold limiting behavior of $K^{(n)}(E,E')$ is [32]

$$K^{(n\to\infty)}(E,E') \to \delta(E-E')/[\sigma(E)/\langle\Phi|\Phi\rangle], \tag{3.23}$$

in accordance with Dirac normalization of continuum states, where

$$\sigma(E) \equiv \sigma_{i\to t}(E-E_i,\ \hat{\varepsilon}_k) \tag{3.24}$$

is the partial-channel photoionization cross section for principal-axis polarization direction $\hat{\varepsilon}_k$ associated with the channel of interest [Eq. (3.1)]. As a consequence, the states $(E>E_t)$

$$\Phi_E^{(n)} = [\sigma(E)^{\frac{1}{2}}/\langle\Phi|\Phi\rangle] \sum_{j=1}^{n} q_j(E) q_j[A(H)]\Phi/\langle\Phi|\Phi\rangle^{\frac{1}{2}}, \tag{3.25}$$

converge to the dipole-prepared states of Eq. (3.3) in the limit $n\to\infty$ [29] The convergence of Eq. (3.25) is demonstrated by inserting the complete set of eigenchannel states, giving,

$$\Phi_E^{(n)} = [\sigma(E)/\langle\Phi|\Phi\rangle]^{\frac{1}{2}} \int_{E_t}^{\infty} dE' \sum_{\alpha_t}^{\infty} \Phi_{E'}^{(\alpha_t)} [\sum_{j=1}^{n} q_j(E) q_j(E')] \langle\Phi_{E'}^{(\alpha_t)}|\Phi\rangle/\langle\Phi|\Phi\rangle^{\frac{1}{2}}. \tag{3.26}$$

Now taking the limit $n\to\infty$ and using Eq. (3.23) gives

$$\begin{aligned} \Phi_E^{(n\to\infty)} &\to \sum_{\alpha_t}^{\infty} \Phi_E^{(\alpha_t)} \langle\Phi_E^{(\alpha_t)}|\Phi\rangle/[\sum_{\alpha_t}^{\infty}|\langle\Phi_E^{(\alpha_t)}|\Phi\rangle|^2]^{\frac{1}{2}} \\ &= \Phi_E^{(d)} \end{aligned} \tag{3.27}$$

from the definition of Eq. (3.3). Thus, when the limit points of L_Φ are included, the subspace $H_s(\Phi)$ corresponds to all discrete states dipole connected to the ground state for the given polarization direction, as well as the dipole-prepared states above threshold.

It remains to investigate the nature of convergence of the Hilbert-space approximations to Schrödinger states of Eqs. (3.15), (3.20), and (3.26). It should also be noted that the development does not contain a prescription for construction of the cross section of Eq. (3.24), which is required as the appropriate normalization factor in Eq. (3.25). Moment theoretical methods for constructing the cross section from the polynomial recurrence coefficients of Eqs. (3.8) and (3.9) are described in the following section [29-32].

4. Moment-theoretical construction of spectral density functions - applications of the classical theory to photoionization cross sections. Previously described procedures for calculating distributions and densities are briefly reviewed and applied to cross sectional determinations [29-32], and recent refinements of these methods for resonances are indicated [36].

4.1 Stieltjes development for photoionization cross sections - nonresonant situations. The cross section $\sigma(E)$ of Eq. (3.24) associated with a particular energetic channel t is seen to provide the appropriate (delta-function) normalization factor required in the dipole interaction-prepared states of Eq. (3.26). It is the case that the polynomial recurrence coefficients of Eqs. (3.8) and (3.9) provide the information necessary for determination of convergent approximations to $\sigma(E)$ employing the theory of moments [15], much as they provide for normalization of the bound states of Eq. (3.19) through the weight factor of Eq. (3.18). To clarify the connection of moment methods with the development of Section 3, consider the (moment) metric

$$(4.1) \qquad \mu_{k+l} = \langle A^k\Phi | A^l\Phi\rangle$$

$$= \int_{-\infty}^{\infty} \epsilon^{-(k+l)} df(\epsilon)$$

with

$$(4.2) \qquad f(\epsilon) = \sum_{f=1}^{\infty} |\langle\Phi_f|\Phi\rangle|^2 \theta(\epsilon_f - \epsilon) + \int_{\epsilon_t}^{\epsilon} d\epsilon' |\langle\Phi_{\epsilon'}^{(d)}|\Phi\rangle|^2,$$

where the transition energy $\epsilon \equiv E-E_i$ has been adopted, and $\theta(\epsilon_f-\epsilon)$ is the unit step function at $\epsilon_f \equiv E_f-E_i$. Evidently, the norms of the (Cauchy) vectors $A^k\Phi$ generate a set of moments [29], with $f(\epsilon)$ corresponding to the appropriate distribution and $\sigma(\epsilon)$ for $\epsilon>\epsilon_t \equiv E_t-E_i$ the associated density [15]. The moments of Eq. (4.1) can be calculated in Hilbert space from a matrix representation of H [29]. However, the Lanczos development of Eqs. (3.8) to (3.14) largely circumvents the instabilities that can arise in dealing explicitly with power moments, with the α_j and β_j providing an alternative computationally stable tridiagonal representation of the metric matrix of Eq. (4.1) [32]. The polynomials of Eqs. (3.12) and (3.13), which follow from the calculated α_j and β_j values, are orthonormal with respect to the distribution $f(\epsilon)$ [15],

$$(4.3) \qquad \int_{-\infty}^{+\infty} df(\epsilon)\; q_j(\epsilon)\; q_k(\epsilon) \;=\; \delta_{jk}.$$

They furthermore contain all the information required to construct bounding approximation to $f(\epsilon)$ from the so-called Tchebycheff inequalities [15]. These provide, in turn, approximations to the density $\sigma(\epsilon)$, which completes the determination of the dipole interaction-prepared states of Eq. (3.23) to (3.27).

Bounds on $f(\epsilon)$ are obtained at any ϵ $(-\infty<\epsilon<+\infty)$ in the form of Tchebycheff inequalities [15]

$$(4.4) \qquad f^{(n)}(\epsilon-o) < f^{(n+1)}(\epsilon-o) \le f(\epsilon) \le f^{(n+1)}(\epsilon+o) < f^{(n)}(\epsilon+o),$$

where,

$$(4.5) \qquad f^{(n)}(\epsilon) = \sum_i \Gamma^{(n)}(\tilde{\epsilon}_i^{(n)}) + \Gamma^{(n)}(\epsilon)$$

is a histogram having step heights

$$(4.6) \qquad \Gamma^{(n)}(\epsilon) = \Big(\sum_{j=1}^{n} q_j(\epsilon)^2\Big)^{-1} \langle\Phi|\Phi\rangle,$$

where the sum over i in Eq. (4.5) is limited to values for which $\tilde{\epsilon}_i^{(n)}<\epsilon$. The n-1 points $\tilde{\epsilon}_i^{(n)}$, which are implicit functions of ϵ, are obtained as the roots of the quasi-orthogonal polynomial [15]

$$(4.7) \qquad \tilde{q}_{n+1}(\tilde{\epsilon}_i^{(n)}, \epsilon) = q_{n+1}(\tilde{\epsilon}_i^{(n)}) - q_{n+1}(\epsilon)\; q_n(\tilde{\epsilon}_i^{(n)}) \;/\; q_n(\epsilon)$$

having one root fixed at the point ϵ. These points ϵ and $\tilde{\epsilon}_i^{(n)}$, and the corresponding weights of Eq. (4.6), are the Radau quadratures associated with the moments of Eq. (4.1), and are seen to be determined entirely by the $q_j(\epsilon)$ [15]. Approximations to the density $\sigma(\epsilon)$ are obtained from the bounding distributions of Eqs. (4.4) to (4.6) in the form

$$(4.8) \qquad \sigma^{(n)}(\epsilon) = \frac{d}{d\epsilon}\,[f^{(n)}(\epsilon+o) + f^{(n)}(\epsilon-o)]/2.$$

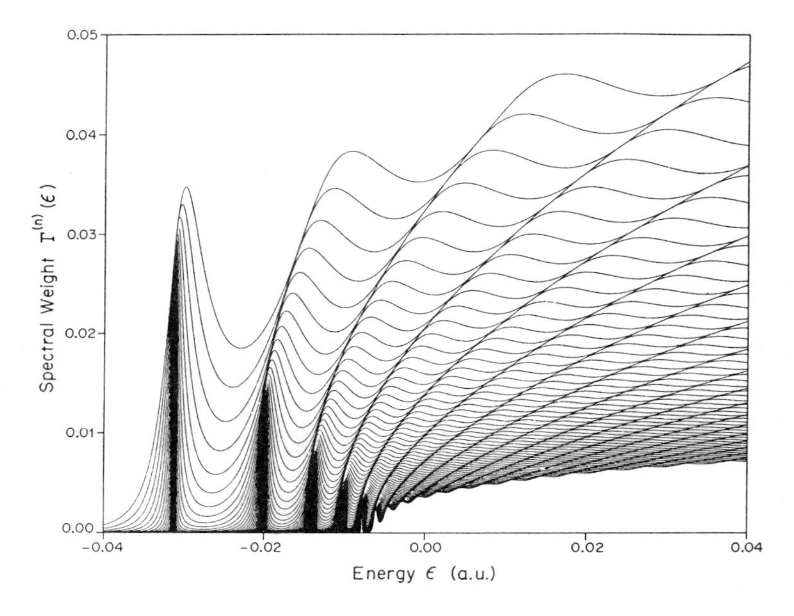

Figure 3. Quadrature weight function of Eq. (4.7) (n = 20 to 100) for the 1s→np/kp spectrum (n>3) in atomic hydrogen [67-69]. (1 a.u. = 27.212 eV).

Here, the derivative is evaluated in the standard fashion, since the average value of the bounds on $f(\epsilon)$ - the Stieltjes value - is a continuous function of ϵ [66]. As an alternative to Eq. (4.8), finite-difference approximations to $\sigma(\epsilon)$ can also be constructed from the points and weights of Eqs. (4.6) and (4.7), both approximations converging to $\sigma(\epsilon)$ in the limit [66].

Rather good cross sectional approximations can be obtained from the foregoing development when $\sigma(\epsilon)$ is a sufficiently slowly varying function of ϵ, or the index n is made sufficiently large. The ground-state p-wave excitation and ionization spectrum of atomic hydrogen provides a useful illustrative example [67-69]. In Figure 3 is shown the quadrature weight of Eq. (4.6) in this case. In accordance with Eqs. (3.17) and (3.18), eigenvalue behavior is evident in the discrete spectral interval ($\epsilon<0$), whereas the weights are seen to decrease monotonically with increasing n above threshold ($\epsilon>0$). The photoionization cross sections obtained for atomic hydrogen from the foregoing development using the weight functions of Figure 3 are found to converge rapidly to the known values, as do the corresponding discrete and continuum states of Eqs. (3.18) and (3.27) [67-69].

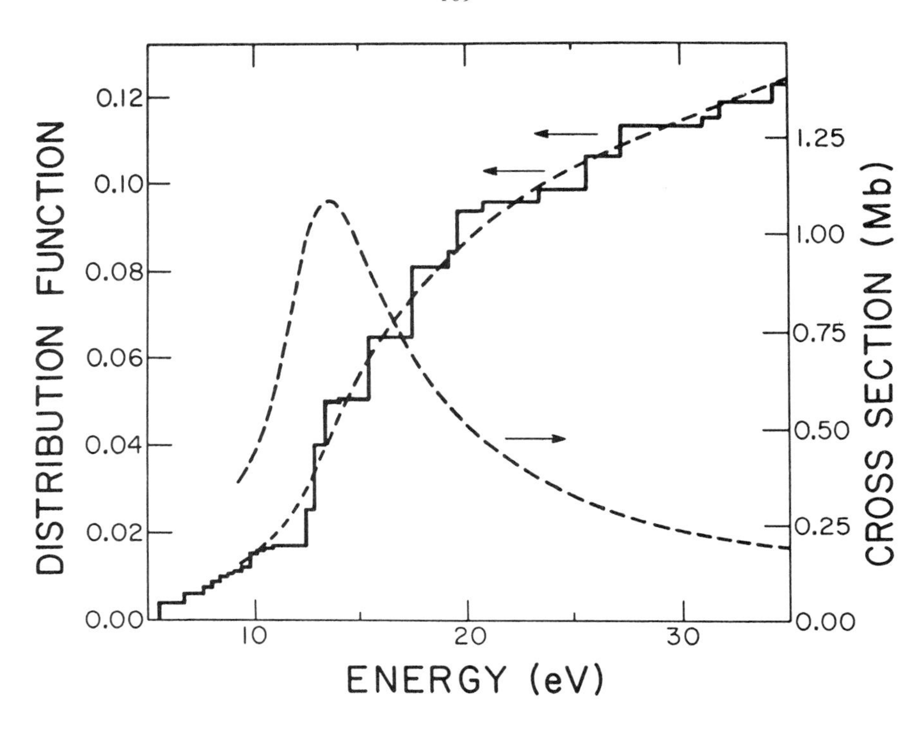

Figure 4. Stieltjes distribution and density of Eqs. (4.5) to (4.9) for $2\pi \rightarrow k\sigma$ ionization of nitric oxide (---), obtained employing n = 10 and a pseudospectrum of transition energies and weights (—) constructed in a Gaussian basis set [70].

As a second example, Figure 4 reports the Stieltjes distribution and density of Eqs. (4.4) to (4.8) for $2\pi \rightarrow k\sigma$ ionization in nitric oxide (NO), calculated employing a pseudostate basis of Gaussian functions in static-exchange approximation [70]. Also shown in the figure is the approximate cumulative histrogram obtained directly from the calculated pseudo transition energies and f numbers or weights (solid line). The Stieltjes distribution of Eq. (4.5) for n = 10 is seen to provide a smooth averaging of the step-function data, the derivative of Eq. (4.8) providing a correspondingly smooth approximation to the photoionization cross section.

Some indication of the resolution achieved from the Stieltjes development in a given order n is obtained from the lineshape functions

$$(4.9)\quad L^{(n)}(\epsilon',\epsilon) = \sigma^{(n)}(\epsilon)\ \Gamma^{(n)}(\epsilon')\sum_{j=1}^{n} q_j(\epsilon)\ q_j(\epsilon'),$$

which provide a decomposition of the cross section in the form

$$(4.10)\quad \sigma^{(n)}(\epsilon) = \sum_{j=1}^{n} L^{(n)}(\tilde{\epsilon}_j^{(n)},\ \epsilon)\ .$$

where the $\tilde{\epsilon}_j^{(n)}$ are nth-order Stieltjes quadrature points [71].

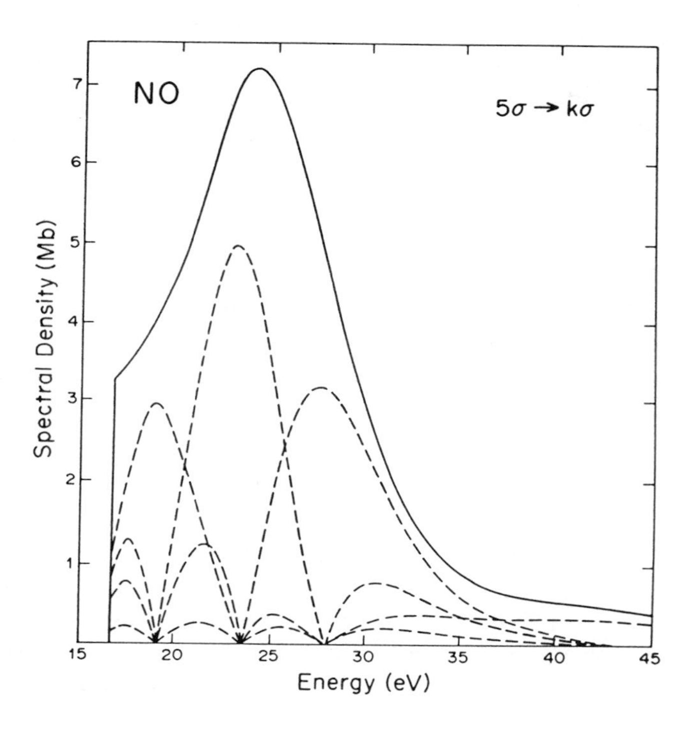

Figure 5. Spectral lineshape functions (squared) and cross section of Eqs. (4.9) and (4.10), respectively, for $(5\sigma \rightarrow k\sigma)b^3\Pi$ ionization of nitric oxide (NO) in 10th order, obtained employing a Gaussian basis set of functions and the development of the text [70].

Typical results are shown in Figure 5 for n = 10 in the case of ($5\sigma \rightarrow k\sigma$)$b^3\Pi$ ionization of the nitric oxide molecule (NO), obtained in a Gaussian basis set [70]. Evidently, the individual lineshape functions are rather broad in this case, suggesting that higher orders (n) of the development may be required to achieve complete resolution of the cross section in this case. Alternatively, Feshbach-Fano methods can be employed in conjunction with the Stieltjes development in cases of highly structured profiles, as described immediately below [36].

4.2 Stieltjes methods for Feshbach-Fano situations - applications to shape resonances. Many of the molecular partial-channel photoionization cross sections [24] investigated to date exhibit resonance behaviors with changing incident photon energy which can be understood on the basis of a single-channel picture, in which the sum of eigenchannel phases in the appropriate channel increases by approximately π radians in the neighborhood of the resonance [25,26,28]. Although the foregoing Stieltjes development can accommodate such shape-resonance behaviors in cross sections quite generally, it is in some ways advantageous to explicitly incorporate spatially compact "zeroth-order" resonance states into the theoretical study of molecular photoionization, following the lead of Feshbach [72,73] and Fano [61,62]. In this way, an underlying nonresonant continuum can be defined which is amenable to computational treatment, the resonance behavior of the cross section being incorporated into a lineshape containing slowly varying functions of the energy [72-74]. In most molecules studied to date the number and types of so-called virtual valence orbitals present correctly indicate the appearance of resonance features in particular ionization continua [27]. Moreover, the spectral locations of such states correlate with bond lengths in a well-understood and predictive fashion [75]. Most important in the present context, however, is the observation that the explicit adoption of such zeroth-order states into the theoretical development can provide a significant improvement in the resolution of calculations of resonant cross sections employing the explicit Hilbert-space methods described in the preceding Sections.

Rigorous formulations of the resonance problem in terms of the interaction of a zeroth-order state with a background continuum have been reported in a number of contexts [72-74]. Feshbach, in particular, showed how the projection-operator formalism could be used to partition the continuum problem in this way beginning with an arbitrary zeroth-order resonance state

[72,73]. Although the state in his development is indeed arbitrary, a good choice may nevertheless favorably affect the ease of calculation and the quality of the results. The present choice of zeroth-order state and discussion of its particular attributes is provided below [36,76]. In the context of the Fano parameterization [61,62], this choice is seen to give a vanishing background cross section [$\sigma_b(E) \to 0$], a diverging profile index [$q(E) \to \infty$], and, in the presence of degenerate continua, a unity correlation coefficient [$\rho(E) \to 1$], providing a Lorentzian cross section with energy-dependent shift [$F(E)$] and width [$\Gamma(E)$].

The classic situation of a single nondegenerate continuum is treated here for simplicity, with attention restricted to the presence of a single zeroth-order state ϕ [61]. In this case, a set of background continuum states ϕ_E is first constructed by diagonalizing PHP, where H is the Hamiltonian and the projector P is $1 - |\phi\rangle\langle\phi|$ [72,73]. The coupling between ϕ and the background continuum is then contained in the off-diagonal matrix elements $V(E) = \langle\phi_E|H|\phi\rangle$; these and the transition amplitudes, $\langle\phi_E|\boldsymbol{\mu}\cdot\hat{\boldsymbol{\varepsilon}}|\Phi_i\rangle$ and $\langle\phi|\boldsymbol{\mu}\cdot\hat{\boldsymbol{\varepsilon}}|\Phi_i\rangle$, completely determine the full wave function Φ_E and hence the cross section $\sigma(E)$. The particular parameterization of Fano gives for the cross section [61,62]

$$\sigma(E) = \sigma_b(E)\,\frac{[q(E) + \epsilon(E)]^2}{1 + \epsilon(E)^2}\,, \tag{4.11}$$

where

$$\sigma_b(E) = |\langle\phi_E|\boldsymbol{\mu}\cdot\hat{\boldsymbol{\varepsilon}}|\Phi_i\rangle|^2 \tag{4.12}$$

is the background cross section, and the dimensionless energy variable $\epsilon(E)$,

$$\epsilon(E) = \frac{E - F(E) - E_o}{\frac{1}{2}\Gamma(E)}\,, \tag{4.13}$$

is expressed in terms of the width,

$$\Gamma(E) = 2\pi|V(E)|^2, \tag{4.14}$$

shift,

$$F(E) = P_c\int_{-\infty}^{+\infty} dE'\,\frac{|V(E')|^2}{E - E'}\,, \tag{4.15}$$

and zeroth-order energy

$$E_o = \langle\phi|H|\phi\rangle. \tag{4.16}$$

All singular integrals in the development, such as that of Eq. (4.15) are regarded as Cauchy principal values (P_c). As can be seen from Eqs. (4.11) and

(4.13), in the neighborhood of a resonance the width and shift functions will correspond to the actual width of the resonance and to the shift in its position from the zeroth-order energy E_o of Eq. (4.16), provided these functions are sufficiently slowly varying with E. The dimensionless profile index q(E), which governs the lineshape, is given by [61,62]

$$(4.17) \qquad q(E) = \frac{\langle\phi|\mu\cdot\hat{\varepsilon}|\Phi_i\rangle + P_c \int dE' \langle\phi|H|\phi_{E'}\rangle\langle\phi_{E'}|\mu\cdot\hat{\varepsilon}|\Phi_i\rangle/(E-E')}{\pi\langle\phi|H|\phi_E\rangle\langle\phi_E|\mu\cdot\hat{\varepsilon}|\Phi_i\rangle} .$$

Finally, the full wave function is given by the expression

$$(4.18) \qquad \Phi_E = a(E)\,(\phi + P_c \int dE' \frac{V(E')}{E-E'} \phi_{E'} + \frac{E-E_o-F(E)}{V(E)} \phi_E) ,$$

where

$$(4.19) \qquad |a(E)|^2 = \frac{|V(E)|^2}{[\Gamma(E)/2]^2 + [E-F(E)-E_o]^2} .$$

In the present treatment, all spatial functions appearing in Eqs. (4.11) to (4.19) are represented in a finite L^2 basis. Diagonalizing PHP in this basis yields background pseudostates $\tilde{\phi}_i$ which define a set of quadrature points $\tilde{E}_i$ and weights

$$(4.20) \qquad \tilde{\Gamma}_i = 2\pi|\langle\tilde{\phi}_i|H|\phi\rangle|^2$$

from which the width function Γ(E) of Eq. (4.14), approximations to associated background wave functions, and all other quantities may be determined following the Stieltjes development for nonresonant continua described in Section 4.1.

As mentioned above, the choice of zeroth-order state is in principle arbitrary, but in practice a mathematically and physically sensible choice will help to approximate the full wave function at resonance. A good choice might be one for which the parameters characterizing the resonance [Γ(E), F(E), and q(E)] vary little with energy, at least over the width of the resonance [61,62]. Alternatively, the shift, F(E), could be minimized or made to vanish [77]. Any reasonable choice of zeroth-order state should also leave a background cross section that is smooth through the region of the resonance; otherwise there is little point in partitioning the problem this way. For shape resonances arising in a nondegenerate photoionization channel, when a single strong resonance dominates the spectrum, the choice

$$(4.21) \qquad \phi = \frac{\int dE\ \Phi_E \langle \Phi_E | \mu \cdot \hat{\varepsilon} | \Phi_i \rangle}{[\int dE\ \sigma(E)]^{\frac{1}{2}}} ,$$

written in terms of the eigenstates Φ_E of H, proves to be particularly convenient for a variety of reasons. First, it should be noted that in an L^2 representation of the operator H the choice of Eq. (4.21) becomes

$$(4.22) \qquad \phi = \frac{\Sigma_j \tilde{\Phi}_j \langle \tilde{\Phi}_j | \mu \cdot \hat{\varepsilon} | \Phi_i \rangle}{[\Sigma_j \tilde{f}_j]^{\frac{1}{2}}}$$

where the $\tilde{f}_j$ are pseudostate oscillator strengths, so that ϕ is easily calculated and normalized. Second, the energy of this ϕ is a cross-section-weighted average over all energies,

$$(4.23) \qquad E_o = \langle \phi | H | \phi \rangle = \frac{\int dE\ E\ \sigma(E)}{\int dE\ \sigma(E)} .$$

A small shift is therefore guaranteed when the cross section contains a single strong resonance. Third, with the choice of Eq. (4.21) for the zeroth-order state, the background is not merely nonresonant, but has zero cross section [Eq. (4.12)],

$$(4.24) \qquad \sigma_b(E) = |\langle \phi_E | \mu \cdot \hat{\varepsilon} | \Phi_i \rangle|^2 = 0.$$

Fourth, although the Fano parameterization of Eq. (4.11) breaks down when the zeroth-order state is defined by Eq. (4.21), as $\sigma_b(E)$ vanishes and q(E) becomes infinite, it is nevertheless easy to rearrange Eq. (4.11) to remove the singularity, leading to

$$(4.25) \qquad \sigma(E) = |\langle \phi | \hat{\mu} \cdot \varepsilon | \Phi_i \rangle|^2 |a(E)|^2$$

$$= \frac{|\langle \phi | \mu \cdot \hat{\varepsilon} | \Phi_i \rangle|^2 |V(E)|^2}{[\Gamma(E)/2]^2 + [E - F(E) - E_o]^2} .$$

The form of Eq. (4.25) is that of a pure Lorentzian lineshape with energy dependent width Γ(E) and shift F(E). For the case of a single strong resonance this gives a very natural description of the cross section, although Eqs. (4.21) and (4.25) can be used for any type of cross-sectional profile [36]. It is to be expected, however, that large variations in Γ(E) may be necessary to describe non-resonant or multi-resonant spectra, in which cases there is perhaps no advantage over the nonresonant development of Section 4.1.

Finally, although attention focuses here on study of a single (lowest) shape resonance, the choice of Eq. (4.21) proves useful for treatment of multiple (higher) resonances, as well, in which cases multiple projection procedures must generally be employed following conventional developments [78].

In the presence of directionally degenerate continua in a single energetic channel, the zeroth-order state of Eq. (4.21) becomes

$$(4.26) \qquad \phi = \frac{\int dE \sum_{\alpha} \Phi_E^{(\alpha)} \langle \Phi_E^{(\alpha)} | \mu \cdot \hat{\varepsilon} | \Phi_i \rangle}{[\int dE \sum_{\alpha} \sigma_{\alpha}(E)]^{\frac{1}{2}}},$$

where the label α distinguishes among the degenerate eigenchannel states of energy E. The dipole interaction with the background orthogonal to Eq. (4.26) vanishes, and only that particular linear combination of background states which contains the entire coupling to ϕ, the dipole-prepared state of Section 3.1, needs to be considered. Thus, the molecular problem is, in effect, reduced to one involving a nondegenerate continuum, even in the presence of directional degeneracy, and the entire development of Eqs. (4.11) to (4.26) is consequently applicable to molecular shape-resonance problems [36].

To illustrate the present approach, applications are reported in the case of the lowest p-wave resonance arising from photoionization of the ground s state of a three-dimensional spherical square well surrounded by a square barrier [79]. Specifically, the potential studied is of the form

$$(4.27) \qquad \begin{aligned} V(r) &= -V_o, && 0 \le r \le 1\ a_o \\ &= V_h, && 1\ a_o \le r \le 1.2\ a_o \\ &= 0, && 1.2\ a_o < r, \end{aligned}$$

where the well depth, V_o, and the barrier height, V_h, are arbitrary. The simplicity of the problem allows an exact solution to be obtained for the cross section [36], whereas by varying the height of the barrier the width of the resonance may be made arbitrarily small. In Figure 6 are shown the exact cross sections for a well depth (V_o) of 5 atomic units (a.u.) and various barrier heights (V_h = 5,10,20,40,80 a.u.). Evidently, both broad and narrow resonances are produced by the range of barrier heights employed. In Figure 7 are shown the width and shift functions for these cases, calculated employing the Stieltjes techniques of Section 4.1 and basis sets of 250 sine waves which are made to vanish at a radius (25 a_o) large compared to the well dimensions.

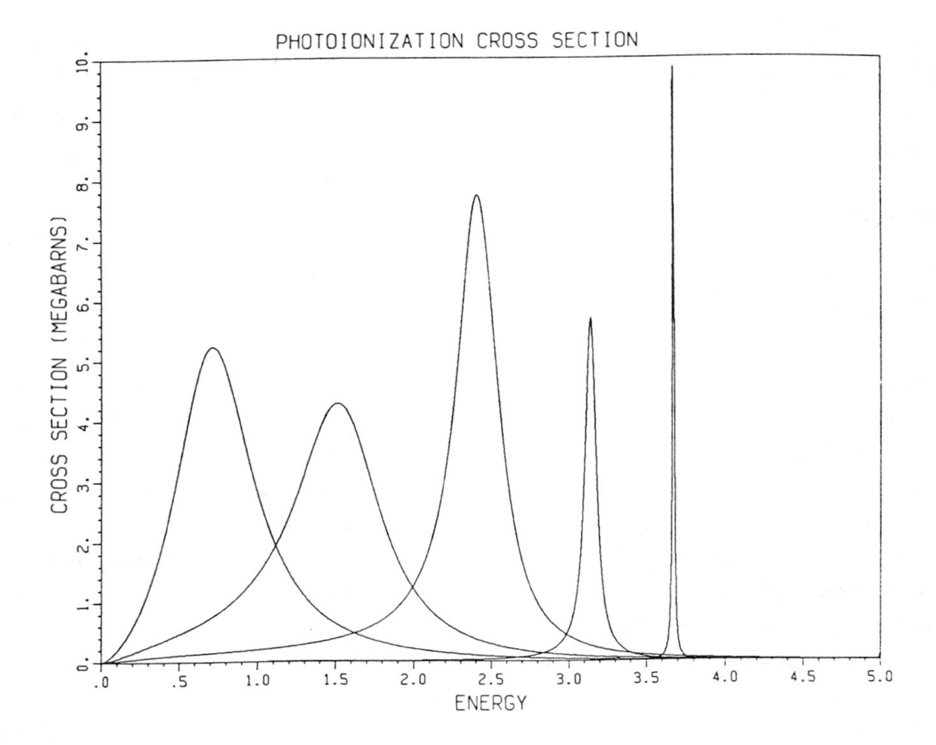

Figure 6. Cross sections for photoionization of the ground state of the potential of Eqs. (4.27) for V_o = 5 a.u. and V_h = 5, 10, 20, 40, and 80 a.u., reading from left to right, calculated from analytic solution of the Schrödinger equation [36]. Values in the cases V_h = 40 and 80 a.u. are reduced by factors of 5 and 20, respectively, for clarity of presentation.

The smallest width function of Figure 7 corresponds to the very narrow cross section (V_h = 80 a.u.) of Figure 6. In spite of the rather different natures of the cross sections, which range from rather broad with peak height of 4.2 Mb to very narrow with peak height of 198 Mb, the width and shift functions of Figure 7 are seen to be generally similar in overall shape and in magnitude. Moreover, although the cross sectional maxima are all at rather different energies, the width functions are seen to all peak in a narrow energy range.

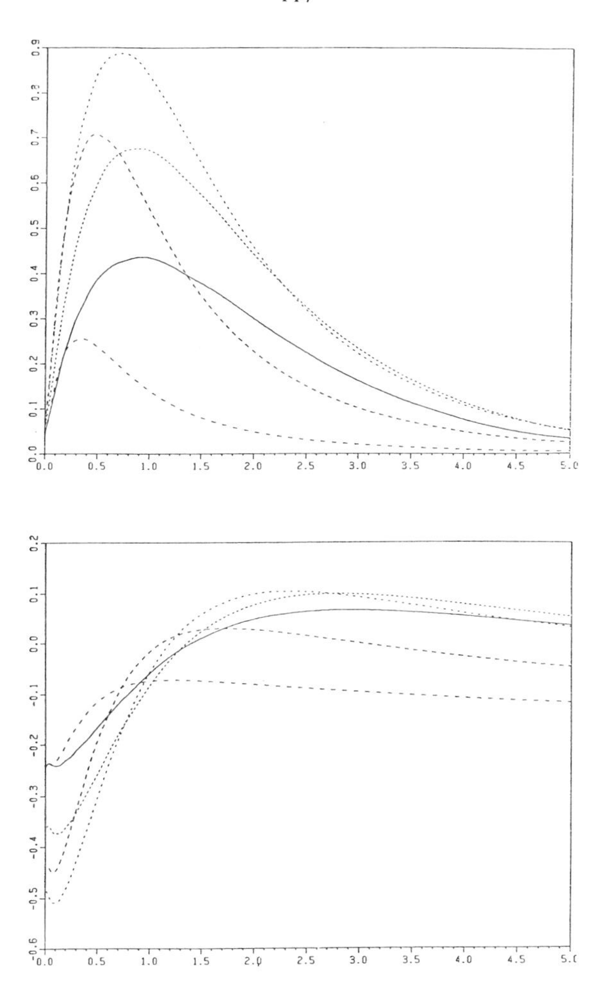

Figure 7. Width (top) and shift (bottom) functions of Eqs. (4.14) and (4.15), respectively, describing the interaction of the zeroth-order state of Eq. (4.21) with the associated background for the barrier problems of Eqs. (4.27) and Figure 6. All values are in atomic units (a.u.). The calculations employ L^2 basis sets of 250 sine functions and the Stieltjes development in tenth-to-fifteenth order to achieve convergence [36], providing an energy resolution of approximately 0.2 a. u. in all cases. The solid lines refer to the cross section at the center of Figure 6 (V_h = 20 a.u.).

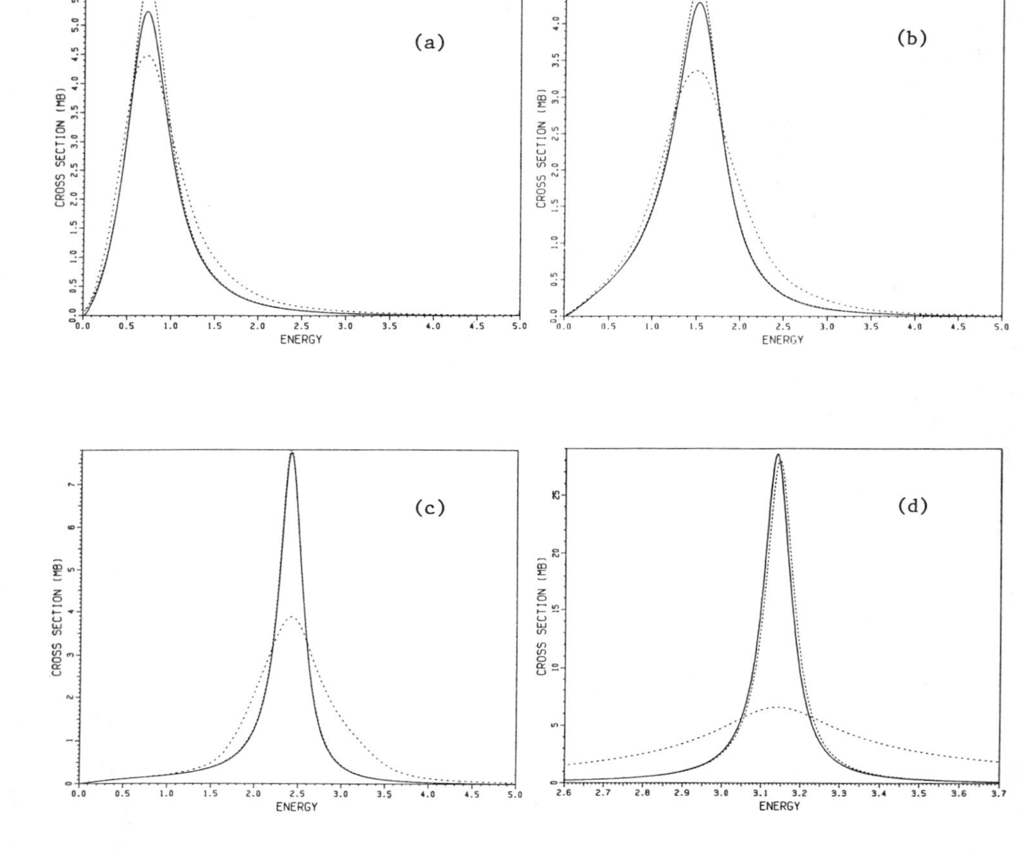

Figure 8. Cross sections for the potential of Eqs. (4.27) obtained from exact solution (——), nonresonant Stieltjes methods (----), and the present resonant Stieltjes development (••••) for barrier values (a) V_h = 5, (b) = 10, (c) = 20, and (d) = 40 a.u. The energy resolution intervals or spacings of the pseudospectra employed in the Stieltjes calculations of the width functions of Figure 7 are approximately 0.2 a.u. in each case. Similar results are also obtained for the case V_h = 80 a.u. (not shown).

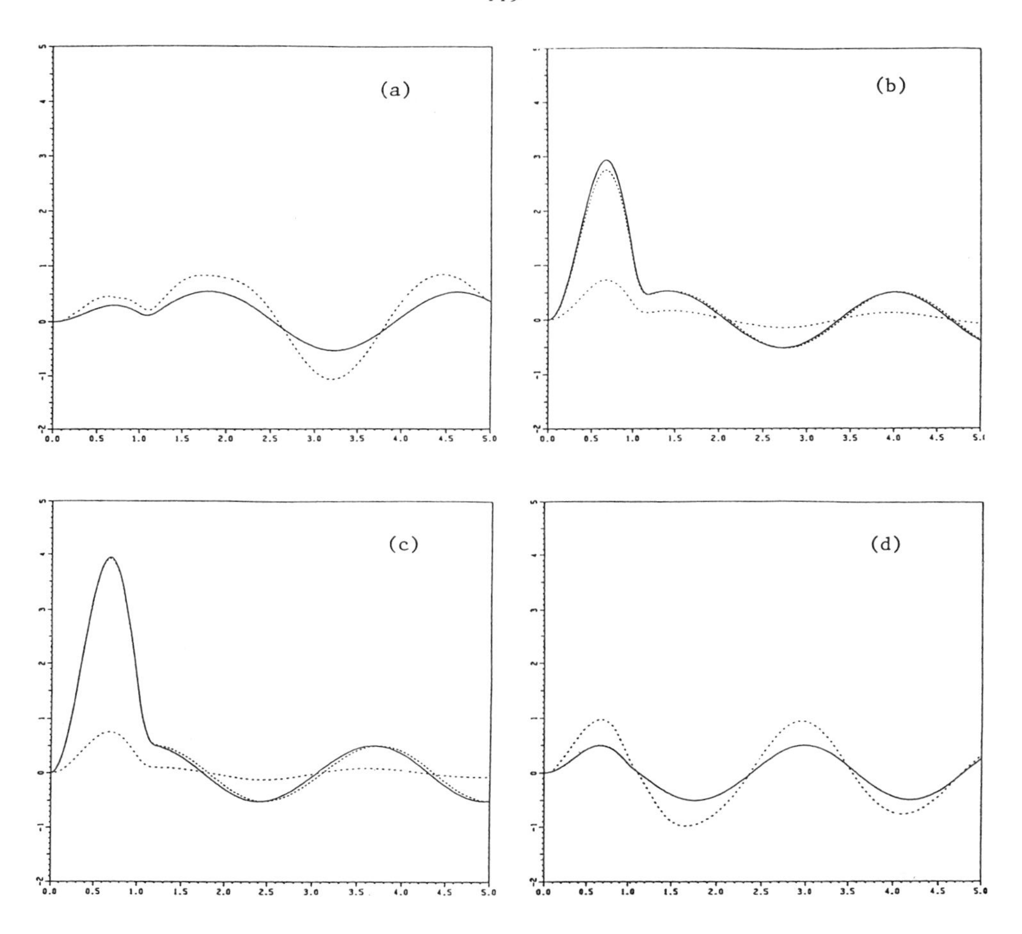

Figure 9. Wave functions for the potential well problem of Eq. (4.27) (V_o = 5 a.u., V_h = 40 a.u.) obtained from exact solution (———), the nonresonant Stieltjes development (-----) and the Stieltjes Feshbach-Fano approach (••••), employing identical Stieltjes orders (n = 15) in each case. The energies of the functions shown are, (a) below resonance - 2.6 a.u.,(b) immediately below resonance - 3.1 a.u, (c) on resonance - 3.14 a.u., and (d) above the resonance energy - 3.5 a.u. [36,76].

As anticipated, the converged widths are generally smooth, slowly varying functions of the energy, and the shift functions are small. In Figure 8 are shown the cross sections calculated by the present formalism in comparison with the exact results and with those obtained from direct Stieltjes construction of the dipole spectrum. Evidently, the present development gives excellent results for both broad and narrow resonances, whereas the direct dipole calculations employing the same numbers of pseudostates required to converge the width functions are unsatisfactory for narrow resonances. Accurate imaging of such narrow resonances employing the nonresonant development would require very fine spacing of pseudostates. By contrast, the width functions of Figure 7 do not show resonant structure, so that the relatively coarse pseudostate spacings achieved ($\Delta\epsilon \sim 0.2$ a.u.) suffice to give an accurate approximation. All of the rapid variation with energy of the cross sections, and of the associated wave functions shown in Figure 9, is incorporated analytically via the formulas of Eqs. (4.11) to (4.25). Numerical applications to broad and narrow molecular resonances indicate the development can provide accurate cross sections from L^2 calculations of nonresonant background continua employing pseudostates with energy spacings that are significantly greater than the resonance width [36,76].

5. Computational applications of Stieltjes methods to the photoionization cross sections of diatomic and polyatomic molecules. A recent data review reports calculations of molecular partial-channel photoionization cross sections employing Stieltjes and other methods [24]. Illustrative results for selected diatomic and polyatomic molecules are presented here to give some idea of the degree of clarification of measured values achieved by such calculations. As indicated in Section 2 above, calculated results reported refer to the fixed-nuclei and static-exchange approximations unless otherwise indicated.

5.1 Calculations on diatomic molecules. The ground-state hydrogen molecule (H_2) provides a particularly simple example of molecular Stieltjes calculations [68]. Moreover, it is a system for which the corresponding experimental values are available [24]. In Figure 10 are shown the weight functions of Eq. (4.6) (n = 5 to 10) for $1\sigma_g \rightarrow k\sigma_u$ and $\rightarrow k\pi_u$ ionization, calculated in static-exchange approximation, as well as the pseudo f numbers obtained from the L^2 Gaussian basis sets employed in the development [68].

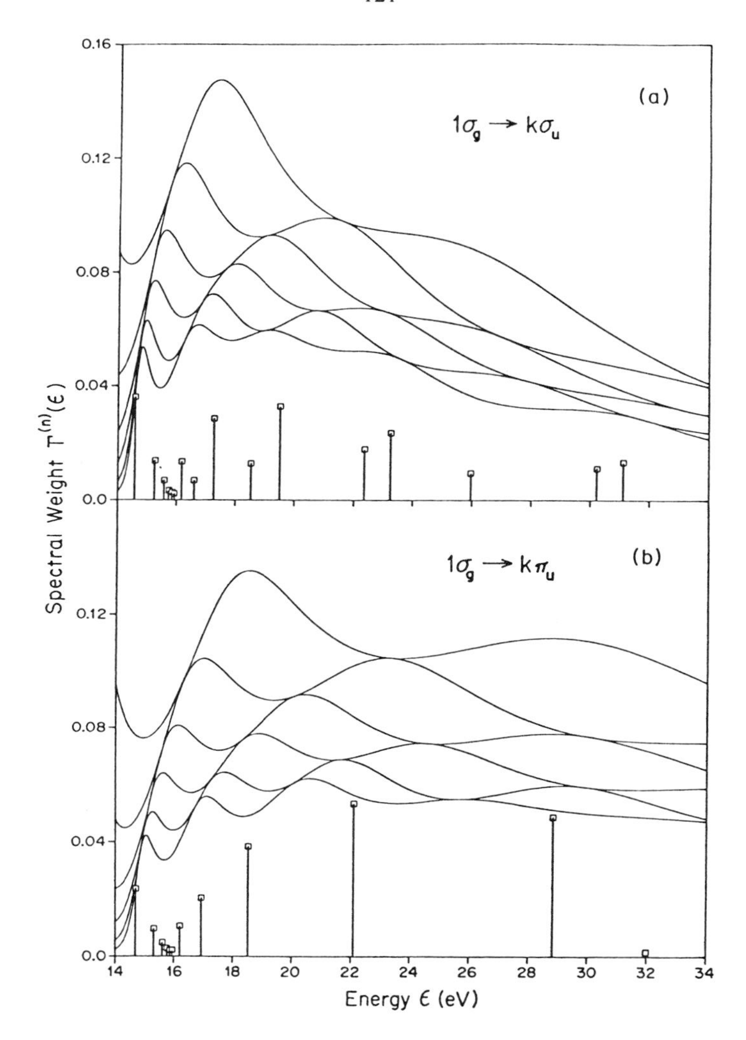

Figure 10. Quadrature weight functions of Eq. (4.6) (n = 5 to 10) for $1\sigma_g \rightarrow k\sigma_u$ and $\rightarrow k\pi_u$ ionization of H_2 in static-exchange approximation, obtained from pseudospectral f numbers shown as vertical lines of appropriate length, calculated employing Cartesian Gaussian basis sets [68].

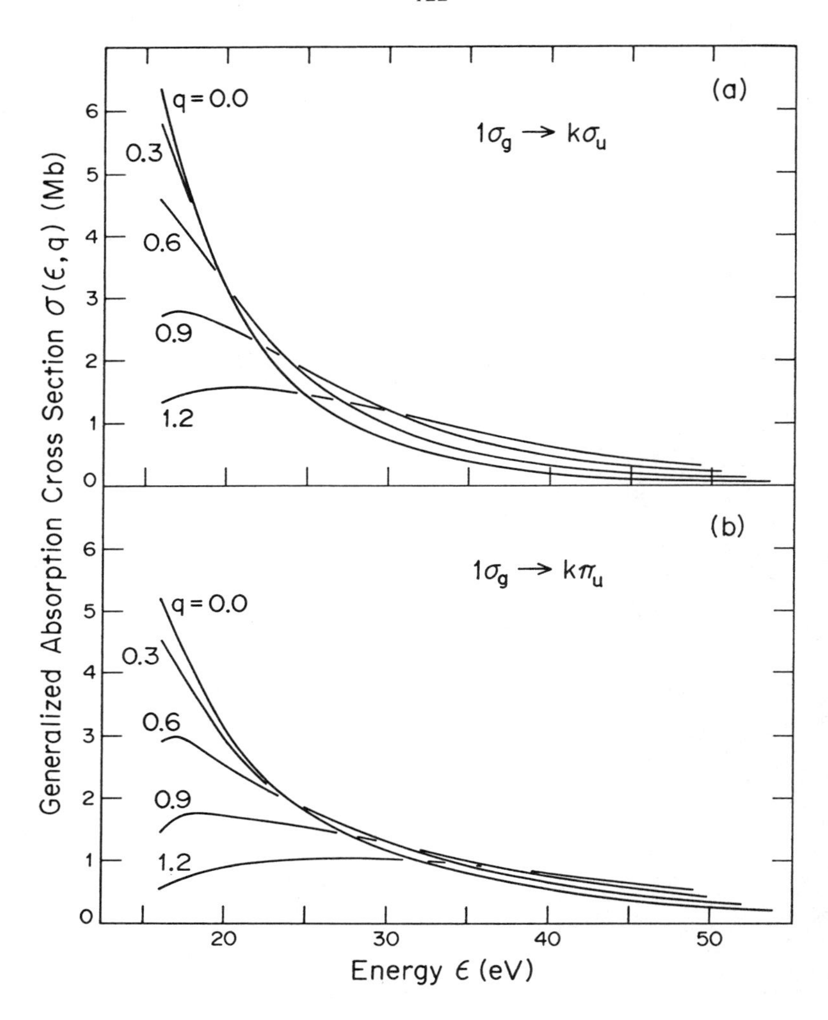

Figure 11. Generalized $1\sigma_g \rightarrow k\sigma_u$ and $\rightarrow k\pi_u$ photoionization cross sections of H_2 obtained in static-exchange approximation employing the weight functions (n = 10) of Figure 10 and corresponding Stieltjes functions in evaluation of the required matrix elements [68]. The results depicted indicate the photoionization anisotropy of the H_2 molecule for polarization perpendicular and parallel to the internuclear axis.

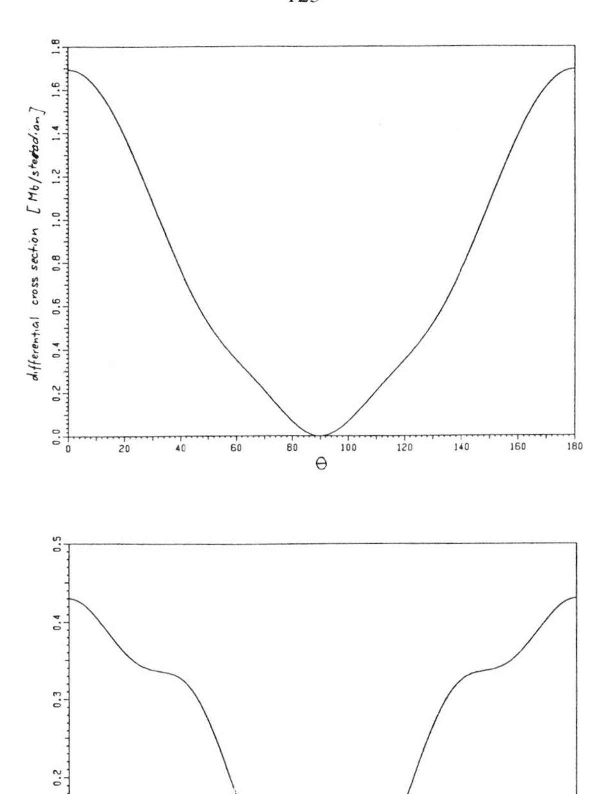

Figure 12. Differential photoionization cross section of Eq. (2.7) for $1\sigma_g \to k\sigma_u$ ionization of oriented H_2 in static-exchange approximation, obtained from Eq. (3.7) and the Stieltjes functions of Eq. (3.25) (n = 10) at the photon energies of 20 eV (top) and 30 eV (bottom) [81]. The angle θ is measure off the light polarization direction, which is fixed along the internuclear axis.

These results indicate that the basis set will "show through" in the weight-function plots for Stieltjes orders (n) much larger than ~ 10. In Figure 11 are shown corresponding polarization cross sections, obtained in this case employing the operator $[(1/q)e^{i\mathbf{g}\cdot\mathbf{r}}]$ in place of the dipole operator of Eq. (2.3), with $\mathbf{g}$ pointing along the internuclear axis for $1\sigma_g \rightarrow k\sigma_u$ ionization and perpendicular to it for $1\sigma_g \rightarrow k\pi_u$ ionization. Note that the results of Figure 11 for $q = 0$ correspond to the dipole photoionization cross section of interest here. When the $q = 0$ results of Figure 11, calculated in the fixed-nuclei approximation, are combined with appropriate Franck-Condon factors for the ionic vibrational states produced, good agreement obtains with conventional Schrödinger calculations and measured values [24,80]. In Figure 12 are shown the differential photoionization cross sections of oriented ground-state H_2 for light polarized along the internuclear axis at two incident photon energies, obtained in static-exchange approximations from Eqs. (2.7), (3.7), and (3.25) ($n = 10$) [81]. The ejection pattern is seen to peak in the direction of the polarization vector ($\theta = 0°$, $180°$) at the lower photon energy, whereas there are inflections at $\theta \simeq 40°$ and $140°$ at higher energy, in accord with results obtained from the conventional Schrödinger development.

Open-shell diatomic molecules provide a significant challenge to cross sectional calculations. In Figure 13 are shown experimental data points and Stieltjes values calculated in the multiplet-specific, static-exchange approximation for the $(5\sigma^{-1})b^3\Pi$ channel in nitric oxide (NO) [70,82]. This channel includes a prominent maximum in the $5\sigma \rightarrow k\sigma$ polarization component of the cross section approximately 10 eV above threshold. There is a small difference between the nonresonant Stieltjes calculations shown in Figure 13 and the resonant Stieltjes calculation (not shown) [36], in general accordance with the higher resolution achieved by the development of Section 4.2 relative to that of 4.1 in a given order ($n = 10$). Agreement with the experimental data in both cases is generally satisfatory, however. In Figure 14 are shown the Stieltjes approximations of Eq. (3.25) to the dipole-prepared $k\sigma$ states for the $(5\sigma \rightarrow k\sigma)b^3\Pi$ channel of Figure 13, plotted in a plane that includes that internuclear axis. As the final-state energy approaches the resonance region, the Stieltjes function is seen to include a compact σ^* portion [27], accounting for the increase in value of the relevant dipole transition matrix element of Eq. (3.5) at these energies. Such compact resonance contributions to dipole prepared states are common in diatomic and polyatomic molecules [27], and account generally for the origins of prominent maxima in molecular partial-channel cross sections [24].

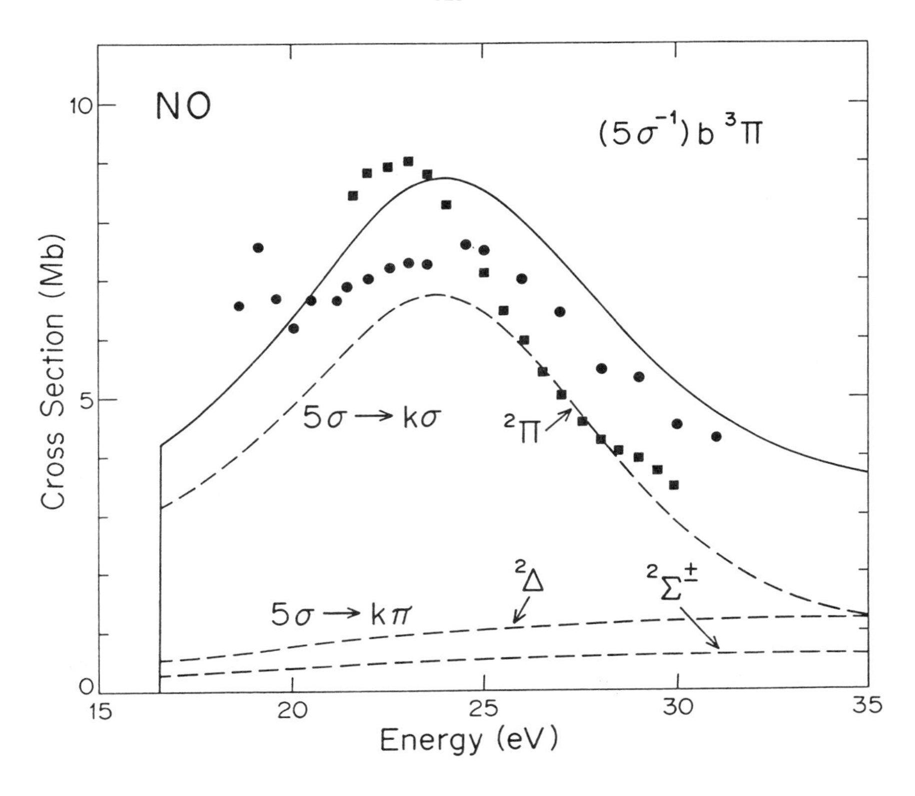

Figure 13. Partial-channel photoionization cross section in nitric oxide (NO) for production of $(5\sigma^{-1})b^3\Pi$ ionic states, obtained in multiplet-specific, static-exchange approximation from the nonresonant Stieltjes development of Section 4.1 in 10th order [70,82]. The polarization components of the cross section are shown as dashed lines. Values obtained from the resonant Stieltjes development of Section 4.2 (not shown) differ somewhat from the nonresonant results shown here, particularly with repect to peak height [36]. Data points refer to experimental values [24].

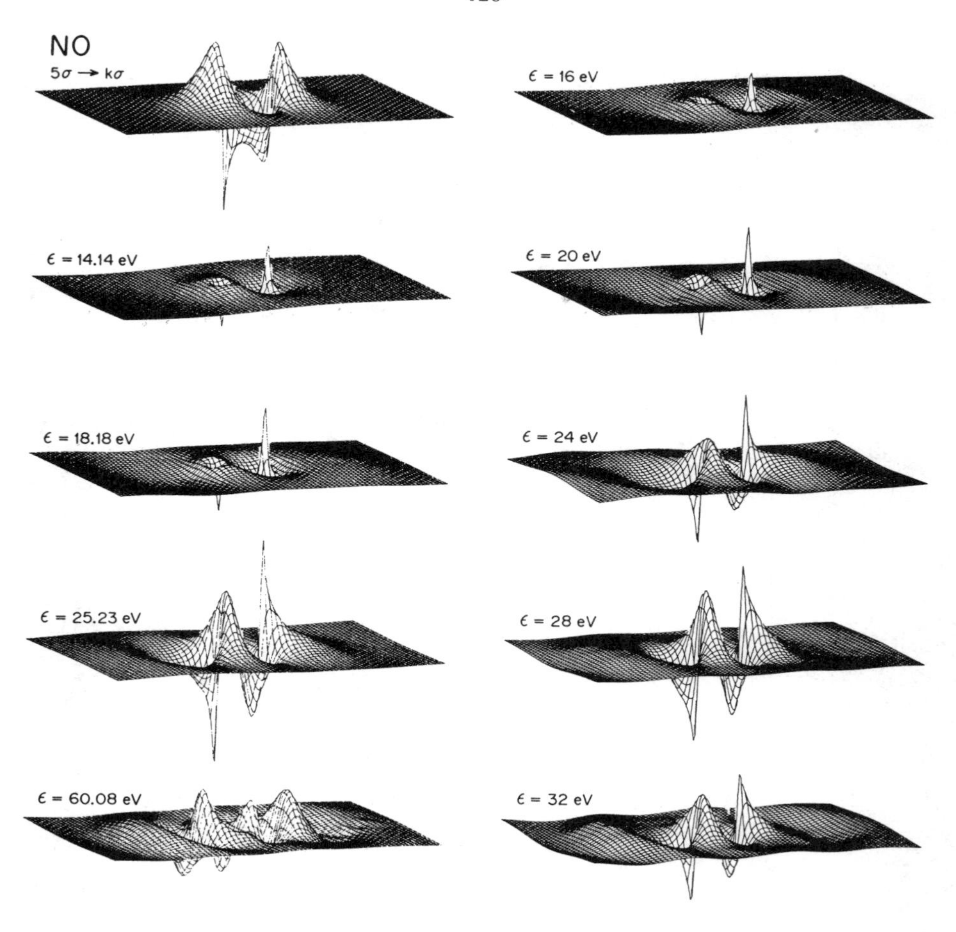

Figure 14. Stieltjes approximations of Eq. (3.25) to the dipole interaction-prepared $k\sigma$ states corresponding to the $5\sigma^{-1}$ photoionization cross section of Figure 13. The column on the left depicts the occupied 5σ and the first four 5th-order $k\sigma$ Stieltjes functions ($n = 5$), whereas the column on the right shows 10th-order functions ($n = 10$) at energies selected to span the resonance region. All values shown refer to orbital functions of Eq. (3.25) evaluated in a plane that includes the internuclear axis.

5.2 Calculations on polyatomic molecules. Large Gaussian basis sets are generally required to obtain convergent Stieltjes approximations to the partial-channel photoionization cross sections of polyatomic molecules. Illustrative results are presented here for N_2O, OCS, and CS_2 molecules, with additional calculations reported separately [24].

In Figure 15 are shown calculated and measured partial-channel cross sections in nitrous oxide (N_2O) for production of the first four $2\pi^{-1}$, $7\sigma^{-1}$, $1\pi^{-1}$, $6\sigma^{-1}$ ionic states. The Stieltjes results obtained in 10th order (n=10)

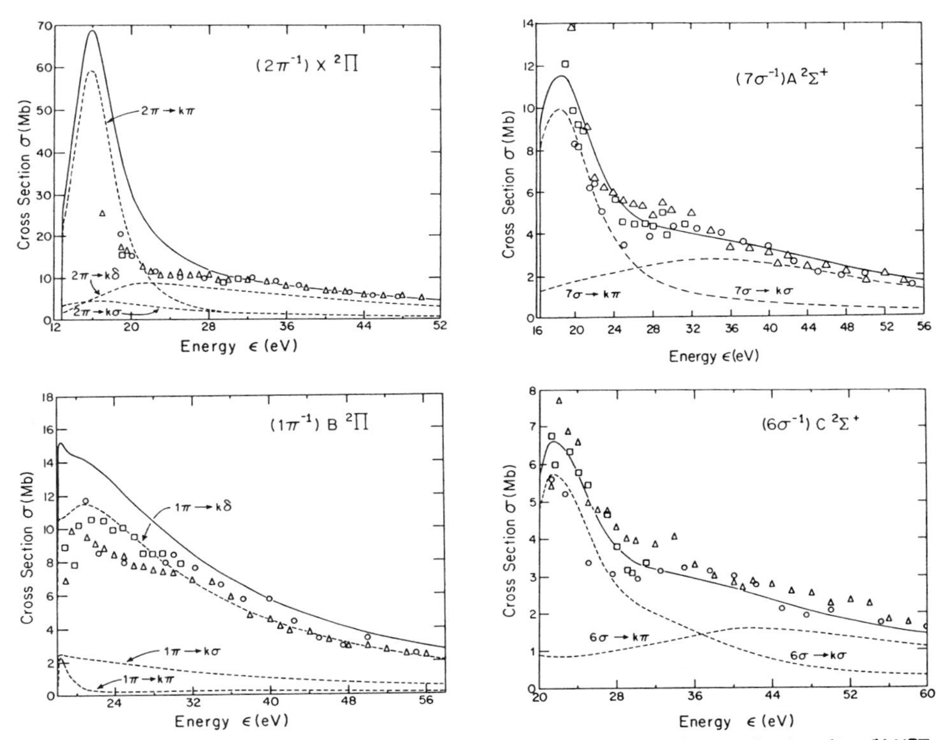

Figure 15. Experimental values and Stieltjes calculations of the $(2\pi^{-1})X^2\Pi$, $(7\sigma^{-1})A^2\Sigma$, $(1\pi^{-1})B^2\Pi$, and $(6\sigma^{-1})C^2\Sigma$ partial-channel cross sections of nitrous oxide (N_2O) [24,84]. The calculated values are obtained in static-exchange approximation employing the nonresonant Stieltjes developments of Section 4.1. Values obtained from the resonant Stieltjes development of Section 4.2 (not shown) are in good agreement with the present results [36]. Data points refer to the experimental values [24].

from both resonant (not shown) and nonresonant developments in these cases are found to be in good agreement, indicating that convergent results are obtained in these cases in the basis sets employed [36]. Agreement with the corresponding measured values is evidently satisfactory, so that the calculations provide a basis for clarifying interpretation. In particular, examination of Stieltjes functions (not shown) in these cases reveals the presence of resonance features in the $7\sigma^{-1}$ and $6\sigma^{-1}$ channels, explaining the near-threshold maxima in the cross sections [36].

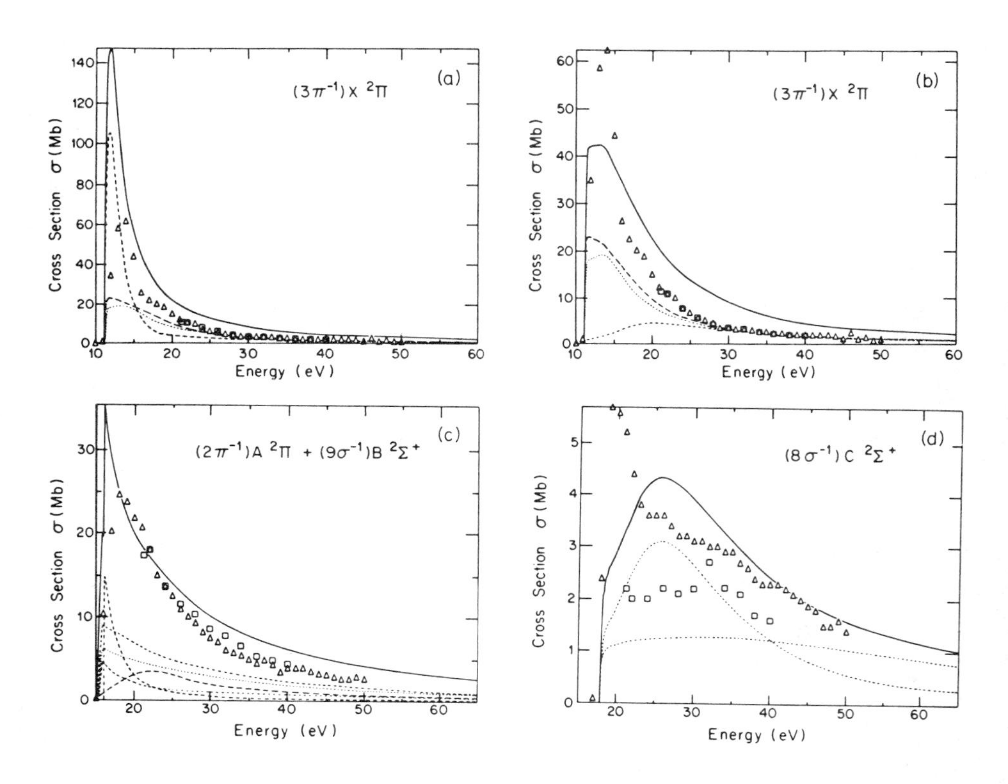

Figure 16. Experimental values and Stieltjes calculations of the $(3\pi^{-1})X^2\Pi$, $(2\pi^{-1})A^2\Pi$, $(9\sigma^{-1})B^2\Sigma$ and $(8\sigma^{-1})C^2\Sigma$ partial-channel cross sections of carbonyl sulfide (OCS) [24,53]. The calculated values are obtained in static-exchange approximation employing the nonresonant Stieltjes development of Section 4.1, whereas the data points are from measurements [24].

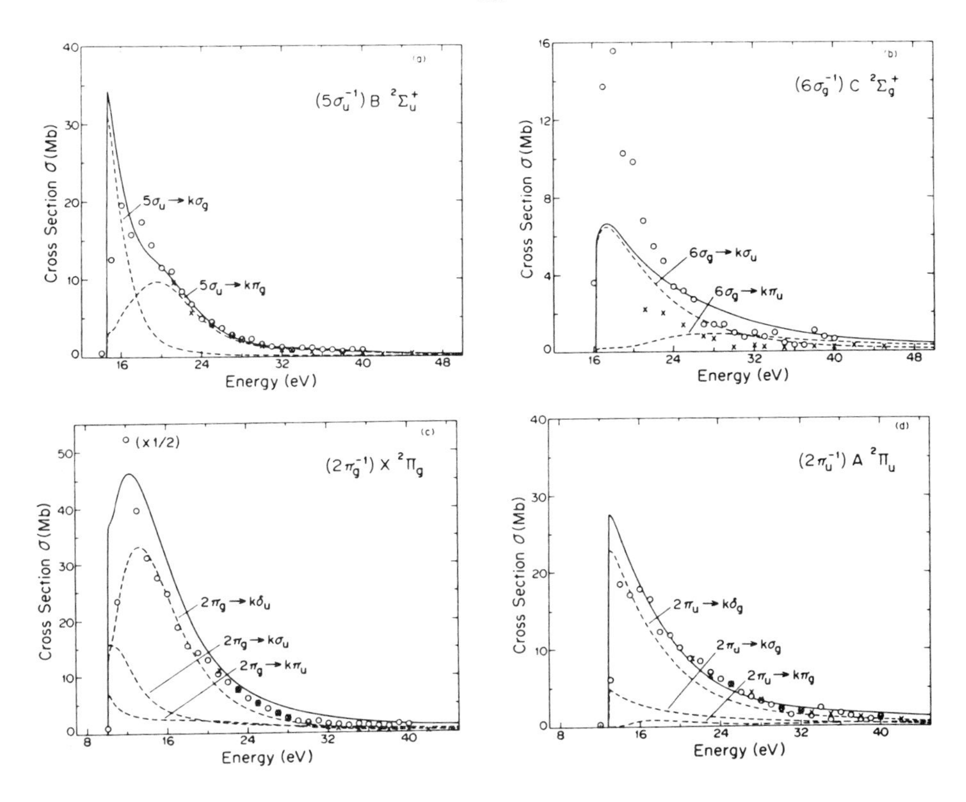

Figure 17. Experimental values and Stieltjes calculations of the $(2\pi_g^{-1})X^2\Pi_g$ $(2\pi_u^{-1})A^2\Pi_u$, $(5\sigma_u^{-1})B^2\Sigma_u$, and $(6\sigma_g^{-1})C^2\Sigma_g$ partial-channel cross sections of carbon disulfide (CS_2) [24,85]. The calculated values are obtained in static-exchange approximation employing the nonresonant Stieltjes development of Section 4.1, whereas the data points are from measurements [24].

Polyatomic molecules containing third-row atoms, in which cases orbital nodes can give rise to Cooper minima [37], have been studied relatively recently. In Figures 16 and 17 are shown partial-channel cross sections for the closely related OCS and CS_2 molecules [53,85]. Referring first to Figure 16, the calculated and measured $3\pi^{-1}$, $2\pi^{-1}$, $9\sigma^{-1}$, and $8\sigma^{-1}$ channel cross sections in OCS are seen to be in generally satisfactory accord, although there is apparently a threshold feature in the $8\sigma^{-1}$ data not present in the Stieltjes calculations. Moreover, the calculated $3\pi^{-1}$ results provide an overestimate at threshold, although the so-called projected results, shown at right, which correspond to removal of a possibly spurious $4\pi(\pi^*)$ contribution, are in better accord with measurement.

Generally satisfactory agreement obtains between theory and experiment for the $5\sigma_u^{-1}$, $6\sigma_g^{-1}$, $2\pi_g^{-1}$ and $2\pi_u^{-1}$ partial channels of CS_2 depicted in Figure 17. A threshold resonance maximum in the $6\sigma_g^{-1}$ channel data is apparently underestimated by the Stieltjes calculations, although this feature may be accounted for on basis of channel-coupling arguments [85]. The degree of agreement shown between theory and experiment in Figures 16 and 17 is typical of most simple linear polyatomic molecules containing third row atoms studied to date [24].

6. Concluding remarks. The present report is designed to provide the reader with a self-contained account of recent progress in the explicit Hilbert-space construction of the electronic Schrödinger states of polyatomic molecules. Much of the development is applicable to anisotropic Hamiltonians more generally. Particular emphasis is placed on devising a treatment that deals with both discrete and continuum states on a common basis, an approach that offers both computational and conceptual advantages. Specifically, working in L^2 basis sets avoids the conventional partial-wave expansion for scattering states, and facilitates the calculation of nondegenerate interaction-prepared states having molecular point-group symmetry, avoiding entirely the customary K-matrix or S-matrix normalized scattering functions. The interaction-prepared states are the natural continuum analogues of the more familiar L^2 bound electronic states of molecules, to which they reduce at energies below threshold without the introduction of separate computational measures in this spectral region. They furthermore provide all of the information necessary for evaluation of photoabsorption cross sections and of photoionization cross sections differential in the direction of ejected electrons. Convergence of the computational procedures employed in construction of the necessary L^2 states follows from aspects of the theory of moments applied in the context of the spectral theory of operators. Feshbach-Fano methods can be incorporated into the general development, avoiding the spectral resolution limitations which can possibly arise in global approachs. In this way narrow spectral profiles can be calculated employing pseudospectra that correspond to relatively coarse energy spacings. Recent calculations on diatomic and polyatomic molecules illustrate the range of applicability of the Hilbert-space development for photoabsorption and ionization processes.

REFERENCES

[1] R. Courant and D. Hilbert, **Methods of Mathematical Physics** (Interscience, New York, 1953, Vol. I.

[2] G. Herzberg, **Molecular Spectra and Molecular Structure, Vol. 3. Electronic Spectra and Electronic Structure of Polyatomic Molecules** (Van Nostrand, Princeton, 1966).

[3] M.B. Robin, **Higher Excited States of Polyatomic Molecules** (Academic, New York, 1974; 1985), Vols. I and II; III.

[4] S.T. Epstein, **The Variational Method in Quantum Chemistry** (Academic, New York, 1974.)

[5] G.H.F. Diercksen and S. Wilson, Editors, **Methods in Computational Molecular Physics** (Reidel, Dordrecht, 1983).

[6] E.A. Hylleraas, Z. Physik, 54, 347 (1929).

[7] L.D. Landau and E.M. Lifschitz, **Quantum Mechanics, NonRelativistic Theory** (Pergamon, Oxford, 1962).

[8] R.G. Newton, **Scattering Theory of Waves and Particles** (Springer, Berlin, 1982).

[9] E. Schrödinger, Ann. Physik 79, 391, 489, 734 (1926); 80, 437 (1926); 81, 109 (1926).

[10] W. Heisenberg, Z. Physik 33, 879 (1925); 38, 411 (1926); 41, 239 (1927); 49, 619 (1928).

[11] M. Born, Z. Physik 38, 803 (1926).

[12] A. Sommerfeld, Ann. d. Phys. 11, 257 (1931).

[13] H.A. Bethe and E.E. Salpeter, **Quantum Mechanics of One- and Two-Electron Atoms and Ions** (Springer, Berlin, 1957).

[14] N.F. Lane, Rev. Mod. Phys. 52, 29 (1980).

[15] J.A. Shohat and J.D. Tamarkin, **The Problem of Moments, Mathematical Surveys 1** (American Mathematical Society, Providence, R.I., 1943, revised 1970).

[16] H.S. Wall, **Analytic Theory of Continued Fractions** (Van Nostrand, New York, 1948).

[17] N.I. Akhiezer, **The Classical Moment Problem** (Oliver and Boyd, London, 1956).

[18] YuV. Vorobyev, **Method of Moments in Applied Mathematics** (Gordon and Breach, New York, 1965).

[19] G.A. Baker, Jr., **Essentials of Padé Approximants** (Academic, New York, 1975).

[20] J. Berkowitz, **Photoabsorption, Photoionization, and Photoelectron Spectroscopy** (Academic, New York, 1979).

[21] G.H.F. Diercksen, W. Huebner, and P.W. Langhoff, Editors, **Molecular Astrophysics** (Reidel, Dordrecht, Holland, 1985).

[22] I. Nenner and J.A. Beswick, "Molecular Photodissociation and Photoionization," in **Handbook on Synchrotron Radiation**, G.W. Marr, Editor (North Holland, Amsterdam, 1986), Vol. II.

[23] C.E. Brion and A. Hamnett, "Continuum Optical Oscillator Strength Measurements by Electron Spectroscopy in the Gas Phase," in **The Excited State in Chemical Physics**, J.W. McGowan, Editor (Wiley, New York, 1981), Part 2; Adv. Chem. Phys. 45, 2 (1981).

[24] J.W. Gallagher, C.E. Brion, J.A.R. Samson, and P.W. Langhoff, "Absolute Cross Sections for Molecular Photoabsorption, Partial Photoionization, and Ionic Photofragmentation Processes," J. Phys. Chem. Ref. Data (to be published).

[25] B.I. Schneider and L.A. Collins, "Resonances in Electron-Molecule Scattering and Photoionization," in **Resonances in Electron-Molecule Scattering, van der Waals Complexes, and Reactive Chemical Dynamics**, D. G. Truhlar, Editor (American Chemical Society, Washington, D.C., 1984), pp. 65-88.

[26] D.L. Lynch, V. McKoy, and R.R. Lucchese, "Dynamics of Molecular Photoionization Processes," in **Resonances in Electron-Molecule Scattering, van der Waals Complexes, and Reactive Chemical Dynamics**, D.G. Truhlar, Editor (American Chemical Society, Washington, D.C., 1984), pp. 89-112.

[27] P.W. Langhoff, "Molecular photoionization resonances, a theoretical chemist's perspective," in **Resonances in Electron-Molecules Scattering, van der Waals Complexes, and Reactive Chemical Dynamics**, D.G. Truhlar, Editor (American Chemical Society, Washington, D.C., 1984), pp. 113-138.

[28] J.L. Dehmer, "Shape Resonances in Molecular Fields," in **Resonances in Electron-Molecule Scattering, van der Waals Complexes, and Reactive Chemical Dynamics**, D.G. Truhlar, Editor (American Chemical Society, Washington, D.C., 1984), pp. 139-163.

[29] P.W. Langhoff, "Stieltjes-Tchebycheff Moment-Theory Approach to Molecular Photoionization Studies," in **Electron-Molecule and Photon-Molecule Scattering**, T. Rescigno, V. McKoy and B. Schneider, Editors (Plenum, New York, 1979), pp. 183-224.

[30] P.W. Langhoff, "Stieltjes-Tchebycheff Moment-Theory Approach to Photoeffect Studies in Hilbert Space," in **Moment Methods in Many-Fermion Systems**, B.J. Dalton, S.M. Grimes, J.P. Vary, and S.A. Williams, Editors (Plenum, New York, 1980), pp. 191-212.

[31] P.W. Langhoff, "Aspects of Electronic Configuration Interaction in Molecular Photoionization," in **Electron-Atom and Electron-Molecule Collisions**, J. Hinze, Editor (Plenum, New York, 1983), pp. 297-314.

[32] P.W. Langhoff, "Schrödinger Spectra," in Methods in Computational **Molecular Physics**, G.H.F. Diercksen and S. Wilson, Editors (Reidel, Dordrecht, Holland, 1983), pp. 299-334.

[33] I.N. Levine, **Quantum Chemistry** (Allyn and Bacon, Boston, 1974), 2nd. edition.

[34] S. Huzinaga, J. Andzelm, M. Klobukowski, E. Radzio-Andzelm, Y. Sakai, and H. Tatewaki, **Gaussian Basis Sets for Molecular Calculations**, Physical Sciences Data (Elsevier, Amsterdam, 1984), Vol. 16.

[35] D. Masson, "Hilbert Space and the Padé Approximant," in The Padé **Approximant in Theoretical Physics**, G.A. Baker, Jr., and J.L. Gammel, Eds. (Academic, New York, 1970), pp.197-218.

[36] C.L. Winstead, Ph.D. Dissertation, Indiana University, Bloomington, 1987, "Projection Operator Formalism for Molecular Photoionization Resonances Using Square-Integrable Basis Sets."

[37] U. Fano and J.W. Cooper, Rev. Mod. Phys. 40, 441 (1968).

[38] F.A. Gianturo and D.G. Thompson, "Computational Models for e - Polyatomic Low-Energy Scattering," in **Electron-Atom and Electron-Molecule Collisions**, J.Hinze, Editor (Plenum, New York, 1983), pp. 231-253.

[39] N. Abusalbi, D.W. Schwenke, C.A. Mead, and D.G. Truhlar, Theor. Chim. Acta. 71, 333 (1987).

[40] G.H.F. Diercksen, W.P. Kraemer, T.N. Rescigno, C.F. Bender, B.V. McKoy, S.R. Langhoff, and P.W. Langhoff, J. Chem. Phys. 76, 1043 (1983).

[41] G. Breit and H.A. Bethe, Phys. Rev. 93, 888 (1954).

[42] C.N. Yang, Phys. Rev. 74, 764 (1948).

[43] J.C. Tully, R.S. Berry and B.J. Dalton, Phys. Rev. 176, 95 (1968).

[44] A.D. Buckingham, B.J. Orr, and J.M. Sichel, Philos. Trans. Roy. Soc. London A268, 147 (1970).

[45] U. Fano and D. Dill, Phys. Rev. A 6, 185 (1972).

[46] D. Dill and J.L. Dehmer, J. Chem. Phys. 61, 692 (1974).

[47] S. Wallace and D. Dill, Phys. Rev. B 17, 692 (1978).

[48] M. White, Phys. Rev. A 26, 1907 (1982).

[49] R.A. Bonham and M.L. Lively, Phys. Rev. A29, 1224 (1984).

[50] P.W. Langhoff, "Molecular Photoionization Processes of Astrophysical and Aeronomical Interest," in **Molecular Astrophysics**, G.H.F. Diercksen, W. Huebner, and P.W. Langhoff, Editors. (Reidel, Dordrecht, Holland, 1985), pp. 551-573.

[51] C.M. Reeves, J. Chem. Phys. 39, 1 (1963).

[52] P.O. Löwdin, Adv. Quantum Chem. 5, 185 (1970).

[53] J.A. Sheehy and P.W. Langhoff, Chem. Phys. Letters 135, 109 (1987).

[54] A. Messiah, **Quantum Mechanics** (Wiley, New York, 1966), Vol. I.

[55] R.F. Barrett, L.C. Biedenharn, M. Danos, P.P. Delsanto, W. Greiner, and H.G. Washweiler, Rev. Mod. Phys. 45, 44 (1973).

[56] D. Loomba, S. Wallace, D. Dill, and J.L. Dehmer, J. Chem. Phys. 75, 4596 (1981).

[57] Z.H. Levine and P. Sovan, Phys. Rev. A 29, 625 (1984).

[58] Y. Itikawa, Chem. Phys. 28, 461 (1978).

[59] W. Thiel, Chem. Phys. 57, 227 (1981).

[60] T.A. Carlson, M.O. Krause, and F.A. Grimm, J. Chem. Phys. 77, 1701 (1982).

[61] U. Fano, Phys. Rev. 124, 1866 (1961).

[62] U. Fano and J.W. Cooper, Phys. Rev. 137 A1364 (1965).

[63] R. Haydock, "Solid State Physics Without Block's Theorem," in **Computational Methods in Classical and Quantum Physics**, M.B. Hooper, Editor (Hemisphere Publishing, Washington, D.C., 1975), pp. 268-283.

[64] C. Lanczos, J. Res. Nat. Bur. Stand. 45, 367 (1950).

[65] J.H. Wilkinson, **The Algebraic Eigenvalue Problem** (Clarenden, Oxford, 1965).

[66] C.T. Corcoran and P.W. Langhoff, J. Math. Phys. 18, 651 (1977).

[67] M.R. Hermann and P.W. Langhoff, J. Math. Phys. 24, 541 (1983).

[68] M.R. Hermann and P.W. Langhoff, Phys. Rev. A 28, 1957 (1983).

[69] M.R. Hermann and P.W. Langhoff, Int. J. Quantum Chem. 23, 135 (1983).

[70] M.R. Hermann, C.W. Bauschlicher, Jr., W.M. Huo, S.R. Langhoff, and P.W. Langhoff, Chem. Phys. 109, 1 (1986).

[71] M.R. Hermann, G.H.F. Diercksen, B.W. Fatyga, and P.W. Langhoff, Int. J. Quantum Chem. S18, 719 (1984).

[72] H. Feshbach, Ann. Phys. (New York) 5, 357 (1958).

[73] H. Feshbach, Ann. Phys. (New York) 19, 287 (1962).

[74] B.W. Shore, Rev. Mod. Phys. 39, 439 (1967).

[75] P.W. Langhoff, N. Padial, G. Csanak, and B.V. McKoy, J. de Chemie 77, 589 (1980).

[76] C.L. Winstead and P.W. Langhoff, Chem. Phys. Letters (to be published).

[77] B.A. Lippmann and T.F. O'Malley, Phys. Rev. A 2 2115 (1970).

[78] W. Domcke, Phys. Rev. A 28, 2777 (1983).

[79] C.E. Woodward, S. Nordholm, and G. Bacskay, Chem. Phys. 69, 267 (1982).

[80] C.M. Dutta, F.M. Chapman, Jr., and E.F. Hayes, J. Chem. Phys. 67, 1904 (1977).

[81] B. Fatyga, C.L. Winstead, T.J. Gil, and P.W. Langhoff, Chem. Phys. Letters (to be published).

[82] M.R. Hermann, S.R. Langhoff, T.J. Gil, and P.W. Langhoff, Chem. Phys. Letters 125, 336 (1986).

[83] P.W. Langhoff, A. Gerwer, C. Asaro, and B.V. McKoy, Int. J. of Quantum Chem. S13, 643 (1979).

[84] C.L. Winstead, M.R. Hermann, and P.W. Langhoff, Chem. Phys. (to be published).

[85] G.H.F. Diercksen, M.R. Hermann, B.W. Fatyga, and P.W. Langhoff, Chem. Phys. Letters 123, 345 (1986).

CLASSICAL MOLECULAR DYNAMICS SIMULATIONS: ALGORITHMS AND APPLICATIONS, STOCHASTIC DYNAMICS, AND FREE ENERGIES

WILFRED F. VAN GUNSTEREN†

Abstract. Methods and algorithms for molecular dynamics (MD) simulations are described with an eye on application to complex molecular systems, such as crystals or solutions of (bio)macromolecules. Those techniques are emphasized that in the experience of the author are reliable and easy to implement. In addition there is a separate section on stochastic dynamics The accuracy of MD simulations of biomolecules is briefly discussed and applications of MD simulation techniques in chemistry are mentioned. The final topic included is the calculation of free energy. One method for this, the so-called thermodynamic cycle integration technique, which combines well known results from statistical thermodynamics with powerful computer simulation methods, is very promising with respect to practical applications in chemistry. Its basic formulae and some applications in the areas of drug design, protein engineering, and conformational analysis are discussed.

1. Introduction. Theoretical description in chemistry involves the physics of complex molecular systems; however, the relative complexity of chemical and biochemical systems limits the ability of physical theory to treat them properly. However, molecular systems are also simple. From a practical chemical point of view they are built up from just two types of particles: nuclei, with negligible size and irrelevant internal structure, and electrons surrounding them. For most practical purposes the Born- Oppenheimer approximation is applicable, separating electronic and nuclear motion. Thus a complex molecular system can be described as a *system of point masses moving in an effective potential field.* It further simplifies things that in the majority of applications this effective potential field is also conservative, that is, it is a function of the atomic coordinates only. Finally, for applications at normal temperatures and not involving details of the behaviour of hydrogen atoms, the laws of *classical mechanics* apply with a sufficient degree of accuracy.

Let us assume that a sufficient reliable description of the potential field describing the atomic interaction is available. We are interested to derive macroscopic properties of a complex system on the basis of this interaction at the atomic level. Such macroscopic properties either concern equilibrium properties of static nature, including all thermodynamic quantities, or they concern dynamic properties, such as various transport coefficients measureable in non-equilibrium systems. To derive such properties is the task of *statistical mechanics.* Unfortunately, complex systems are almost never amenable to analytical treatment in the framework of statistical mechanics, with the rather dull exceptions of dilute gases, very simple liquids and almost perfect crystals. So we have to resort to *simulation* of macroscopic behaviour by constructing a representative statistical ensemble of a size large enough to provide us with sufficient statistical accuracy.

---

†Laboratory of Physical Chemistry, University of Groningen,
Nijenborgh 16, 9747 AG Groningen, The Netherlands

Molecular dynamics (MD) computer simulations have considerably added to our understanding at the atomic level of the properties of molecular systems, such as liquids or solutions, of (bio)molecules during the past two decades. A rather static picture of molecular conformation has been gradually transformed into a more dynamic one, where the molecular properties are dynamic averages over an ensemble of molecular configurations. The development of computer simulation techniques has been made possible by the continuous and rapid development of computer hardware. Every six to seven years the ratio of performance to price has increased by a factor ten and due to the emerging parallel computing techniques the end of this increase is not yet in sight. At present small macromolecules, like proteins, in aqueous solution, involving many thousands of atoms, can be simulated over periods of about 10–100 ps. For reviews we refer to [1-4].

Section 2 forms a very brief introduction to molecular dynamics simulation methods and applications [1, 5–10]. Section 3 considers the extension to stochastic dynamics, and section 4 considers a special topic, namely the calculation of free energies [11-14], that is currently receiving considerable attention.

2. Molecular dynamics simulations: Algorithms and applications. This section is based on a number of more extensive introductory papers [1, 5–10]. For references to the various techniques that are described, we refer to the references quoted in Refs. [1, 5–10].

2.1. Simulation methods. In the field of computer simulation of many-particle systems, several different methods can be distinguished. In order to classify these we consider a system consisting of N atoms with masses $m_i (i = 1, 2, ...N)$ and Cartesian position vectors $\{\vec{r}_i\}$. The interaction between the atoms of this system is described by an *interaction function* or *force field*

$$V(\vec{r}_1, \vec{r}_2, ..., \vec{r}_N). \tag{1}$$

The various simulation techniques are the following ones.

1. Systematic search (SS). If the system contains only a few degrees of freedom, the complete configuration space can be scanned for low-energy molecular configurations and an *ensemble* can be generated using the *Boltzmann factor*

$$\exp[-V(\vec{r}_1, \vec{r}_2, ..., \vec{r}_N)/kT] \tag{2}$$

as a weight function, where k denotes Boltzmann's constant and T refers to the absolute temperature of the system.

2. Monte Carlo (MC). If a system contains many degrees of freedom, straightforward scanning of the complete configuration space is impossible. In that case, an *ensemble* of configurations can be generated by a combination of *random sampling* and use of the *Boltzmann factor* (2). Given a starting configuration a new configuration is generated by random displacement of one (or more) atoms. The displacements should be such that in the limit of a large number of successive displacements the available Cartesian space of all atoms is uniformly sampled. The newly generated configuration is either accepted or rejected on the basis of an energy criterion involving the change ΔE of the potential energy (1) with respect to the previous configuration. In the case of rejection, the previous configuration is counted again and used as a starting point for another random displacement. The criterion is the following: accept if $\Delta E \leq 0$, or for $\Delta E > 0$, accept if $\exp(-\Delta E/kT) > R$, where R is a random number, taken from a uniform distribution over the interval $(0, 1)$. In this way each configuration occurs with a probability proportional to its Boltzmann factor (2). In order to obtain high computational efficiency, one would like to combine a large (random) step size with a high acceptance ratio. For complex systems involving many covalently bound atoms, a reasonable acceptance ratio can only be obtained for very small step size. This makes MC much less efficient than MD for (macro)-molecular systems.

3. Molecular Dynamics (MD). By application of MD a *trajectory* (configurations as a function of time) of the system is generated by simultaneous integration of *Newton's equations of motion*

$$d^2\vec{r}_i/dt^2 = m_i^{-1}\vec{F}_i \tag{3}$$

$$\vec{F}_i = -\frac{\partial}{\partial \vec{r}_i} V(\vec{r}_1, \vec{r}_2, ..., \vec{r}_N) \tag{4}$$

for all the atoms in the system. Here, the force on atom i is denoted by $\vec{F}_i$.

4. Stochastic Dynamics (SD). The technique of stochastic dynamics is an extension of MD. A *trajectory* of the system is generated by integration of the stochastic *Langevin equation of motion*

$$d^2\vec{r}_i/dt^2 = m_i^{-1}[\vec{F}_i + \vec{R}_i] - \gamma_i\, d\vec{r}_i/dt. \tag{5}$$

Here, a stochastic force $\vec{R}_i$ and a frictional force propertional to a friction coefficient γ_i are added to Eq. (3). An introduction to SD techniques is given in Section 3.

5. Energy minimization (EM). Applying *EM* one searches for a *minimum energy configuration* by *moving along the gradient*

$$\Delta\vec{r}_i \sim -\frac{\partial}{\partial \vec{r}_i} V(\vec{r}_1, \vec{r}_2, ..., \vec{r}_N) \tag{6}$$

through configuration space. Since in this way one moves basically only downhill over the energy hypersurface, EM yields only a local minimum energy configuration, which is generally not far from the initial one. In MD the available kinetic energy makes it possible to cross over energy barriers of the order of kT. This feature makes MD a more efficient technique to locate low energy regions of the interaction function (1) than regular EM techniques.

6. Normal mode analysis (NMA). Having obtained a minimum energy configuration one may *find the eigenvectors of the second derivative matrix*

$$\frac{\partial^2}{\partial \vec{r}_i \partial \vec{r}_j} V(\vec{r}_1, \vec{r}_2, ..., \vec{r}_N) \tag{7}$$

at that point in configuration space, from which the *harmonic motion of the system around that configuration* may be obtained. Although the technique is elegant, its value is rather limited due to the observation that the dynamics of a molecular system at room temperature cannot be satisfactorily be characterized by pure harmonic motion around one molecular configuration.

Since the MD method is a much richer method than SS, MC, EM and NMA, we shall concentrate entirely on MD.

2.2. Molecular dynamics in practice. When applying MD to simulate a system a number of questions have to be answered and choices related to the level of accuracy must be made. Below we list and briefly discuss these basic questions and choices.

1. Is a classical description appropriate? When considering a molecular system, such as a liquid or a solution, at room temperature quantum effects generally play a minor role. However, there are many exceptions to classical behaviour. For example, high frequency motions in covalently bonded molecules do not behave classically. But, a quantum correction to a classical treatment is generally sufficiently accurate.

2. Is the time scale of the simulation adequate? With use of modern computers the length of a MD run extends from a few tens of picoseconds up till nanoseconds, depending on the size of the system. This means that the time scale of the process that can be simulated is limited. For the simulation of activated processes special techniques have been developed, which require the pathway of the process to be known. The time span of a simulation should be chosen as short as possible while retaining sufficient statistical accuracy and without running into ergodicity problems. The latter consideration is serious: many systems of interest have the awkward habit of residing for fairly long periods of time in a certain region of configuration space before traversing to another region.

3. How many atoms are essentially involved in the phenomenon to be studied? Only a limited number of atoms can be simulated. The largest macromolecular systems which have been simulated, contain about 10^4 atoms. Atomic degrees of freedom that are not essential for an adequate description of the phenomenon to be studied, may be removed by applying constraints, or stochastic techniques in combination with potentials of mean force, or the extended wall region boundary condition (see below).

4. Is the interaction function that is used, sufficiently accurate? The formulation of an effective potential force field could be based on the Schrödinger equation, *e.g.*, on the Born-Oppenheimer approximation in which the effective potential for unclear motion is obtained by solving the electronic Schrödinger equation with fixed nuclei. In practice this is not simple, since the number of electrons is usually large and the extent of configuration space is too vast to allow accurate *ab initio* determination of the effective force field (1). One has to resort to simplifications and semi-empirical or empirical adjustments of the potential field.

A *typical* molecular *force field* looks as follows:

$$
(8) \qquad V(\vec{r}_1, \vec{r}_2, ..., \vec{r}_N) = \sum_{\text{bonds}} 1/2\, K_b\, (b - b_0)^2 + \sum_{\text{angles}} 1/2\, K_\theta\, (\theta - \theta_0)^2 + \sum_{\text{torsions}} 1/2\, K_\xi\, (\xi - \xi_0))^2 + \sum_{\text{dihedrals}} K_\phi(1 + \cos(n\phi - \delta)) + \sum_{\text{pairs}(i,j)} [C_{12}/r_{ij}^{12} - C_6/r_{ij}^6 + q_i q_j / 4\pi\, \varepsilon_0 \varepsilon_r\, r_{ij}]
$$

the first term represents the covalent bond (b) interaction and the second one that of the bond angles (θ). Two forms are used for the dihedral angle interactions: a harmonic term for dihedrals (torsions ξ) that do not undergo transitons, e.g. dihedral angles within aromatic rings, and a sinusoidal term for the other dihedrals (ϕ), which may make 360° turns. The last term represents the nonbonded interaction between pairs of atoms i and j at a distance r_{ij} with charges q_i and q_j .

There exists a large number of *variations* on the form (8). Some force fields contain mixed terms like $K_{b\theta}(b - b_0))(\theta - \theta_0))$ which directly couple bond-length and bond-angle vibrations. Others use more complex dihedral terms. The choice of the relative dielectric constant ranges from $\varepsilon_r = 1$ to $\varepsilon_r = 8$, while others take it proportional to the distance r_{ij}. Another possibility is to allow for virtual charges, that is, charges not located on atoms.

When *determining the parameters* of the interaction function (8) there are basically two ways to go. The most elegant procedure is to fit them to results of ab-initio quantum calculations on small molecular clusters. However, due to various serious approximations that have to be made in this type of procedure, the resulting force fields are in general not very satisfactory. The alternative is to fit the force field parameters to experimental data (crystal structure, energy and lattice dynamics, infrared, X-ray data on small molecules, liquid properties, free energies of hydration, nuclear magnetic resonance data, etc.). In our opinion the best results have been obtained by this semi-empirical method.

Finally we note that the choice of a particular force field should depend on the system properties one is interested in. Some applications require more refined force fields than others. Moreover, there should be a balance between the level of accuracy or refinement of different parts of a molecular model. Otherwise the computing effort put into a very detailed and accurate part of the calculation may easily be wasted due to the distorting effects of the crude parts of the model.

5. Treatment of boundaries. Most macromolecular simulations have been performed *in vacuo*, that is, without any wall or boundary. In this case the pressure is approximately correct (0 atm), but other properties, especially those of atoms near or at the macromolecular surface will be distorted by the vacuum boundary condition.

The classical way to minimize edge effects is to use *periodic boundary conditions.* The atoms of the system that is to be simulated are put into a cubic (or any periodically space filling shaped) box, which is surrounded by 26 identical translated images of itself. When calculating the forces of an atom in the central box, all interactions with atoms in the central box or images in the surrounding boxes that lie within a spherical cut-off radius R_c are taken into account. Thus in fact a crystal is simulated. For a molecule in solution the periodicity is an artifact of the computation, so the effects of periodicity on the forces on the atoms should not be significant. This means that an atom should not simultaneously interact with another atom and its image. Consequently, the length R_{box} of the edge of the computational box should exceed twice the cut-off radius R_c : $R_{box} > 2R_c$. Possible distorting effects of the periodic boundary condition may be traced by simulating systems of different size.

For large macromolecular systems application of periodic boundary conditions is far too expensive. In that case the number of atoms in the simulation can be limited by simulating only part of the molecular system, the reduced system. Edge effects can be minimized by restraining the motion of the atoms in the outer shell of the reduced system viz. the *extended wall region*. The atoms in this region can be kept fixed or harmonically restrained to stationary positons. In this way the reduced system will not deform. Atoms in the inner region are simulated without such restraints. For macromolecular systems a reduction by a factor of 10 may be obtained by applying the extended wall region technique instead of periodic boundary conditions. It still remains to be tested to what extent the properties of the atoms in the inner region are distorted by restraining the dynamics of the atoms in the wall region.

6. Types of molecular dynamics. Normally, MD is performed at constant volume and constant total energy, yielding a microcanonical ensemble. For various reasons this is not very convenient. When MD is performed in non-equilibrium situations in order to study irreversible processes, catalytic events or transport properties, the need to impress external constraints or restraints is apparent. In such cases the temperature should be controlled as well in order to absorbe the dissipative heat produced by the irreversible process. But also in equilibrium cases the automatic control of temperature and pressure as independent variables is very convenient. We will briefly describe a method that we have found to be very reliable and convenient, which is easy to implement and which produces a stable algorithm even under adverse non-equilibrium conditions. The method achieves a coupling to an external bath with *constant temperature and/or pressure* with adjustable time constant of the coupling. This has the advantage that the coupling can be chosen sufficiently weak to avoid disturbance of the system and sufficiently strong to achieve the desired restraint. The basic idea is to modify the equations of motion in such a way that the net result on the system is a first-order relaxation of temperature T and pressure P toward given reference values T_0 and P_0:

$$dT/dt = (T_0 - T)/\tau_T \tag{9}$$

and

$$dP/dt = (P_0 - P)/\tau_p \tag{10}$$

The modification of the equations of motion is such that the local disturbances are minimized while the global effects of Eqs. (9, 10) are conserved. This is effected by scaling the velocities in every MD time step with a factor λ:

$$\vec{v}_i \leftarrow \lambda \vec{v}_i \tag{11}$$

$$\lambda = [1 + (T_0)/T - 1)\Delta t/\tau_T]^{-1/2} \tag{12}$$

where

$$T = E_{\text{kin}}/(3Nk/2) \tag{13}$$

$$E_{\text{kin}} = \sum_{i=1}^{N} 1/2\, m_i\, \vec{v}_i^{\,2} \tag{14}$$

and scaling the coordinates $\vec{r}_i$ and size $\vec{R}_{\text{box}}$ of the periodic box in every step with a factor μ:

$$\vec{r}_i \leftarrow \mu \vec{r}_i \tag{15}$$

$$\vec{R}_{\text{box}} \leftarrow \mu \vec{R}_{\text{box}} \tag{16}$$

$$\mu = [1 - \beta(P_0 - P)\Delta t/\tau_P]^{1/3} \tag{17}$$

where β denotes the isothermal compressibility of the system and the pressure P is computed from the relation

$$P = (2/3V)[E_{\text{kin}} - \Xi] \tag{18}$$

Here, V denotes the volume of the box, E_{kin} the kinetic energy and the virial Ξ is defined as

$$\Xi = -1/2 \sum_{i<j}^{N} \vec{r}_{ij} \cdot \vec{F}_{ij} \tag{19}$$

In the calculation of pressure, forces within a molecule can be omitted, together with kinetic contributions of internal degrees of freedom. The equations can be modified to include non-isotropic systems. The coupling time contstants τ_T and τ_P can be arbitrarily chosen, but should exceed ten time steps to ensure stability of the algorithm. The value of the compressibility β used in (17) is not critical, since it only influences the accuracy of the pressure coupling time constant τ_P.

7. Algorithms for MD. We now return to the solution of Eq. (3) assuming that the force calculation can be carried out for any configuration of the atoms. Various algorithms for integrating Eq. (3) have been proposed, of which the simplest is the so-called leap-frog scheme:

$$\vec{v}_i(t + \Delta t/2) = \vec{v}_i(t - \Delta t/2) + m_i^{-1} \vec{F}_i(t)\Delta t \tag{20}$$

$$\vec{r}_i(t + \Delta t) = \vec{r}_i(t) + \vec{v}_i(t + \Delta t/2)\Delta t \tag{21}$$

It belongs to the class of open integration methods and possesses the useful property of being invariant upon time-reversal. This implies that the total energy which is a constant of motion, shows essentially no drift. Although it is of *third order* in the integration time step Wt and so less accurate than higher-order Gear algorithms, it is much easier to use, requires less storage and is more stable and efficient for large values of Δt. In practice it only makes sense to use higher-order algorithms when accurate dynamics is required based on accurately known forces and using a relatively small time step Δt, or when the forces are highly harmonic, which makes them predictable by higher-order derivatives. Algorithms that are of lower than third order are not efficient in MD of molecular systems. This is due to the fact that molecular potential functions V (Eq. (1)) have generally positive second derivatives, requiring a third or higher order algorithm.

8. Simplifications and time-saving techniques. Various simplifications can be applied to the basic procedures sketched above, with the aim of saving computing effort. One may think of *constraining the bond lengths and/or the bond angles* in the molecules of the system. The length of the time step Δt in a MD simulation is limited by the highest frequency $\nu_{\max}$ in the system: $\Delta t << \nu_{\max}^{-1}$. By constraining the degrees of freedom with the highest frequencies, the time step can generally be lengthened, which reduces the computer time required for a MD run of a specific length. The application of constrained dynamics makes sense only when:

a) the frequencies of the frozen degrees of freedom are (considerably) higher than those of the remaining ones, thereby allowing a (considerable) increase of Δt, and when
b) the frozen degrees of freedom are only weakly coupled to the remaining ones, viz. when the motion of the molecules is not affected by application of the constraints.

For macromolecular system the application of bond length constraints saves about a factor of 2 in computer time when hydrogen atoms are explicitly treated and a factor of 3 when the united-atom model is used. The use of bond angle constraints is not allowed, since it affects the macromolecular dynamics considerably.

Another way to reduce the computing effort is the application of *stochastic dynamics* (SD). Less relevant degrees of freedom are ignored and their influence is approximated by a combination of mean force interactions, stochastic forces and frictional forces, as discussed in Sect. 3.

9. Derivation of properties from MD. Given the ability to simulate and store long dynamic trajectories of a complex system, how can the derived properties be derived? We must realize that the precise details of any trajectory are totally irrelevant: not only are all interesting properties related to statistical averages, but the precise course of a trajectory is even unpredictable. Any error, however slight, will cause at some later time a significant deviation of the exact trajectory. But that need not bother us since also in real physical systems the predctability of trajectories in phase space is destroyed by noise from outside.

On the basis of a trajectory ensemble averages of any *directly measurable* quantity can be obtained by simple averaging: the kinetic and potential energy, pressure, structural properties such as radial distribution functions, structure factors, etc. Properties like the specific heat, compressiblity and other derivatives of directly measureable quantities can be obtained from fluctuations. Auto and cross correlation functions of time dependent quantities yield transport properties like diffusion coefficients, viscosity, conductance and variety of relaxation properties.

A large class of very relevant thermodynamic quantities is ***not directly measurable***, since they depend on an integral over phase space, rather than an ensemble average. These quantities are the entropy S, the Helmholtz free energy A, the Gibbs free energy G, the thermodynamic potential μ, and all quantities derived from these, such as binding constants association and dissociation constants, solubilities, etc. These quantities are of more fundamental and practial interest than the directly measureable ones. In recent years techniques have become available to determine free energies, be it at the expense of considerable computing effort; these are discussed in Sect. 4.

2.3. Experimental basis. In order to provide a firm basis for the application of simulation methods the results should be compared to experimental data whenever possible. Here, one may think of X-ray or neutron diffraction data, spectroscopic information from nuclear magnetic resonance and other (optical) techniques and from thermodynamic measurements. With current force fields, crystal structures of small molecules are reproduced withing 0.2–0.3 K. Even dynamic effects, like the flip-flopping of hydrogen bonds, which was observed experimentally from neutron diffraction, are reproduced in MD simulations [15]. However, a continuous testing of techniques and force fields is required in order to improve the accuracy of the simulations.

2.4. Applications in chemistry and physics. Of the many possible applications of MD simulation techniques we list a few of the most interesting ones:

(1) prediction of the dependence of a molecular conformation on the type of environment (e.g. water, apolar solvent, crystal, etc.);
(2) calculation of relative binding constants by evaluating free energy differences between various molecular complexes;
(3) prediction of energetic and structural changes caused by modification of amino acids in enzymes or of base pairs in DNA;
(4) derivation of three-dimensional (3D) molecular structure on the basis of 2D-NMR data by using restrained MD techniques;
(5) dynamic modelling of molecular complexes by searching configuration space using MD;
(6) prediction of properties of materials under extreme conditions of temperature and pressure, which may be experimentally inaccessible.

These applications can be carried out using the GROningen MOlecular Simulation program package (GROMOS), which is available at nominal cost. The package contains FORTRAN routines and vectorized versions of some routines for Cray, Cyber 205, and Fujitsu as well as Convex computers. It handles molecular dynamics and energy minimzation for any molecular system, with topologies and potential functions included for molecules of biological interest. A large number of analysis programs is included.

3. Stochastic dynamic simulations: Algorithms and applications. When modelling macromolecular systems using molecular dynamics (MD), practical limitations of computational effort form a strong incentive to reduce the number of degrees of freedom as much as possible, while retaining a truthful simulation of the physical characteristics of interest. In principle there are two ways to achieve this reduction.

a. Hard degrees of freedom, that is, those corresponding to high-frequency normal modes, can be treated as if they were completely constrained and thus be eliminated from the system. This approach leads to constrained dynamics, which increases the computational efficiency by allowing for a larger time step in the simulation [5].
b. Less relevant degrees of freedom can be ignored and their influence approximated by a combination of mean force interactions, stochastic and frictional forces. This approach leads to *stochastic dynamics* (SD).

When simulating a macromolecule in solution, one is generally not interested in the details of the motion of the solvent molecules. One would like to get rid of explicit treatment of the solvent molecules in the simulation, but would like to maintain their influence on the dynamics of the solute. This influence can be divided into three types.

a. The solvent affects the average interaction between solute atoms.
b. The solute atoms experience a frictional force due to the solvent.
c. The solvent molecules collide with the solute atoms, inducing a randomly fluctuating motion of the latter.

The average effect of the solvent atoms a on the interaction between solute atoms denoted by i can be incorporated in the interaction function $V_s(\{x_i\})$ for the solute by averaging over the solvent (α) degrees of freedom in an equilibrium ensemble:

$$V_s(\{x_i\}) = \langle V(\{x - i\}, \{x_\alpha\})\rangle_\alpha \tag{22}$$

Here the indices i or a label atoms and Cartesian components. The functions V_s is called a *potential of mean force* [16]. The two other types of solvent influence can be mimicked by the inclusion of a frictional force and a random force in Newton's equations of motion.

The simplest case is a frictional force proportional to the velocity of the atom to which it applies, and a random force with white-noise character, uncorrelated between the various degrees of freedom:

$$m_i\dot{v}_i(t) = F_i(\{x_i(t)\}) - m_i\gamma_i v_i(t) + R_i(t) \tag{23}$$

This is called the *simple Langevin equation.* The atomic mass is denoted by m_i , the velocity by v_i , the friction coefficient by γ_i . The systematic force F_i is to be derived from the potential of mean force and the random force is denoted by R_i. It is assumed to be a stationary gaussian random variable with zero mean and to have no correlation with prior velocities nor with the systematic force:

$$\langle R_i(0)R_j(t)\rangle = 2m_i\gamma_i kT_{\text{ref}}\delta_{ij}\delta(t), \tag{24}$$

$$W(R_i) = [2\pi\langle R_i^2\rangle]^{-1/2}\exp\{-R_i^2/(2\langle R_i^2\rangle)\}, \tag{25}$$

$$\langle R_i\rangle = 0, \tag{26}$$

$$\langle v_i(0)R_j(t)\rangle = 0,\ t \geq 0, \tag{27}$$

$$\langle F_i(0)R_j(t)\rangle = 0,\ t \geq 0. \tag{28}$$

Here $\langle \dots \rangle$ denotes averaging over an equilibrium ensemble, k is Boltzmann's constant, T_{ref} is the reference temperature and $W(R_i)$ is the (gaussian) probability distribution of the stochastic force. Formula (24) relates the random force with the velocity-dependent friction force by the second fluctuation-dissipation theorem. A minor correction to (24) has been discussed in [17–19].

The choice of the values for the friction coefficients depends on the nature of the molecular system. For a solute in aqueous solution they depend on the hydration of the solute atoms and on the viscosity of the solvent. For example, they can be taken equal to the friction coefficient of pure water at the given temperature times the fraction of the atomic surface area which is accessible to water.

3.1. Types of stochastic dynamics. More sophisticated forms of SD can be obtained by incorporating time and/or spatial correlations into the dynamics.

a. Time correlations are taken into account in the generalized Langevin equation [20]:

$$m_i\dot{v}_i(t) = F_i(\{x_i(t)\}) - m_i\int_0^t \gamma_i(t')v_i(t-t')dt' + R_i(t) \tag{29}$$

where the firction coefficient γ_i has become time-dependent. It has been transformed into the memory function $\gamma_i(t)$. Instead of (24) the second fluctuation-dissipation theorem now gives [20]

$$\langle R_i(0)R_j(t)\rangle = m_i kT_{\text{ref}}\delta_{ij}\gamma_i(t) \tag{30}$$

When the correlation properties of the random force are poorly described by (24), or equivalently, when the assumption of a time independent friction coefficient is too crude, one may try to improve the description of the dynamics of the system by using (29) and (30) [21, 22].

b. Spatial correlations can be taken into account by making the friction coefficient γ_i space dependent, for example by writing [23]

$$m_i\dot{v}_i(t) = F_i(\{x_i(t)\}) - m_i\sum_{j=1}^{3N}\gamma_{ij}v_j(t) + \sum_{j=1}^{3N}\beta_{ij}R_j(t). \tag{31}$$

Instead of (24) we now have the relations

$$\langle R_i(0)R_j(t)\rangle = 2\delta_{ij}\delta(t) \tag{32}$$

$$\gamma_{ij} = (m_i k T_{\text{ref}})^{-1} \sum_{k=1}^{3N} \beta_{ik}\beta_{jk} \tag{33}$$

This stochastic dynamics with hydrodynamic interaction has been used to simulate simple dimers and trimers [23].

When modelling heterogeneous macromolecules, the application of time and/or spatial correlations in SD becomes cumbersome. Therefore we restrict the further discussion to simple Langevin dynamics.

3.2. Algorithms for stochastic dynamics. The Langevin equation (23) differs from Newton's equation of motion which is solved in MD, by the occurrence of the random term in the force. This makes the force in SD less predictable than in MD. One might suppose that a very low order algorithm would suffice, but this is not the case due to the fact that molecular potential functions generally have consistently positive second derivatives. This requires a third- or higher order algorithm. When integrating (23) the relation between the time step Δt and the accuracy of the trajectory is entirely determined by the systematic force F_i and not by the random force. Without the systematic force the Langevin equation is exactly solvable by integration of the force to obtain the velocity and subsequent integration of the velocity to obtain the coordinates. These steps involve integration of random variables, sometimes multiplied by known functions of time, over a time interval. Such integrals are again stochastic variables, the statistical properties of which follow from their derivation. So, they can directly be sampled from the appropriate probability distributions. A slight complication lies in the fact that some random variables are obtained by integration over the same interval, but with different time-dependent weight functions, which introduces correlation between them. Therefore, sampling of two correlated random variables from bivariate gaussian distributions turns out to be required [21, 24, 25].

3.3. Leap-frog algorithm for SD. One of the most simple, efficient and stable algorithms for performing MD simulations is the leap-frog scheme [26]:

$$v(t_n + \Delta t/2) = v(t_n - \Delta t/2) + m^{-1}F(t_n)\Delta t \tag{34}$$

$$x(t_n + \Delta t) = x(t_n) + v(t_n + \Delta t/2)\Delta t \tag{35}$$

Here we have omitted the indices denoting atoms and Cartesian components and the index n denotes the n-th integration step of step size Δt. We will briefly derive a SD equivalent of the leap-frog scheme, which will reduce to (34) and (35) in the limit of zero friction ($\gamma = 0$) [27].

Solving the linear, inhomogeneous first order differential equation (23) one finds

$$\begin{aligned} v(t) &= v(t_n)\exp[-\gamma(t - t_n)] \\ &\quad + m^{-1}\exp[-\gamma(t - t_n)] \int_{t_n}^{t} \exp[-\gamma(t_n - t')].\{F(t') + R(t')\}dt' \end{aligned} \tag{36}$$

Since the stochastic properties of $R(t')$ are known (3-7), the integral over $R(t')$ can directly be obtained. The integral over $F(t')$ is obtained by expanding $F(t')$ in a power series

$$F(t) = F(t_n) + \dot{F}(t_n)(t - t_n) + O[(t - t_n)^2] \tag{37}$$

The derivative of the systematic force with respect to time at $t = t_n$ is denoted by $\dot{F}(t_n)$. Since the leap-frog scheme is a third-order algorithm, an expansion of $F(t)$ up to second order suffices. Expansion (37) is only valid in the limit of small friction coefficients γ. This can be understood by noting that the systematic force $F(\{x(t)\})$ depends on the coordinates, which must satisfy the Langevin equation (23), which includes apart from the systematic force two extra terms, the friction force and the random force, which will influence the dependence of $x(t)$ on time. In the high friction limit the latter terms will dominate in (23), in which case inclusion of the derivative of the systematic force would make no sense. Therefore, we will omit from the formulae below all terms beyond third order in Δt in the coordinates and beyond second order in the velocities.

Performing the integration in (36) we find

$$\begin{aligned} v(t) &= v(t_n)\exp[-\gamma(t - t_n)] \\ &+ (m\gamma)^{-1}F(t_n)[1 - \exp[-\gamma(t - t_n)]] \\ &+ (m\gamma^2)^{-1}\dot{F}(t_n)[\gamma(t - t_n) - [1 - \exp[-\gamma(t - t_n)]]] \\ &+ m^{-1}\exp[-\gamma(t - t_n)]\int_{t_n}^{t}\exp[-\gamma(t_n - t')]R(t')dt' \end{aligned} \tag{38}$$

If we take (17) with $t = t_n - \Delta t/2$, multiply it by $\exp[-\gamma\Delta t]$ and subsequently subtract it from (17) with $t = t_n + \Delta t/2$, we find

$$\begin{aligned} v(t_{n+1/2}) &= v(t_{n-1/2})\exp[-\gamma\Delta t] \\ &+ m^{-1}F(t_n)\Delta t(\gamma\Delta t)^{-1}[1 - \exp[-\gamma\Delta t]] \\ &+ V_n(\Delta t/2) - \exp[-\gamma\Delta t]V_n(-\Delta t/2) \end{aligned} \tag{39}$$

where the term involving $\dot{F}(t_n)$ has been omitted, since it is of third order in Δt. Here $t_{n+1/2} \equiv t_n + \Delta t/2$, and

$$V_n(\Delta t/2) \equiv m^{-1}\exp[-\gamma\Delta t/2]\int_{t_n}^{t_n+\Delta t/2}\exp[-\gamma(t_n - t')]R(t')dt' \tag{40}$$

When $\gamma \to 0$, formula (39) reduces to (34), which is the velocity formula of the leap-frog algorithm.

The positions can be obtained from

$$x(t) = x(t_{n+1/2}) + \int_{t_n+1/2}^{t} v(t')dt'. \tag{41}$$

If we use in this equation formula (38) for $v(t')$ with t_n replaced by $t_{n+1/2}$, perform the integrations and reduce the double integral over $R(t')$ using partial integration, we find for $t = t_{n+1/2} + \Delta t/2$:

$$
\begin{aligned}
(42)\quad x(t_{n+1/2}&+\Delta t/2) \\
&= x(t_{n+1/2}) + v(t_{n+1/2})\gamma^{-1}[1 - \exp[-\gamma\Delta t/2]] \\
&+ (m\gamma^2)^{-1} F(t_{n+1/2})[\gamma\Delta t/2 - [1 - \exp[-\gamma\Delta t/2]]] \\
&+ (m\gamma^3) F(t_{n+1/2}[1/2(\gamma\Delta t/2)2 - [\gamma\Delta t/2 - [1 - \exp[-\gamma\Delta t/2]]]] \\
&+ X_{n+1/2}(\Delta t/2)
\end{aligned}
$$

where

$$
(43)\quad X_{n+1/2}(\Delta t/2) = (m\gamma)^{-1} \int_{t_{n+1/2}}^{t_{n+1/2}+\Delta t} [1 - \exp[-\gamma(t_{n+1/2} + \Delta t/2 - t')]] R(t')dt'
$$

If we write (21) with $-\Delta t$ instead of Δt and subtract it from (42), the term involving $x(t_{n+1/2})$ is eliminated and we find

$$
\begin{aligned}
(44)\qquad x(t_{n+1}) &= x(t_n) \\
&+ v(t_{n+1/2})\Delta t(\gamma\Delta t)^{-1}[\exp(+\gamma\Delta t/2) - \exp(-\gamma\Delta t/2)] \\
&+ X_{n+1/2}(\Delta t/2) - X_{n+1/2}(-\Delta t/2),
\end{aligned}
$$

where terms involving $F(t_{n+1/2})$ and $F(t_{n+1/2})$ have been omitted since they are of third order in Δt. When $\gamma \to 0$, this formula reduces to (35), which is the coordinate formula of the leap-frog algorithm.

When using the SD leap-frog algorithm, (39) and (44), it must be noted that $V_n(-\Delta t/2)$ is correlated with $X_{n-1/2}(\Delta t/2)$, since they are different integrals of $R(t)$ over the time interval $(t_{n-1/2}, t_n)$. The same observation holds for $X_{n+1/2}(-\Delta t/2)$ and $V_n(\Delta t/2)$ occurring in (39) and (44). These are different integrals of $R(t)$ over the time interval $(t_n, t_{n+1/2})$. In both cases the pair of random variables obeys a bivariate gaussian distribution, the parameters of which can be obtained by evaluating the quantities $\langle X^2_{n-1/2}(\Delta t/2)\rangle, \langle V^2_n(-\Delta t/2)\rangle$ and $\langle X_{n-1/2}(\Delta t/2)V_n(-\Delta t/2)\rangle$ for the first pair and the quantities $\langle V^2_n(\Delta t/2)\rangle$, $\langle X^2_{-n+1/2}(-\Delta t/2)\rangle$ and $\langle V_n(\Delta t/2)X_{n+1/2}(-\Delta t/2)\rangle$ for the second pair of random variables using formula (3) [27, 28].

3.4. Applications of stochastic dynamics. For a small polyatomic molecule it has been checked whether the SD approach yields a reasonable approximation of the solvent effects [29]. It appears that the conformations and motion of one decane molecule that is simulated using SD agree very well with the results of a full MD simulation of liquid decane. For a protein in aqueous solution a corresponding test is still to be performed. Application of SD instead of MD in simulations of protein in solution reduces the computational effort considerably. Simulation of a small protein like BPTI (bovine pancreatic trypsin inhibitor) consisting of about 500 atoms in a box (truncated octahedron) filled with water molecules, requires a minimum of about 1500 water molecules to fill the remaining volume of the box. So, the MD simulation includes about 15000 degrees of freedom or coupled equations to be integrated. A SD simulation of this protein involves only the solute degrees of freedom, which amount to only 1500, a reduction by a factor of 10.

SD can also be used instead of MD or energy minimization techniques in order to search the conformational space of a molecule for low energy conformations. Finally, the application of SD allows one to vary the strength of the coupling of the atoms (through the friction coefficients γ) to the heat bath. This yields the possibility to study the process of transfer of energy through a molecule.

4. Calculation of free energies in biomolecular systems. From a molecular dynamics trajectory the statistical equilibrium averages can be obtained for any desired directly measurable property *of the molecular system*, that is, for which a value can be computed at each point of the trajectory. Examples of such properties are the potential or kinetic energy of relevant parts of the system, structural properties, electric fields, etc. From such averages a number of thermodynamic properties can be derived. However, two important thermodynamic quantities, the entropy and the (Helmholtz or Gibbs) free energy can generally not be derived from a statistical average. These are global properties that depend on the extent of phase (or configuration) space accessible to the molecular system. Therefore, computation of the absolute free energy of a molecular system is virtually impossible. Yet, the most interesting chemical quantities like binding constants of donor-acceptor complexes or molecular solubilities are directly related to the free energy. Over the past few years several statistical mechanical procedures have evolved for evaluating relative free energies [8–11]. They are rather demanding as far as computer time is concerned, but will open up a wide area of most interesting applications in chemistry.

4.1. Methods to determine free energy or entropy. In a canonical ensemble the fundamental formula for the Helmholtz free energy A is:

$$A = -kT \ln Z \tag{45}$$

where k is Boltzmann's constant, T refers to the absolute temperature and Z is the partition function, determined by the Hamiltonian $H(\vec{p}, \vec{q})$ that describes the total energy of the system in terms of (generalized) coordinates $\vec{q}$ and momenta $\vec{p}$. For a system consisting of N particles one has

$$Z = (h^{3N} N!)^{-1} \iint \exp[-H(\vec{p}, \vec{q})/kT] d\vec{p}\, d\vec{q} \tag{46}$$

where h is Planck's constant. The entropy S is related to the free energy by

$$A = U - TS \tag{47}$$

where U is the internal energy, the ensemble average $\langle H \rangle$ of $H(\vec{p}, \vec{q})$. In terms of the probability

$$\pi(\vec{p}, \vec{q}) = \exp[-H(\vec{p}, \vec{q})/kT]/(h^{3N} N!\, Z) \tag{48}$$

one has:

$$U = \iint H(\vec{p}, \vec{q}) \pi(\vec{p}, \vec{q}) d\vec{p}\, d\vec{q} \tag{49}$$

Using (1)-(5) the entropy S can be expressed in terms of the probabilities $\pi(\vec{p}, \vec{q})$ in phase space:

$$S = -k \iint \pi(\vec{p}, \vec{q}) \ln[\pi(\vec{p}, \vec{q}) hN!] d\vec{p}\, d\vec{q} \tag{50}$$

In order to discuss the various techniques for evaluating A or S we must distinguish between diffusive and non-diffusive systems [9].

Diffusive systems, such as liquids or solutions, are characterized by the eventual diffusion of particles over all of the available space. Non-diffusive systems, such as solids, glasses and macromolecules with a definite average structure are characterized by time independent average positions around which the atoms are fluctuating. For these systems the configuration space is limited and the integral (50) can be obtained by a *direct determination of the probability ditribution function* $\pi(\vec{p}, \vec{q})$, e.g. from a simulation. For diffusive systems this road is closed, since the multidimensional configuration space is so vast that it can never be integrated using simulation techniques. For these systems two other techniques are available, *probe methods*, like the particle insertion methods, and the very powerful *thermodynamic perturbation and integration methods.*

4.2. Direct determination of the probability distribution. [30–32] If the potential function is not dependent on the momenta $\vec{p}$ of the atoms, the kinetic contribution to the entropy can be split off and analytically determined. The configurational part S_c depends on the multidimensional probability distribution $\pi(\vec{q})$, which is difficult to determine with sufficient accuracy from a simulation. However, if it is assumed to be a multivariate Gaussian distribution, it is completely characterized by its covariance matrix

$$\sigma_{ij} = \langle [q_i - \langle q_i \rangle][q_j - \langle q_j \rangle] \rangle, \tag{51}$$

and the configurational entropy becomes only a function of its determinant:

$$S_c = 1/2k[3N(1 + \ln 2\pi) + \ln \det \underline{\sigma}]. \tag{52}$$

When treating a macromolecule, the computation of this determinant of a matrix of thousands degrees of freedom is quite computer time consuming. The application of this method is also restricted by the fact that the inclusion of diffusive solvent degrees of freedom is not possible.

4.3. Probe methods. [33–36] These methods insert a test particle in the system at a very large number of uniformly distributed random test positions. The interaction energy of the test particle with the particles in the system is computed and the average Boltzmann factor of the obtained energy yields the free energy. The method works well for homogeneous low and medium density systems. At high density, or in heterogeneous systems, poor statistics is obtained when averaging the Boltzmann factor, due to the generally large insertion energies.

4.4. Thermodynamic perturbation and integration methods. [37–49] These methods make use of the fact that free energy changes related to small perturbations of a molecular system can be determined in the course of a simulation. The free energy difference between two states A and B of a system can be determined from a MD simulation in which the potential energy function V is slowly changed such that the system reversibly changes from state A to state B. In principle the free energy is determined as the work necessary to change the system from A to B over a reversible path.

The method works as follows. Firstly, the Hamiltonian $H(\vec{p},\vec{q})$ is made a function of a parameter γ, such that $H(\vec{p},\vec{q},\lambda_A)$ characterizes state A of the system and $H(\vec{p},\vec{q},\lambda_B)$ state B. Then the (Helmholtz) free energy of the system is also a function of λ:

(53) $$A(\lambda) = -kT \ln Z(\lambda)$$

and the partition function Z becomes also dependent on λ:

(54) $$Z(\lambda) = (h^{3N}N!)^{-1} \iint \exp[-H(\vec{p},\vec{q},\lambda)/kT] d\vec{p}\, d\vec{q}.$$

The free energy difference ΔA_{BA} then reads

(55) $$\Delta Ad_{AB} = A(\lambda_B) - A(\lambda_A) = -kT \ln[Z(\lambda_B)/Z(\lambda_A)]$$

which can be expressed as an ensemble average

(56)
$$\begin{aligned}\Delta A_{AB} = -kT \ln &\Big\{ \iint \exp[-(H(p,q,\lambda_B) - H(p,q,\lambda_A))/kT] \\ &. \exp[-H(p,q,\lambda_A)/kT] dp\, dq \Big/ \iint \exp[-(H(p,q,\lambda_A)/kT] dp\, dq \Big\} \\ = -kT &\ln \{\langle \exp[-(H(p,q,\lambda_B) - H(p,q,\lambda_A))/kT]\rangle\}\end{aligned}$$

where the brackets $\langle \ldots \rangle_\lambda$ denote an ensemble average over $\vec{p}$ and $\vec{q}$, generated at a specific value of λ. Formula (56) is called the *perturbation formula*, since it will only yield accurate results when state B is close to state A. If this difference is large, the change from A to B must be split up in a number of steps between intermediate states that are close enough to allow for the use of formula (56) and then ΔA_{BA} is just the sum of the ΔA for all intermediate steps. The thermodynamic *integration formula* is obtained by straightforward differentiation of (53) with respect to λ:

$$dA(\lambda)/d\lambda = \langle \partial H(\vec{p}, \vec{q}, \lambda)/\partial\lambda \rangle_\lambda. \tag{57}$$

In that case

$$\Delta A_{BA} = \int_{\lambda_A}^{\lambda_B} \langle \partial H(\vec{p}, \vec{q}, \lambda)/\partial\lambda \rangle d\lambda. \tag{58}$$

If λ is being changed very slowly during a MD simulation, the integration (58) can be carried out during the MD run. Then ΔA_{BA} can be directly obtained for rather different states A and B. The continuous change in λ should be so slow that the system remains essentially in equilibrium for each intermediate value of λ.

Various parametrizations of the Hamiltonian are possible. The choice should be such that the change of $A(\lambda)$ as a function of λ is as smooth as possible. We use

$$H(p, q, \lambda) = \sum_{i=1}^{N} \vec{p}_i^{\,2}/(2m_i(\lambda)) + V(r, ..., r_N, \lambda). \tag{59}$$

For the masses we choose

$$m_i(\lambda) = (1-\lambda)m_i^A + \lambda m_i^B \tag{60}$$

For the terms in the potential energy function $V(r_1, ..., r_N, \lambda)$ representing bond-stretching, bond-angle bending, dihedral torsion and non-bonded interactions we choose

$$V(b, \lambda) = 1/2[(1-\lambda)K_b^A + \lambda K_b^B].[b - [(1-\lambda)b_0^A + \lambda b_0^B]]^2, \tag{61}$$

$$V(\theta, \lambda) = 1/2[(1-\lambda)K_\theta^A + \lambda K_\theta^B].[\theta - [(1-\lambda)\theta_0^A + \lambda \theta_0^B]]^2, \tag{62}$$

$$V(\phi, \lambda) = (1-\lambda)K_\phi^A[1 + \cos(n^A\phi - \delta^A)] + \lambda K_\phi^B[1 + \cos(n^B\phi - \delta^B)], \tag{62}$$

$$\begin{aligned} V(r_{ij}, \lambda) = {} & (1-\lambda)[C_{12}^A(i,j)/r_{ij}^{12} - C_6^A(i,j)/r_{ij}^6 + q_i^A q_j^A/(4\pi\varepsilon_0 r_{ij})] \\ & + \lambda[C_{12}^B(i,j)/r_{ij}^{12} - C_6^B(i,j)/r^6 + q_i^B q_j^B/(4\pi\varepsilon_0 r_{ij})]. \end{aligned} \tag{64}$$

4.5. Thermodynamic cycles. The basis on which the thermodynamic cycle approach rests is the fact that the (Helmholtz) free energy A is a thermodynamic state function. This means that as long as a system is changed in a reversible way the change in free energy WA will be independent of the path. Therefore, along a closed path or cycle one has $\Delta A = 0$. This result implies that there are two possibilities of obtaining ΔA for a specific process; one may calculate it directly using the techniques discussed above along a path corresponding to the process, or, one may design a cycle of which the specific process is only a part and calculate the ΔA of the remaining part of the cycle. The power of this thermodynamic cycle technique lies in the fact that on the computer also non-chemical processes such as the conversion of one type of atom into another type may be performed.

In order to visualize the method we consider the relative binding of a repressor protein R to two different DNA operator fragments O_A and O_B. The appropriate thermodynamic cycle for obtaining the relative binding constant is

$$
\begin{array}{ccc}
R + O_A & \xrightarrow{1(\text{exp})} & (R : O_A) \\
{\scriptstyle 3(\text{sim})}\downarrow & & \downarrow{\scriptstyle 4(\text{sim})} \\
R + O_B & \xrightarrow[2(\text{exp})]{} & (R : O_A)
\end{array}
\tag{65}
$$

where the symbol : denotes complex formation. The relative binding constant of processes 1 and 2 equals

$$K_2/K_1 = \exp[-(\Delta A_2 - \Delta A_1)/RT]. \tag{66}$$

However, simulation of processes 1 and 2 is virtually impossible, since it would involve the displacement of many solvent molecules from the region on the operator O where the repressor will be bound. But, since (21) is a cycle, we have

$$\Delta A_2 - \Delta A_1 = \Delta A_4 - \Delta A_3 \tag{67}$$

and, if the operator O_B is not too different (one or a few base pairs) from operator O_A, the desired result can be obtained by simulation of the non-chemical processes 3 and 4.

4.6. Applications and limitations. It will be clear that the applications of the thermodynamic cycle integration technique are manifold in chemistry. One may think of studying free energy of solvation, of binding different inhibitors to enzymes, of binding mutant enzymes to substrates or inhibitors, of studying phenomena like adsorption or phase separation, etc. A major limitation to the application of the technique may reside in the requirement of sufficiently dense sampling along a reversible path. This may involve very long MD simulations. The length of such a simulation must be much longer than the relaxation times characterizing the relevant degrees of freedom in the molecular system. When studying hydration this conditions can be met, but conformational relaxation in macromolecules may require much longer than 10–100ps simulations.

REFERENCES

[1] W. F. van Gunsteren and H. J. C. Berendsen, *Molecular dynamics: Perspectives for complex systems*, Biochem. Soc. Trans., 10 (1982), pp. 301–305.

[2] J. Hermans, ed., *Molecular Dynamics and Protein Structure*, Polycrystal Book Service, Western Springs, IL, 1985.

[3] G. Ciccotti and W. G. Hoover, eds., *Molecular-Dynamics Simulation of Statistical Mechanical Systems*, Proceedings of the International School of Physics "Enrico Fermi", course 97, North-Holland, Amsterdam, 1986.

[4] J. A. McCammon and S. C. Harvey, *Dynamics of Proteins and Nucleic Acids*, Cambridge University Press, London, 1987.

[5] H. J. C. BERENDSEN AND W. F. VAN GUNSTEREN, *Molecular dynamics simulations: Techniques and approaches*, in *Molecular Liquids-Dynamics and Interactions*, A.J. Barnes et al., eds., NATO ASI Series C135, Reidel, Dordrecht, 1984, pp. 475-500.

[6] W. F. VAN GUNSTEREN AND H. J. C. BERENDSEN, *Molecular dynamics simulations: Techniques and approaches to proteins*, in *Molecular Dynamics and Protein Structure, J. Hermans, ed.*, Polycrystal Book Service, Western Springs, IL, 1985, pp. 5-14.

[7] H. J. C. BERENDSEN AND W. F. VAN GUNSTEREN, *Practical algorithms for dynamics simulations*, in *Molecular Dynamics Simulation of Statistical-Mechanical Systems*, Proceedings of the International School of Physics "Enrico Fermi", course 97, G. Ciccotti and W. G. Hoover, eds., North-Holland, Amsterdam, 1986, pp. 43–65.

[8] H. J. C. BERENDSEN, *Biological molecules and membranes*, in *Molecular-Dynamics Simulation of Statistical-Mechanical Systems*, Proceedings of the International School of Physics "Enrico Fermi", course 97, G. Ciccotti and W. G. Hoover, eds., North-Holland, Amsterdam, 1986, pp. 496–519.

[9] P. KOLLMAN AND W. F. VAN GUNSTEREN, *Molecular mechanics and dynamics in protein design*, in *Methods in Enzymology: Recombinant DNA, Section VIII, Site Specific Mutagenesis and Protein Engineering*, R. Wu and L. Grossman, eds., 1987 (to appear).

[10] H. J. C. BERENDSEN, W. F. VAN GUNSTEREN, E. EGBERTS AND J. DE VLIEG, *Dynamic simulation of complex molecular systems*, in *ACS Symposium Series, No. 353*, D. G. Truhlar ed. (American Chemical Society, Washington, 1987), pp. 106-122.

[11] N. QUIRKE AND G. JACUCCI, *Energy difference functions in Monte Carlo simulations. Application to (1) the calculation of free energy of liquid nitrogen, (2) the fluctuation of Monte Carlo averages*, Molec. Phys., 45 (1982), pp. 823–838.

[12] H. J. C. BERENDSEN, J. P. M. POSTMA AND W. F. VAN GUNSTEREN, *Statistical mechanics and molecular dynamics: The calculation of free energy*, in *Molecular Dynamics and Protein Structure*, Polycrystal Book Service, Western Springs, IL, 1985, pp. 43–46.

[13] D. FRENKEL, *Free energy computation and first-order phase transitions*, in *Molecular Dynamics Simulation of Statistical-Mechanical Systems*, Proceedings of the International School of Physics "Enrico Fermi", course 97, North-Holland, 1986.

[14] M. MEZEI AND D. L. BEVERIDGE, *Free energy simulations*, Ann. New York Acad. Sci., 482 (1986), pp. 1–23.

[15] J. E. H. KOEHLER, *On Structure and Dynamics of Cyclodextrins in the Crystal and in Solution*, thesis, Freie Universität Berlin, pp. 1–45.

[16] C. PANGALI, M. RAO AND B. J. BERNE, *Hydrophobic hydration around a pair of apolar species in water*, J. Chem. Phys., 71 (1979), pp. 2982-2990.

[17] G. CICCOTTI AND J. P. RYCKAERT, *On the derivation of the generalized Langevin equation for interacting Brownian particles*, J. Statist. Phys., 26 (1981), pp. 73–82.

[18] G. BOSSIS, B. QUENTREC AND J. P. BOON, *Brownian dynamics and the fluctuation-dissipation theorem*, Molec. Phys., 45 (1982), pp. 191–196.

[19] W. F. VAN GUNSTEREN AND H. J. C. BERENDSEN, *On the fluctuation-dissipation theorem for interacting brownian particles*, Molec. Phys., 47 (1982), pp. 721–723.

[20] R. KUBO, *The fluctuation-dissipation theorem*, Rep. Progr. Phys., 29 (1966), pp. 255–284.

[21] F. J. VESELY, *Algorithms for Mori chain dynamics: a unified treatment*, Molec. Phys., 53 (1984), pp. 505–524.

[22] B. CICCOTTI, M. FERRARIO AND J. P. RYCKAERT, *Computer simulation of the generalized Brownian motion II. An argon particle in argon fluid*, Molec. Phys., 46 (1982), pp. 875-899.

[23] D. L. ERMAK AND J. A. MCCAMMON, *Brownian dynamics.*

[24] M. P. ALLEN, *Brownian dynamics simulation of a chemical reaction in solution*, Molec. Phys., 40 (1980), pp. 1073-1087.

[25] W. F. VAN GUNSTERN AND H. J. C. BERENDSEN, *Algorithms for Brownian dynamics*, Molec. Phys., 45 (1982), pp. 637-647.

[26] R. W. HOCKNEY AND J. W. EASTWOOD, *Computer Simulation using Particles*, McGraw-Hill, New York, 1981.

[27] W. F. VAN GUNSTEREN AND H. J. C. BERENDSEN, *A leap-frog algorithm for stochastic dynamics*, Molecular Simulation (1987) (to appear).

[28] A. PAPOULIS, *Random Variables and Stochastic Processes*, McGraw-Hill, New York, 1965.

[29] W. F. VAN GENSTEREN, H. J. C. BERENDSEN AND J. A. C. RULLMANN, *Stochastic dynamics for molecules with constraints Brownian dynamics of n-alkanes*, Molec. Phys., 44 (1981), pp. 69–95.

[30] M. Karplus and J. N. Kushick, *Method for estimating the configurational entropy of macromolecules*, Macromolecules, 14 (1981), pp. 325–332.

[31] O. Edholm and H. J. C. Berendsen, *Entropy estimation from simulations of non-diffusive systems*, Molec. Phys., 51 (1984), pp. 1011–1028.

[32] A. DiNola, H. J. C. Berendsen and O. Edholm, *Free energy determination of polypeptide conformations generated by molecular dynamics*, Macromolecules, 17 (1984), pp. 2044–2050.

[33] B. Widom, *Some topics in the theory of fluids*, J. Chem. Phys., 39 (1963), pp. 2808–2812.

[34] J. G. Powles, W. A. B. Evans and N. Quirke, *Non-destructive molecular dynamics simulation of the chemical potential of a fluid*, Molec. Phys., 46 (1982), pp. 1347-1370.

[35] B. Guillot and Y. Guissani, *Investigation of the chemical potential by molecular dynamics simulation*, Molec. Phys., 54 (1985), pp. 455–465.

[36] K. S. Shing and K. E. Gubbins, *The chemical potential in dense fluids and fluid mixtures via computer simulation*, Molec. Phys., 46 (1982), pp. 1109-1128.

[37] J. G. Kirkwood, *Statistical mechanics of fluid mixtures*, J. Chem. Phys., 3 (1935), pp. 300–313.

[38] R. W. Zwanzig, *High-temperature equation of state by a perturbation method. I. Nonpolar gases*, J. Chem. Phys., 22 (1954), pp. 1420–1426.

[39] G. M. Torrie and J. P. Valleau, *Monte Carlo free energy estimates using non-Boltzmann sampling: Application to the sub-critical Lennard-Jones fluid*, Chem. Phys. Letters, 28 (1974), pp. 578–581.

[40] S. Okazaki, K. Nakanishi, H. Touhara and Y. Adachi, *Monte Carlo studies on the hydrophobic hydration in dilute aqueous solutions of nonpolar molecules*, J. Chem. Phys., 71 (1979), pp. 2421-2429.

[41] J. P. M. Postma, H. J. C. Berendsen and J. R. Haak, *Thermodynamics of cavity formation in water. A molecular dynamics study*, Faraday Symp. Chem. Soc., 17 (1982), pp. 55-67.

[42] B. L. Tembe and J. A. McCammon, *Ligand-Receptor Interactions*, Computers & Chemistry, 8 (1984), pp. 281–283.

[43] W. L. Jorgensen and C. Ravimohan, *Monte Carlo simulation of differences in free energies of hydration*, J. Chem. Phys., 83 (1985), pp. 3050–3054.

[44] P. H. Berens, D.H. J. Mackay, G. M. White and K. R. Wilson, *A Monte Carlo study of ion-water clusters*, J. Chem. Phys., 64 (1976), pp. 481-491.

[45] P. H. Berens, D. H. J. Mackay, G. M. White and K. R. Wilson, *Thermodynamics and quantum corrections for molecular dynamics of liquid water*, J. Chem. Phys., 79 (1983), pp. 2375–2389.

[46] M. Mezei, S. Swaminathan and D. L. Beveridge, *Ab Initio calculation of the free energy of liquid water*, J. Am. Chem. Soc., 100 (1978), pp. 3255–3256.

[47] J. P. M. Postma, *Molecular Dynamics of* H_2O, thesis, University of Groningen, 1985.

[48] W. F. van Gunsteren and H. J. C. Berendsen, *The power of dynamic modelling*, in *Proceedings of the Symposium on Computational Methods in Chemical Design: Molecular Modelling and Graphics*, edited by J. Stozowski, Oxford University Press, London, 1986 (to appear).

[49] T. P. Straatsma, H. J. C. Berendsen and J. P. M. Postma, *Free energy of hydrophobic hydration: A molecular dynamics study of noble gases in water*, J. Chem. Phys., 85 (1986), pp. 6720–6727.

COLLECTIVE PHENOMENA IN STATISTICAL MECHANICS AND THE GEOMETRY OF POTENTIAL ENERGY HYPERSURFACES

FRANK H. STILLINGER†

Abstract. Static and dynamic properties in dense many-body systems are considered from a unified point of view that emphasizes geometric features of the potential energy function Φ for interparticle interactions. Steepest descent mapping on the Φ hypersurface generates a natural division of the multidimensional configuration space into "basins" whose boundaries contain the fundamental transition states for the system as a whole. The resulting statistical representation (a) identifies an inherent structure for the liquid state, (b) leads to a simple description of phase transitions, (c) offers a natural formalism for supercooled liquids, and (d) provides an analysis of relaxation phenomena in glasses. Classical mechanics is used throughout the development.

1. Introduction. The condensed phases of matter (crystalline, liquid, amorphous solid, liquid-crystalline, quasicrystalline, etc.) display a fascinating variety of "collective" properties whose existence stems from the strong and often complicated intermolecular forces that are present. Predicting and understanding these collective phenomena from just a knowledge of intermolecular forces has been an enduring challenge to statistical mechanics. In spite of some remarkable successes, the subject remains open. The object of these three lectures is to develop a comprehensive framework for studying collective phenomena in condensed matter, to demonstrate some of the new insights it produces, and most importantly to identify some outstanding mathematical questions that it generates.

2. Condensed systems. By definition, the constituent particles (atoms, molecules) of a sample of dense matter are in continuing interaction with one another. The interactions are comprised in a potential energy function $\Phi(\mathbf{r}_1 \ldots \mathbf{r}_N)$, whose variables $\mathbf{r}_i$ represent the position coordinates of the N particles. Generally, Φ is bounded below by $-KN$ for some $K > 0$, is at least twice differentiable in all its variables when no pair of particles overlap, and vanishes when all particles recede infinitely far from one another.

A simple (but still challenging!) class of Φ's is frequently used in statistical mechanical modeling, namely a sum of particle pair interactions. When all N particles are identical, one then has

$$\phi(\mathbf{r}_1 \ldots \mathbf{r}_N) = \sum_{i<j=1}^{N} v(\mathbf{r}_i, \mathbf{r}_j). \qquad (2.1)$$

If the particles furthermore are just atoms or ions, the pair potential v would depend only on distances r_ij.

The temporal evolution of the N-particle system is determined by its dynamical equations, and these depend on Φ. For simplicity it will be assumed that classical mechanics applies, *i.e.*, Newton's equations of motion are relevant.

†AT&T Bell Laboratories, Murray Hill, New Jersey 07974, USA

If the total system energy is large enough it is normally expected that the N-body system would be able to achieve thermal equilibrium at some absolute temperature T. If this is so, thermodynamic properties can be derived from the canonical partition function Z_N which in classical statistical mechanics has the following form [1]:

$$Z_N(\beta) = (\lambda_T^{3N} N!)^{-1} \int d\mathbf{r}_1 \cdots \int d\mathbf{r}_N \exp(-\beta\Phi), \tag{2.2}$$

where $\beta = (k_B T)^{-1}$, k_B is Boltzmann's constant, and λ_T is the mean thermal de Broglie wave length (independent of Φ, and irrelevant in most of the following). Integration limits on the $\mathbf{r}_i$ are determined by the containment vessel for the material sample which we assume has volume V. The logarithm of Z_N gives $-\beta$ times the Helmholtz free energy F for the system, from which standard thermodynamic formulas allow pressure, energy, entropy, $\cdots$, to be obtained. It is usually the case (as in this presentation) that primary interest concerns the large-N limit with temperature and density held constant.

No generating function analogous to Z_N is available to produce properties in the general nonequilibrium regime. Instead one may be obliged to consider the full details of Newtonian dynamical orbits to extract nonequilibrium behavior.

2.1 Steepest descent mapping. No general exact procedures are available to evaluate multiple integrals of the type in Z_N, Eq. (2.2), even for pairwise additive potentials. However, a series of transformations can be applied that creates formal simplifications and that has important conceptual advantages [2]. These transformations are generated by Φ itself.

For rotational simplicity let $\mathbf{R} \equiv (\mathbf{r}-1 \cdots \mathbf{r}_N)$, $\Phi \equiv \Phi(\mathbf{R})$, and suppose that we are dealing with structureless point particles in ordinary 3-space so that $\mathbf{R}$ has $3N$ components. Steepest-descent trajectories on the Φ hypersurface in $(3N+1)$-space are generated by solutions to

$$\partial\mathbf{R}/\partial\tau = -\nabla\Phi(\mathbf{R}), \tag{2.3}$$

where τ is a "virtual time" variable. These trajectories "relax" the collection of particles, moving each one at a velocity proportional to the force it experiences, until (as $\tau \to +\infty$) the steepest-descent trajectory lodges at a local minimum of Φ. Thus the solutions to Eq. (2.3) generate a mapping of (almost) all particle configurations $\mathbf{R}$ onto a discrete set of configurations $\mathbf{R}_\alpha$ which are the local minima of Φ. Here we use α as an index for local minima.

The connected set of particle configurations all of which map onto minimum α defines a "basin" B_α. The collection of such basins essentially covers the entire $3N$-dimensional configuration space available to the particles. Basin boundaries are exceptional since Eq. (2.3) does not unambiguously map them to minima, but they constitute a zero-measure set with no consequence in the following.

It has been argued on simple intuitive grounds [2] that if all N particles are identical and are confined to a container with volume proportional to N, then in the large-N limit the number $\Omega_0(N)$ of Φ minima should behave as

$$\Omega_0(N) = N! \exp[\nu N + o(N)], \tag{2.4}$$

where ν is positive and depends on container volume V. The $N!$ factor simply acknowledges that permuting particle positions converts any munimum to an essentially equivalent one.

There is a substantial need to validate Eq. (2.4) on a mathematically sound basis. It would be useful if bounds on ν could be supplied that are sharp enough to show how this quantity depends on number density N/V and on details of Φ. Another mathematical challenge concerns whether basins B_α are always simply connected, or whether some physically sensible Φ's produce multiple connectivity.

2.2 Partition function transformations. Steepest-descent basins can usefully be classified according to ϕ, their depths on a per-particle basis:

$$\phi = \Phi(\mathbf{R}_\alpha)/N. \tag{2.5}$$

Since the number of minima is large, and since the Φ's should be bounded between finite limits, it should be possible to define a distribution G of minima versus ϕ. In the large-N limit this distribution must have a form consistent with Eq. (2.4), namely

$$G(\phi) \sim N! \exp[N\sigma(\phi)] \tag{2.6}$$

where $\sigma(\phi)$ is independent of N but generally will vary with particle density.

An intrabasin configurational integral I_α at inverse temperature β may be defined for B_α as follows:

$$I_\alpha(\beta) = \int_{B_\alpha} d\mathbf{R} \exp[-\beta \Delta_\alpha \Phi(\mathbf{R})], \tag{2.7}$$

$$\Delta_\alpha \Phi(\mathbf{R}) = \Phi(\mathbf{R}) - \Phi(\mathbf{R}_\alpha). \tag{2.8}$$

It includes, with appropriate Boltzmann-factor weighting, all possible "vibrational" excursions away from the basin bottom. Not all basins of a given depth have the same shapes or I_α's, but it makes sense to lump those in a narrow depth interval $\phi \pm \epsilon$ together and to use the resulting average I_α to define a "vibrational free energy per particle" $f_v(\beta, \phi)$:

$$f_v(\beta, \phi) = -\lim_{\epsilon \to 0} \lim_{N \to \infty} (N\beta)^{-1} \ln \langle I_\alpha(\beta) \rangle_{\phi \pm \epsilon}. \tag{2.9}$$

Armed with definitions (2.6) and (2.9) for σ and f_v, one finds that for large N the partition function Z_N has the following asymptotic representation:

$$Z_N(\beta) \sim \lambda_T^{-3N} \int d\phi \exp\{N[\sigma(\phi) - \beta\phi - \beta f_v(\beta, \phi)]\}. \tag{2.10}$$

The original multiple integral of high order, Eq.(2.2), has been replaced by a simple quadrature. In addition, the problem has been clearly separated into a purely enumerative part (σ), and a thermal vibration part (f_v).

Since N is large, the dominant contribution to the integral in Eq. (2.10) is provided by the maximum of the integrand, *i.e.* $-\beta$ times the Helmoholtz free energy per particle is given by

$$\begin{aligned} -\beta F/N &= (\ln Z_N)/N \\ &\cong -3\ln\lambda_T + \sigma(\phi_m) - \beta\phi_m - \beta f_v(\beta,\phi_m) \end{aligned} \tag{2.11}$$

(with an error that vanishes in the large-N limit), where $\phi_m(\beta)$ is determined by the criterion:

$$\sigma(\phi_m) - \beta\phi_m - \beta f_v(\beta,\phi_m) = maximum. \tag{2.12}$$

The significance of $\phi_m(\beta)$ is that it gives the depth of those basins that dominate equilibrium properties at the given temperature and density. It seems clear that ϕ_m must be a decreasing function of $\beta = 1/k_\beta T$, and as $\beta \to +\infty \phi_m$ approaches the potential energy per particle for the global ϕ minimum (normally expected to correspond to a crystalline arrangement of the N particles).

Although it may be difficult generally to determine the enumeration function σ for a given potential function Φ, the inverse problem might be amenable: Given $\sigma(\phi)$, is it possible to construct a potential which has this as its local minimum enumeration function?

Generalizations of the foregoing to include the case of several distinguishable species are conceptually trivial.

2.3 Particle pair distribution functions. The distribution of distances between particle centers has experimental significance, since it can be measured by X-ray or neutron diffraction. The theoretical significance is that when Φ is pairwise additive and the particles are spherically symmetric, the thermodynamic energy and pressure can be expressed as simple quadratures of the distance distribution.

Conventionally the distance distribution is presented as the pair correlation function $g^{(2)}(s)$. At equilibrium in a single-species system it is given by [1]:

$$g^{(2)}(s) = \frac{\int d\mathbf{r}_1 \cdots \int d\mathbf{r}_N \delta(s - r_{12}) \exp(-\beta\Phi)}{B(s)\int d\mathbf{r}_1 \cdots \int d\mathbf{r}_N \exp(-\beta\Phi)}, \tag{2.13}$$

where

$$r_{12} = |\mathbf{r}_2 - \mathbf{r}_1|, \tag{2.14}$$

$$B(s) = B(s) = V^{-2}\int d\mathbf{r}_1 \int d\mathbf{r}_2 \delta(s - r_{12}), \tag{2.15}$$

and δ is the Dirac-delta function. When s is large compared to the range of intermolecular forces, $g^{(2)}$ is essentially independent of s. For small s it reflects the way that interactions produce preferred spatial arrangements of particles. Generally, s variations of $g^{(2)}$ have greater amplitude at low temperature then at high temperature.

Steepest-descent mapping is applicable to the distribution of pair distances, and produces a remarkable "image enhancement" effect. The corresponding "quenched pair correlation function" $g_q^{(2)}(s)$ can be written in a form analogous to that of Eq. (2.13):

$$g_q^{(2)}(s) = \frac{\int d\mathbf{r}_1 \cdots \int d\mathbf{r}_N \delta[s - r_{12q}(\mathbf{r}_1 \cdots \mathbf{r}_N)] \exp(-\beta\Phi)}{B(s) \int d\mathbf{r}_1 \cdots \int d\mathbf{r}_N \exp(-\beta\Phi)}. \tag{2.16}$$

Here r_{12q} is the scalar distance between particles 1 and 2 *after* the steepest descent mapping has been applied to configuration $\mathbf{r}_1 \cdots \mathbf{r}_N$. Consistent with previous remarks, $g_q^{(2)}$ is the result of removing intra-basin vibrational deformations from $g^{(2)}$.

Molecular dynamics computer simulation results are now aviailable, giving both $g^{(2)}$ and $g_q^{(2)}$ for some simple atomic models [3]. Figure 1 schematically presents the principal result, namely that virtually all of the temperature dependence of the conventiaonal $g^{(2)}$ for liquids stems from intra-basin processes. More precisely, under constant density conditions, $g_q^{(2)}$ is essentially independent of the temperature of the starting liquid, provided that the temperature is at or above the thermodynamic melting temperature. Hence $g_q^{(2)}$ represents the inherent structure present in the liquid state. That such a temperature-independent inherent structure should exist seems (from computer simulation studies) to be connected with the narrowness of the distribution of basin depths, *i.e.*, the curvature of $\sigma(\phi)$ near its maximum. But why the vast majority of minima should be so closely spaced in ϕ is still an unanswered question.

In several popular many-body models, Φ has the additive form shown in Eq. (2.1) with a central pair potential that diverges at the origin thus:

$$v(r) \sim Ar^{-p} \quad (r \to 0), \tag{2.17}$$

where $A > 0$, $p > 3$. Under equilibrium conditions it is widely believed that the corresponding small-r behavior of the pair correlation function will be ($\beta = 1/k_B T$):

$$\ln g^{(2)}(s) \sim -\beta A s^{-p} \quad (s \to 0). \tag{2.18}$$

The steepest-descent mapping is expected to produce a function $g_q^{(2)}$ which, like $g^{(2)}(s)$, vanishes strongly at the origin. But does $\ln g_q^{(2)}$ behave as $-(const)s^{-p}$, or is a fundamentally different asymptote involved? Once again no answers yet exist.

3. Phase transitions. Phase transitions are dramatic examples of the collective phenomena exhibited by condensed matter. They include melting and freezing, spontaneous changes of crystal symmetry, transitions that form or destroy liquid crystals, and a variety of surface reconstructions on both crystalline and liquid substrates. Understanding and controlling phase transitions is scientifically compelling and technologically urgent.

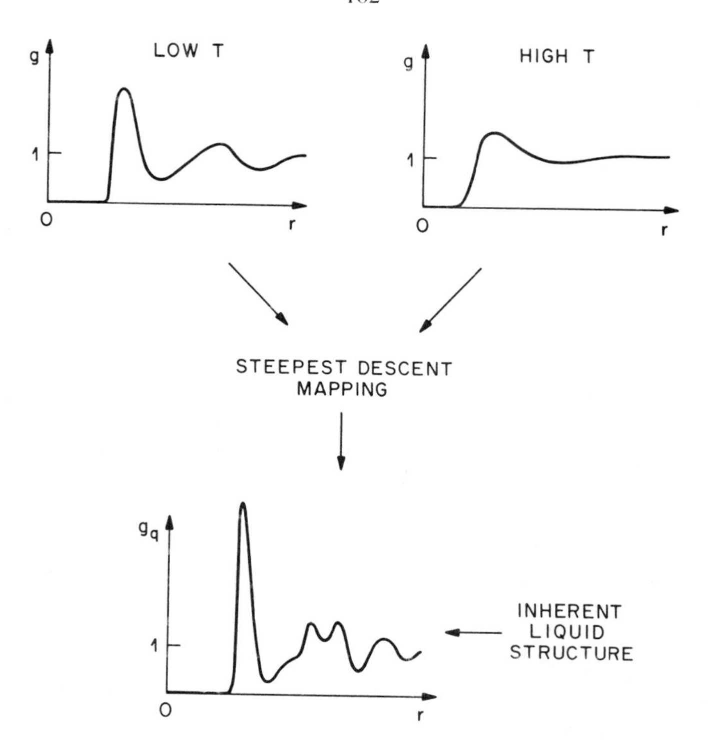

Figure 1. Inherent structure in the liquid state revealed by steepest descent quenching.

The formalism presented in Sect. 2 offers a convenient way to discuss and to analyze phase transitions of the types just mentioned. Since it is fundamentally a classical statistical-mechanical approach, it cannot in its present form describe those transitions (metal-insulator, normal metal-superconductor, helium Bose condensation, ...) whose basis is explicitly quantum mechanical.

For the most part this Section will be concerned with thermodynamic equilibrium aspects of phase change (Sects. 3.1–3.3). However, in Sect. 3.4 Section E we will briefly consider the properties of long-lived metastable states, such as supercooled liquids.

3.1 Order-disorder transition in substitutional alloys. Solid state physics offers several examples of crystalline binary alloys whose component atoms A and B have an alternating periodic arrangement at low temperature, but are disordered among the crystal sites at high temperature (but still below the alloy's melting point). A thermodynamic phase transition separates these two regimes. A well-known case is β-brass, incorporating equal numbers of copper and zinc atoms in a body-centered cubic structure (see Figure 2). Copper atoms are segregated on one of the simple cubic sublattices at low temperature, zinc atoms on the other. At high temperature the sublattices have equal occupancies due to thermally driven random substitutions. A phase transition between ordered and disordered forms occurs at about $460°C$, accompanied by an infinite heat capacity anomaly [4].

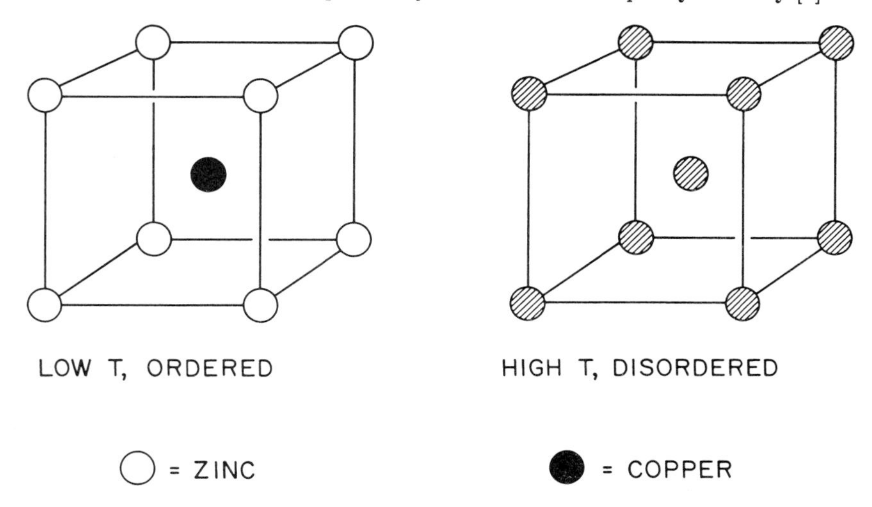

Figure 2. Atomic arrangements in the ordered low temperature and disordered high temperature forms of β-brass.

β-brass melts of course (at about $850°C$), and the system then moves in regions of its configuration space whose basins β_α have minima corresponding to amorphous atom packings. However, these regions of configuration space are not at issue in discussing the order-disorder transition. For simplicity we can restrict attention just to those basins and their minima corresponding to equal numbers (say $N/2$) of copper (A) and zinc (B) atoms arranged on the sites of a perfect body-centered cubic lattice. The distribution of potential energy minima for this two-species system restricted to the crystalline state may be written asymptotically as follows:

$$G(\phi) = [(N/2)!]^2 \exp[N\sigma(\phi)] \tag{3.1}$$

where as before ϕ is the potential energy per atom at the minima of the total potential energy function Φ.

The physics of these substitutional alloys suggests two convenient simplifying assumptions. The first is that the Φ_α, the local minima of Φ, depend only on the interactions of nearest-neighbor atom pairs. This is equivalent to the statement

$$\Phi_\alpha = Na + N_\ell b, \tag{3.2}$$

where a and b are constants and N_ℓ is the number of nearest-neighbor like-atom pairs. We must have

$$b > 0 \tag{3.3}$$

so that the absolute Φ minimum has $N\ell = 0$ (A and B segregated on separate sublattices with each A having eight B's as nearest neighbors, and vice versa).

The second simplifying assumption is that the vibrational properties, specifically $f_v(\beta, \phi)$, are the same regardless of the atomic arrangement over the body-centered cubic lattice. Under this circumstance the dominant basin depth at any given inverse temperature β which we have denoted by $\phi_m(\beta)$, can be obtained from a simplified form of Eq. (2.12),

$$\sigma(\phi_m) - \beta\phi_m = maximum \tag{3.4}$$

or equivalently (since σ'' seems always to be nonpositive),

$$\sigma'(\phi_m) = \beta. \tag{3.5}$$

Equation (3.2) states that ϕ is a linear function of N_ℓ/N, and it is clear that

$$a \le \phi \le a + 4b. \tag{3.6}$$

In the randomly substituted state ϕ is just at the middle of this range, $a + 2b$.

Evaluation of $\sigma(\phi)$ is an enumeration problem, namely "How many ways can $N/2$ A's plus $N/2$ B's be placed on a body-centered-cubic lattice so that there are exactly $N_\ell (\equiv N_\phi/b)$ nearest-neighbors atom pairs of like species?" Exactly the same enumeration problem lies at the heart of the Ising model on the same lattice (atoms A and B are equivalent to up and down spins in the Ising model)[1]. Although two-dimensional Ising models have been solved exactly [5], only approximate results are avilable for three-dimensional cases. Nevertheless a variety of techniques have been applied (series analysis, renormalization group methods, Monte Carlo computer simulations) that seem to reveal the principal features of interest.

A graphical construction for $\phi_m(\beta)$ corresponding to the determing in Eq. (3.5) involves rolling a straight line (slope β) on the curve of σ versus ϕ. The order-disorder transition point ($\beta = \beta_c$) is associated with a vanishing-curvature singularity of $\sigma(\phi)$ in the neighborhood of which we have

$$\begin{aligned} \sigma(\phi) = \sigma(\phi_c) + \beta_c(\phi - \phi_c) - A_\pm|\phi - \phi_c|^{p_\pm} + \dots, \\ A_\pm > 0, \quad p_\pm > 2, \quad \phi_c = \phi_m(\beta_c), \end{aligned} \tag{3.7}$$

where subscripts + and − refer respectively to $\phi > \phi_c$ and $\phi < \phi_c$. Terms beyond those shown in Eq. (3.7) have higher powers of $|\phi - \phi_c|$ than $p_\pm$.

By carrying out the construciton of $\phi_m(\beta)$ near β_c graphically or otherwise, we find

$$\phi_m(\beta) - \phi_c = \left(\frac{\beta_c - \beta}{p_+ A_+}\right)^{1-\alpha_+} + \dots, \quad (\beta < \beta_c), \tag{3.8a}$$

$$\phi_c - \phi_m(\beta) = \left(\frac{\beta - \beta_c}{p_- A_-}\right)^{1-\alpha_+} + \dots, \quad (\beta > \beta_c), \tag{3.8b}$$

where

$$\alpha_{\pm} = \frac{p_{\pm} - 2}{p_{\pm} - 1} > 0. \tag{3.9}$$

A β derivative of these potential energy expressions produces essentially the heat capacity $C_{\pm}$, which we therefore see exhibits the characteristic divergence exponent(s) $\alpha_{\pm}$.

$$C_{\pm}(\beta) = K_{\pm}|\beta - \beta_c|^{-\alpha_{\pm}} + \ldots. \tag{3.10}$$

Experimental measurements, and the various approximate theories of order-disorder critical phenomena for 3-dimensional systems seem to agree that critical exponent $\alpha_{\pm}$ is small and of the order of 0.1. But it remains a challenge to combinatorial mathematics to provide an exact evaluation of heat capacity and other critical exponents, perhaps by obtaining the relevant $\sigma(\phi)$.

3.2 Melting and freezing transitions. The order-disorder phase transition just described is a "continuous" phase transition. Any property evaluated in the low-temperature ordered phase and in the high-temperature disordered phase respectively approaches the same limit as T approaches $T_c = (k_B\beta_c)^{-1}$ from either side. This contrasts vividly with "discontinuous" or "first-order" phase transitions where the two phases involved remain distinctly different at any temperature where both can exist. We will now consider the case of melting in single-component substances, where the low-temperature crystal transforms discontinuously to the liquid; the two phases differ in symmetry, mechanical properties, density, and energy.

The convention we have used is that the container volume V is kept fixed as temperature changes, and as steepest-descent mappings are effected. Under circumstances the singularity structure of $\sigma(\phi)$ will be influenced by the *constant-pressure* densities of the crystal and liquid. If these differ (as is virtually always the case) then at *constant volume* the crystal melting temperature T_m will be less than the liquid freezing temperature T_f. In the intervening interval

$$T_m < T < T_f, \tag{3.11}$$

the system displays coexisting macroscopic regions of crystal and liquid, *i.e.*, a mixed state with proportions of crystal and liquid varying continuously with T.

Although it was reasonable to suppose that the basin vibrational free energy $f_v(\beta, \phi)$ was independent of ϕ in the preceding order-disorder example, that is no longer the case for the melting-freezing transition. The crystal and liquid phases are structurally too unlike to permit such an approximation. Just as does $\sigma(\phi)$, $f_v(\beta, \phi)$ is now expected to exhibit nontrivial ϕ variation and to show a singularity structure consistent with phase coexistence.

Figure 3 indicates the graphical relationsip of the terms in Eq. (2.12) at the endpoints of the coexistence interval (3.11). Matching of slopes for σ and for $\beta(\phi + f_v)$ as before locates $\phi_m(\beta)$. When coexistence obtains, the steepest descent mapping of particle configurations presumably produces mechanically stable particle packings which themselves exhibit coexistence. That is, the dominant packings identified by $\phi_m(\beta)$ in this temperature interval will have large side-by-side regions respectively with ordered and with disordered particle arrangements. To show that such heterogeneous packings exist for physically reasonable potentials is a nontrivial matter, but accumulating computer simulation data demonstrates that it is so [6].

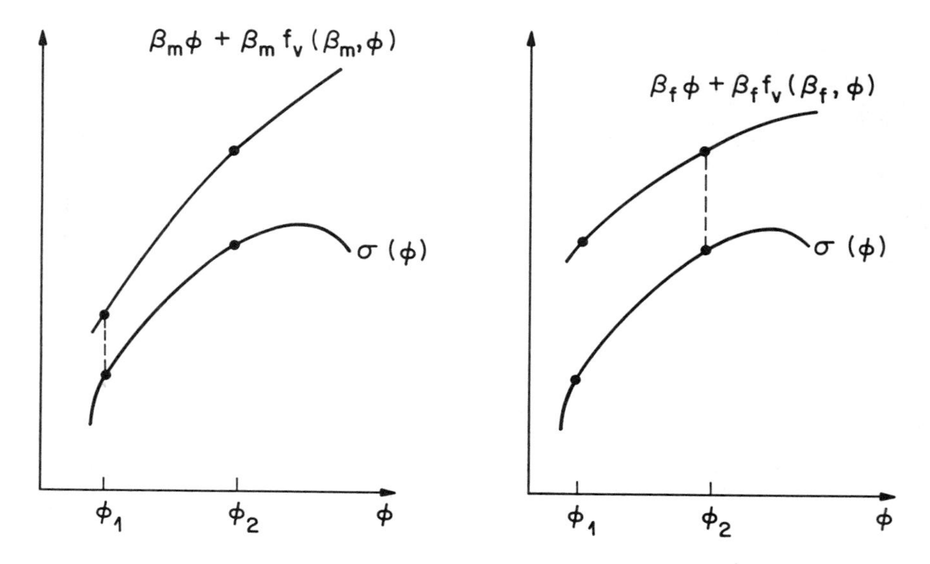

Figure 3. Matching of slopes for terms in Eq. (I.11) for the crystal-liquid first-order phase transition. The left panel corresponds to the fixed-volume melting point of the crystal, the right panel to the freezing point of the liquid.

The most basic unanswered questions concerning this description involve the nature of the singularities, with respect to ϕ, of $\sigma(\phi)$ and of $f_v(\beta, \phi)$ at coexistence endpoints ϕ_1 and ϕ_2. These singularities are unlike that in the order-disorder case since the heat capacity remains finite at constant volume. The best present guess is that essential singularities are involved, and that these arise from "heterophase fluctuations," a distribution of small inclusions of one type of packing within large domains of the other type [7].

3.3 Two examples. We now focus attention on two models that have been extensively studied with computer simulation (molecular dynamics and Monte Carlo), and which exhibit melting and freezing phenomena. They both entail pairwise additive potential energy functions as in Eq (2.1), with structureless point particles. Disregarding trivial constants the respective pair potentials are the inverse power potential

$$v(r) = r^{-p} \quad (p > 3) \tag{3.12}$$

and the Gaussian core potential

$$v(r) = \exp(-r^2). \tag{3.13}$$

Both involve only repulsive forces. The restriction on p in the first of these is necessary to keep energy, entropy, and free energy extensive (*i.e.*, proportional to N).

The inverse power potential is a homogeneous function of distance, the implications of which are significant in the present connection. In particular it means that the density (ρ) and inverse-temperature (β) variations of σ and f_v must have the forms:

$$\sigma(\phi, \rho) \equiv \overline{\sigma}(\rho^{-p/3}\phi) \tag{3.14}$$

and

$$\beta f_v(\beta, \phi, \rho) \equiv y\overline{f}(\rho^{-p/3}\phi, y), \tag{3.15}$$

where

$$y = \beta \rho^{p/3}. \tag{3.16}$$

These follow from the fact that dilation and compression neither create nor destroy relative minima, but simply rescale the coordiunates that yield those minima. The structure of the low-temperature crystal is face-centered-cubic; relations (3.14) and (3.15) imply that the melting and freezing points correspond to fixed values of the parameter y.

The analytical properties of the Gaussian core potential are quite different. First, there is a duality relation between the low-density and high-density absolute minima of ϕ [8]:

$$\rho_f^{-1/2}[1 + 2\phi_f(\rho_f)] = \rho_b^{-1/2}[1 + 2\phi_b(\rho_b)], \tag{3.17a}$$

$$\rho_f \rho_b = \pi^{-3}. \tag{3.17b}$$

Here the subscripts f and b refer to the respective stable crystal forms, face-centered cubic at low density, body-centered cubic at high density. This identity stems from the self-similarity of the Gaussian function under Fourier transformation. The density at which the two crystal forms have equal energy is obvious from (3.17), namely

$$\rho = \pi^{-3/2}. \tag{3.18}$$

A second analytical property of the Gaussian core potential is that in the high density limit the energies of all periodic structures (even with very large unit cells) become asymptotically equal. More precisely the energy differences between distinct periodic structures go to zero (apparently) as $\exp(-C\rho^{2/3})$, $C > 0$. This implies that the melting and freezing temperatures vanish in the limit $\rho \to +\infty$. However, it is by no means clear at present how the basic quantities σ and f_v individually behave in this limit. Input from interested mathematicians on this question would certainly be welcome and illuminating.

3.4 Supercooling. Although the basic criterion (1.12) was formulated to describe strict equilibrium conditions, it is nevertheless also suggestive concerning a specific deviation from equilibrium, namely the supercooling of liquids below their freezing point. Careful experiments show that this is a repeatable and reproducible phenomenon, and a little thought suggests how Eq. (1.12) should be modified to describe that situation.

It seems clear that only a subset of potential minima and their basins should be considered. Supercooled liquids have avoided nucleation. The relevant packings therefore must be devoid of crystal fragments, at least those beyond some critical size. In principle this requires implementing an algorithm to decide if any packing is devoid of such crystallinity. It would clearly be useful to exhibit such an algorithm explicity and to determine if its implementation is a polynomial-time or an NP-complete problem [9].

The "amorphous" subset of packings which meet the noncrystallinity criterion can be enumerated just as before by a function $\sigma_a(\phi)$, where

$$\sigma_a(\phi) \leq \sigma(\phi). \tag{3.19}$$

This amorphous packing subset then will have a well-defined vibrational free energy function $f_{va}(\beta, \phi)$ in analogy to Eq. (1.9). Finally, then, the "quasi-equilibrium" condition for supercooled liquids has a form directly analogous to Eq. (1.12):

$$\sigma_a(\phi_{ma}) - \beta\phi_{ma} - \beta f_{va}(\beta, \phi_{ma}) = \text{maximum}. \tag{3.20}$$

The optimizing $\phi_m a(\beta)$ locates the depth of the dominating potential minima within the noncrystalline subset.

By projecting out of consideration the crystal-containing packings, the singularities present originally in σ and f_v which were associated with melting and freezing necessarily have been eliminated. It will be important to establish eventually whether σ_a and $f_v a$ display new singularities as might be associated, for instance, with a low-temperature glass transition or an instability point for strongly supercooled liquids. In any case it would be important to determine the lowest potential energy in the amorphous subset, what geometrical packing structure is involved, and how these attributes depend on the details of the algorithm employed to identify that subset.

Finally we note that pair correlation functions (both pre-quench and post-quench) can be precisely defined on the amorphous subset. Once again it would be important to determine how algorithm-sensitive are these functions.

4. Rate processes. Many-body systems in a condensed state display a wide variety of time-dependent collective properties. These include self diffusion, viscous flow, fracture, shock-wave propagation, crystal nucleation, and in some cases chemical reactions. Observable time dependence arises from the motion of the system configuration point $R(t)$ which in classical dynamics obeys the multi-dimensional Newton equation (masses taken to be unity):

$$d^2\mathbf{R}(t)/dt^2 = -\nabla\Phi[\mathbf{R}(t)]. \tag{4.1}$$

In virtually all cases of interest the dynamics occurs at sufficiently large (conserved total energy that the system is not trapped in any single basin B_α, but is more or less free to roam through a sequence of basins.

4.1 Collective transition states. With respect to the potential energy hypersurface, the least costly dynamical pathway between two contiguous basins passes through a "transition state" located on the common boundary hypersurface between the basins. This transition state is a horizontal saddle point with a single negative principal curvature, *i.e.*, it is characterized by a vanishing of $\nabla\Phi$ and by the existence of a single negative eigenvalue of the Hessian matrix $\nabla\nabla\Phi$. The eigenvector corresponding to the negative eigenvalue is conventionally identified with the direction of the so-called "reaction coordinate" through the transition state.

One should keep in mind that two or more transition states could occur on the boundary between two contiguous basins. Furthermore, horizontal saddle points that are *not* transition states could occur in the interior of basins. The frequency of such occurrences and how they depend upon interparticle interactions is unknown at present.

The number of transition states, their distribution by height with respect to basin bottoms, their spectrum of curvatures, and other basic geometric characteristics are directly relevant to the understanding of rate processes and how fast they proceed at different temperatures.

Unfortunately no general procedure for locating saddle points in multidimensional surfaces is presently available, analogous to the steepest-descent method [Eq. (1.3)] for local minima. But the importance of transition states has often forced researchers to devise and employ inefficient search routines. One that has been used is the following [10]:

(1) Using molecular dynamics computer simulation with frequent steepest-descent mapping, locate two potential minima at $\mathbf{R}_1$ and $\mathbf{R}_2$ with contiguous basins.

(2) Construct the linear path between $\mathbf{R}_1$ and $\mathbf{R}_2$:

$$\mathbf{R}(\ell) = \ell\mathbf{R}_1 + (1-\ell)\mathbf{R}_2. \tag{4.2}$$

The potential energy $\Phi[\mathbf{R}(\ell)]$ along this path has relative minima at $\ell = 0$ and 1, and must have at least one relative maximum in between. Let $\mathbf{R}_3$ be the position along this connecting line which produces the largest relative Φ maximum. This is a first crude estimate of the transition state position.

(3) Consider the scalar function

$$\Psi(\mathbf{R}) = [\nabla\Phi(\mathbf{R})]^2: \tag{4.3}$$

its zeros are all minima and include all of the extrema of the original function Φ. Starting at $\mathbf{R}_3$, construct a steepest-descent path on the Ψ hypersurface to identify one of its minima at location $\mathbf{R}_4$.

(4) Check to see if the Hessian matrix for Φ has only one negative eigenvalue at $\mathbf{R}_4$. If it does, use the corresponding eigenvector as the direction for

infinitesimal displacements both positive and negative away from the saddle point, followed by Φ-hypersurface steepest descent to verify that the originally-selected minima are the ones for which this saddle point serves as transition state.

Clearly this is an unwieldy process. Improved procedures would be an important advance.

4.2 Localization. The steepest descent mapping of the continuous dynamical trajectory $\mathbf{R}(t)$ for the many-body system produces $\mathbf{R}_q(t)$, a piecewise constant function whose values are positions of the bottoms of basins successively visited by the system. The left and right side limits for $\mathbf{R}_q$ at each time t_i that a discontinuity is encountered are (with unit probability) the bottoms of contiguous basins. It is important to note that the corresponding configurational shifts between those minima:

$$\Delta\mathbf{R}_q(i) = \lim_{\epsilon\to 0}[\mathbf{R}_q(t_i+\epsilon) - \mathbf{R}_q(t_i-\epsilon)] \tag{4.4}$$

appear from computer simulation studies to exhibit a localization property [10]. Specifically this means that most of the change typically is concentrated on a small subset $[O(1)]$ of the N particles, and furthermore the typical displacements of particles in the small subset are comparable to particle diameters. Correspondingly the potential energy changes only by $O(1)$ between these minima, in spite of the fact that it is an $O(N)$ quantity.

This localization property implies that the directions of the $\Delta\mathbf{R}_q(i)$ are far from uniformly distributed over the surface of the $(3N-1)$-sphere. Instead they are concentrated strongly so that all but a small number of direction cosines are essentially zero.

Computer simulation studies reveal another nontrivial feature. Only rarely do the individual (localized) transitions carry the N-body system through a particle permutation. This *could* conceivable take place as a pair of nearest neighbors rotates 180° about their centroid, or more generally as a close loop of particles executes simultaneous jumps around that loop. But such shifts are exeptional.

These observations about localization and rarity of direct permutational transsitions are based only on studies of N-body systems with short-range interactions. The conclusions might well be different for long-range interactions, particularly if oscillatory pair potentials were present. This aspect clearly needs further study.

4.3 Self diffusion. The self-diffusion constant D measures the rate at which Brownian motion moves particles about. In terms of the time-dependent displacement $\Delta\mathbf{r}_i(t)$ for particle i, for large t,

$$D \sim \langle|\Delta\mathbf{r}_i(t)|^2\rangle/6t. \tag{4.5}$$

Here it is assumed that the system is in thermal equilibrium. The diffusion process is simultaneously underway for all N particles, and it reflects the sequence of basins visited in the $3N$-dimensional configuration space. On account of the localization property for elementary transitions, the number of exit channels from any basin to neighboring basins is $O(N)$, so that rate of transition is also expected to be $O(N)$. In other words the mean residence time in any basin is $O(N^{-1})$.

Evidence thus far available from simulations indicates that under steepest descent quenching, individual particles tend only to move by an amount comparable to their diameter. If that is generally the case an alternative expression for D can be written as follows:

$$D \sim \langle |\Delta \mathbf{r}_{iq}(t)|^2 \rangle / 6t. \tag{4.6}$$

The quantity $\Delta \mathbf{r}_{iq}(t)$ is the displacement of particle i from the basin bottom relevant at time 0 to that relevant at time t. It would be desirable to know if there is any difference in the way that expressions (4.5) and (4.6) approach D with increasing t, specifically which converges more rapidly.

Experimental and simulational data for D as a function of absolute temperature T are is often fitted to expressions of the form:

$$D(T) \cong D_0 T^{1/2} \exp[-E(T)/k_B T]. \tag{4.7}$$

where D_0 is a positive constant. The factor $T^{1/2}$ represents mean particle speed, and the exponential Boltzmann factor with positive $E(T)$ supposedly represents the probability of climbing up potential energy barriers. Indeed for many liquids in their thermodynamic stability range, E is approximately independent of temperature. However, it is not at all clear how $E(T)$ in Eq. (4.7) relates specifically to the distribution of transition state barrier heights between basins in the $3N$-dimensional configuration space. This remains an important research topic.

4.4 Glass transitions. We have already seen how a quasi-equilibrium description of supercooled liquids can be based on a subset of the configuration space basins. Its applicability requires that the system of interest (a) be able to explore the allowed basins adequately to achieve a representative sampling, while (b) avoiding nucleation (*i.e.*, penetration of the excluded set of crystalline basins). As a liquid is superecooled more and more its rate of basin exploration declines strongly, so that even if nucleation were to be avoided the quasi-equilibrium expression (3.20) must eventually become inapplicable to real experiments. For many substances this happens over a narrow temperature interval, often identified as a glass transition temperature T_g. Keep in mind that this is a somewhat ill-defined quantity, not on the same par as the precisely defined thermodynamic melting and freezing temperature.

Cooling well below T_g produces an amorphous solid whose properties depend upon the rate of cooling through T_g. In this low-temperature regime the system typically behaves as though it were trapped in a relatively small group of basins. The conserved total energy is so low that it is very unlikely or even impossible to find sufficiently low transition states as escape routes to surrounding basins. The self-diffusion constant becomes immeasurably small, consistent with the trapping.

As a supercooled liquid enters the glass transition regime, its time-dependent behavior (fluctuation, regression, and response to external perturbations) demonstrates the presence of a broad spectrum of relaxation times. Near T_g, various measurable properties vs. time often can be fitted to a "stretched exponential" decay

$$\exp[-(t/\tau)^{\beta_0}]; \quad 0.3 < \beta_0 < 1.0, \tag{4.8}$$

where β_0 can depend on temperature and τ increases rapidly as T decreases through the T_g range. Laplace-transforming (4.8) identifies a wide range of superposed simple exponential decay rates whose presence indicates the complexity of the potential energy hypersurface over which the system is sluggishly diffusing.

Clearly what is required is an appropriate description for the multiple-length-scale topography of the $3N$-dimensional Φ hypersurface in the amorphous region, and how that determines the wide spectrum of relaxation times. One approach, which exhibits some features of a "renormalization group" program, involves combining basins into metabasins [11]. Specifically, let basins B_α and B_β belong to the same metabasin $\overline{B}_\gamma$ if (a) B_α and B_β are inequivalent (not related simply by particle permutation), (b) B_α and B_β are either contiguous or are connected by a sequence of contiguous basins in the same metabasin, and (c) the barrier heights of the lowest transition states between B_α and B_β (either in a shared boundary, or along a chain of basins in the metabasin that connects them) do not exceed $\eta > 0$. Here we take η to measure the lesser elevation of the transition state measured from the two flanking minima.

As η increases, the original basins aggregate into a declining number of ever-larger metabasins. In analogy with Eq. (1.4), the η-dependent number of metabasins asymptotically can be written

$$\Omega(N, \eta) \sim N! \exp[\nu(\eta)N], \tag{4.9}$$

where ν is a strictly decreasing function of η.

Consider just the transitions between metabasins. These should include the longest relaxation times present in the system (they are evidently associated with large-scale topographic features of the Φ-scape), but should exclude rapid relaxation times. The dividing line roughly should occur for times proportional to the Boltzmann factor:

$$\exp(-\eta/k_B T). \tag{4.10}$$

The function $\nu(\eta)$ is a fundamental measure of the multidimensional Φ topography. Unfortunately it is poorly known, but obviously it must vanish as $\eta \to +\infty$. Heuristic arguments suggest that in order to produce relaxation spectra of the type underlying the stretched exponential function, $\nu(\eta)$ would have to possess roughly the following form:

$$\nu(\eta) \sim \exp[-A \exp(a\eta)], \quad A, a > 0 \tag{4.11}$$

wherein a is related to β_0 as follows:

$$\beta_0 = k_B T a/(1 + k_B T a), \qquad (T \approx T_g). \tag{4.12}$$

Quite obviously this loose heuristic connection between topography and relaxation rates needs substantial strengthening with rigorous analysis.

4.5 Dynamical recurrences. As a final matter, we consider briefly the return of the configuration-space dynamical trajectory $\mathbf{R}(t)$ to the basin B_α in which it was located at $t = 0$. This forms a kind of momentum-blind and spatially course-grained version of the venerable Poincaré recurrence problem in dynamics [12].

Let $U_\alpha(\mathbf{R})$ be the characteristic function for B_α, 1 inside and 0 outside. Consider the autocorrelation function

$$u(t) = \langle U_\alpha[\mathbf{R}(0)] U_\alpha[\mathbf{R}(t)] \rangle, \tag{4.13}$$

where the average involved is canonical *i.e.*, fixed temperature, and it includes all basins. Obviously $u(0) = 1$. The initial decay of $u(t)$ is related to the mean escape rate from basins. If the system is ergodic, the long-time limit should be determined essentially by the number of basins with depths given by ϕ_m, Eq. (1.12); that is:

$$\lim_{t \to \infty} u(t) = (N!)^{-1} \exp\big\{ - N\sigma[\phi_m(\beta)] \big\}. \tag{4.14}$$

Here it is assumed that most of the relevant basins for the given temperature are "typical."

The detailed behavior of $u(t)$ should reveal the statistics of return times. Those statistics should be particularly illuminating as temperature declines, raising the possibility of transition from ergodic to nonergodic regimes. Perhaps this could be related to the glass transition in a useful way.

REFERENCES

[1] T. L. HILL, *Statistical Mechanics*, McGraw-Hill, New York, 1956.

[2] F. H. STILLINGER AND T. A. WEBER, *Hidden structure in liquids*, Phys. Rev. A, 25 (1982), pp. 978–988.

[3] F. H. STILLINGER AND T. A. WEBER, *Inherent pair correlation in simple liquids*, J. Chem. Phys., 80 (1984), pp. 4434–4437.

[4] C. KITTEL, *Introduction to Solid State Physics*, 2nd edition, Wiley, New York, 1956, p. 340.

[5] R. J. BAXTER, *Exactly Solved Models in Statistical Mechanics*, Academic Press, New York, 1982.

[6] F. H. STILLINGER AND R. A. LAVIOLETTE, *Local order in quenched states of simple atomic substances*, Phys. Rev. B, 34 (1986), pp. 5136–5144.

[7] A. F. ANDREEV, *Singularity of thermodynamic quantities at a first order phase transition point*, Sov. Phys., JETP (English trans.), 18 (1984), pp. 1415–1416.

[8] F. H. STILLINGER, *Duality relations for the Gaussian core model*, Phys. Rev. B, 20 (1979), pp. 299–302.

[9] M. R. GAREY AND D. S. JOHNSON, *Computers and Intractability: A Guide to the Theory of NP-Completeness*, W. H. Freeman, San Francisco, 1979.

[10] T. A. WEBER AND F. H. STILLINGER, *Interactions, local order, and atomic rearrangement kinetics in amorphous nickel-phosphorous alloys*, Phys. Rev. B, 32 (1985), pp. 5402–5411.

[11] F. H. STILLINGER, *Role of potential energy scaling in the low-temperature relaxation behavior of amorphous materials*, Phys. Rev. B, 32 (1985), pp. 3134–3141.

[12] A. MÜNSTER, *Prinzipien der statistischen Mechanik*, in *Handbuch der Physik, Band III/2, Prinzipien der thermodynamik und statistik*, ed. S. Flügge, Springer-Verlag, Berlin, 1959, pp. 216–218.

QUANTUM DYNAMICS

QUANTUM EVOLUTION IN EXTERNAL ELECTROMAGNETIC FIELDS: EXACT RESULTS AND ASYMPTOTIC APPROXIMATIONS

T.A.OSBORN†

Abstract. The quantum evolution of an N-body system of charged particles that couple to external electromagnetic fields is described. The existence of both operator and kernel valued (propagator) solutions to the evolution problem is established. The singular perturbation behavior of this system as mass becomes large is realized in terms of a gauge-invariant asymptotic expansion. This expansion for the propagator is a nonperturbative approximation with coefficients that are determined from a manifestly gauge invariant transport recurrence relation.[1]

1. Introduction. In quantum mechanics the Hamiltonian (operator) is the generator of time evolution. Consider the physical system composed of N non-relativistic particles, each with mass m and charge q, that mutually interact through scalar fields and couple via the Lorentz force to external electromagnetic fields. The differential form of the Hamiltonian appropriate for this type of system is

$$H(x,p,t,m) = \frac{1}{2m}[p - qa(x,t)]^2 + q\phi(x,t) + V(x,t). \tag{1.1}$$

Here x is the generic coordinate in the Euclidean space $\mathbf{R}^d$ ($d = 3N$) that determines the position of all N particles, p is the momentum operator conjugate to x ($p = -i\hbar\nabla$) and $t \in \mathbf{R}$ is the time variable. The interaction of the charged particles with an external electromagnetic field is described by the vector and scalar potentials, $a : \mathbf{R}^{d+1} \to \mathbf{R}^d$ and $\phi : \mathbf{R}^{d+1} \to \mathbf{R}$ respectively. The additional potential $V(x,t)$ in (1.1) accounts for the mutual interaction energy of all N particles and their interaction with any other non-electromagnetic external potentials. For example the pairwise Coulomb interactions between the charged particles is part of $V(x,t)$ since the energy associated with this interaction is not part of the external electromagnetic field. All the potentials a, ϕ, and V are assumed to be known functions. Obviously Hamiltonians of the form (1.1) are time dependent and in general do not conserve energy.

Observe that there is no loss of generality in the assumption of a common mass m for all N particles since a coordinate scale change transforms a Hamiltonian with different mass particles into one with the form (1.1). Likewise the effect of a different charge coupling constant q for each particle can be absorbed into the definition of a and ϕ.

The state of a quantum system is represented by a vector in Hilbert space. For this system the appropriate Hilbert space is composed of L^2 functions on $\mathbf{R}^d$, $\mathcal{H} = L^2(\mathbf{R}^d)$. Consider the evolution problem in $\mathcal{H}$ on the given interval

†Department of Physics, University of Manitoba, Winnipeg, Manitoba R3T 2N2, Canada

[1]The results discussed in this review paper have been obtained in the past several years in a series of collaborations with Robert Corns, Yoshikazu Fujiwara, Frank Molzahn, and Lech Papiez.

$[0,T]$. Denote by T_Δ and T_Δ^o the closed and open two-dimensional time domains $T_\Delta = \{(t,s) \in \mathbf{R}^2 : 0 \le s \le t \le T\}$ and $T_\Delta^o = \{(t,s) \in \mathbf{R}^2 : 0 < s < t < T\}$. Typically s has the interpretation as the starting time for the evolution and t is the present or running time variable. The vector and scalar potentials that occur in most physical applications are well enough behaved so that the differential operator $H(x, -i\hbar\nabla, t, m)$ has a unique closed extension $H(t,m)$ with domain $D\big(H(t,m)\big) \subset \mathcal{H}$. Of course $H(t,m)$ depends on more parameters than are displayed in its argument list. In particular it also depends on $\hbar$ (the rationalized Planck's constant), q (the charge coupling constant) and all the potentials. The mass parameter is maintained as an explicit variable because it plays an important role in the theory to be developed below.

In the circumstances where the domain of $H(t,m)$ is t-invariant, *i.e.*, $D\big(H(t,m)\big) = D_o$, $t \in [0,T]$, the abstract ($\mathcal{H}$-valued) evolution problem [1] in T_Δ takes the following form. A function $\psi : [s,T] \to \mathcal{H}$ is said to be a solution of

$$i\hbar\dot{\psi}(t) = H(t,m)\psi(t) \tag{1.2a}$$

if ψ takes values in D_o, possesses a strong derivative, $\dot{\psi}$, throughout the interval $[s,T]$, and satisfies (1.2a) for all $t \in [s,T]$. Let f be an arbitrary function on D_o and s be the time at which the initial data condition is imposed. The *Cauchy problem in the triangle* T_Δ is the problem of finding, for each fixed $s \in [0,T]$, a solution $\psi(\cdot;s)$ of (1.2a) on the interval $[s,T]$ that satisfies the initial condition

$$\psi(s;s) = f. \tag{1.2b}$$

Two widely studied descriptions of the evolution process are available. In the first, one seeks a family of bounded operators $U(t,s;m) : \mathcal{H} \to \mathcal{H}$ such that the solution $\psi(t;s)$ of (1.2) is obtained from the relationship

$$\psi(t;s) = U(t,s;m)f, \qquad (t,s) \in T_\Delta, \quad f \in D_o. \tag{1.3}$$

In the time independent case where $H(t,m)$ has the same static value H for all t and where H is self-adjoint,then $U(t,s;m)$ is only a function of the time displacement $\tau = t - s$ and may be written [2] in the well known form

$$U(\tau + s, s; m) = \exp\{-i\tau H/\hbar\}. \tag{1.4}$$

The operators on the right hand side of (1.4) form a one parameter ($\tau \in \mathbf{R}$) unitary group. A basic effect of the t dependence of $H(t,m)$ is to cause $U(t,s;m)$ to depend separately on t and s. For this reason an alternate representation to (1.4) is required for the two-time evolution operator $U(t,s;m)$.

A second description of evolution emerges if one describes all the functions in (1.3) by their pointwise values in $\mathbf{R}^d$. In the circumstance where $U(t,s;m)$ is an integral operator with a complex valued kernel $K(x,t;y,s;m)$, (1.3) is replaced by the statement

$$\psi(x,t;s) = \int_{\mathbf{R}^d} K(x,t;y,s;m) f(y)\,dy. \tag{1.5}$$

In the physics literature, K is called the propagator of the quantum system. The propagator clearly plays a central role in quantum evolution, since once it is known all properties of an evolving system may be determined. As a rule, except for simple examples like the harmonic oscillator problem, the function K is too complicated to be obtained in closed form.

The basic goals of this paper are to:

1°) establish the existence and fundamental properties of the Schrödinger evolution operators $U(t,s;m)$ for a class of potentials wide enough to include most applications of interest in physics;

2°) provide a mathematically precise definition of the propagator and for a class of analytic potentials obtain a convergent series expansion for K that is based on the Dyson series [3];

3°) to obtain the large mass non-perturbative gauge invariant asymptotic approximation for the propagator of 2°) together with uniform bounds for the error and explicit expressions for the expansion coefficients.

The third objective above requires some additional preliminary discussion. Often the propagator K is understood to be the distribution-valued solution of the time-dependent Schrödinger equation

$$i\hbar\frac{\partial}{\partial t}K(x,t;y,s;m) = H(x,-i\hbar\nabla,t,m)K(x,t;y,s;m) \tag{1.6a}$$

that satisfies the delta-function initial condition

$$K(x,s;y,s;m) = \delta(x-y). \tag{1.6b}$$

Clearly $H(x,-i\hbar\nabla,t,m)$ is an elliptic operator whose highest order differential structure is proportional to the Laplacian, *i.e.*, $-[\hbar^2/2m]\Delta$. In looking at the constants in (1.6a) it is seen that there are two singular perturbation problems of interest. The adjustable constants $\hbar^2$ and m^{-1} both multiply $-\Delta$ and the description of the solutions K of (1.6a), as either $\hbar \to 0$ or $m^{-1} \to 0$, define a singular perturbation problem. The limit $\hbar \to 0$ gives one the generalized WKB expansion and the $m^{-1} \to 0$ provides a type of eikonal expansion [4]. In 3°) our aim is to provide a rigorous discussion of the less familiar $m^{-1} \to 0$ limit.

2. Evolution operators. In this section Schrödinger evolution operators will be investigated. We begin by enlarging the framework of the evolution problem. In the case of forward time evolution ($t > s$) the mass parameter m will be allowed to take values in the upper half complex plane. The symbols $\mathbf{C}_>$ and $\mathbf{C}_\geq$ ($\mathbf{C}_<$ and $\mathbf{C}_\leq$) will respectively denote the open and closed upper (lower) half planes, while $\mathbf{C}_+$ is the set $\mathbb{C}_+ = \mathbb{C}_\geq \backslash \{0\}$. The original quantum evolution is realized as the positive real axis $m \in (0, \infty)$ boundary value of the extended evolution in $\mathbb{C}_+$. The existence of solutions for this extended evolution with $\operatorname{Im} m > 0$ is of particular importance in the construction of the propagator.

Consider the evolution problem (1.2) for $m \in \mathbb{C}_+$, where $H(t,m)$ is a closed operator on $\mathcal{H}$ for each $(t,m) \in [0,T] \times \mathbb{C}_+$ having domain D_o for all m and t. Let $\mathcal{B}(\mathcal{H})$ be the Banach space (with norm $\|\cdot\|$) of all bounded operators mapping $\mathcal{H}$ into $\mathcal{H}$. For each mass the evolution map $f \mapsto \psi(t,s)$ defines a linear operator from D_o into $\mathcal{H}$. The extension of this operator from the domain D_o to all of $\mathcal{H}$ gives one the evolution operator $U(t,s;m)$.

DEFINITION 1. Fix $m \in \mathbb{C}_+$. A two-parameter operator valued function $U : T_\Delta \to \mathcal{B}(\mathcal{H})$ is said to be the Schrödinger evolution generated by the family of Hamiltonian operators $\{H(t,m) : t \in [0,T]\}$ if

(1) For $(t,s) \in T_\Delta$, $U(t,s;m)$ maps the domain D_o into itself.

(2) U is uniformly bounded in T_Δ and for some non-negative c,

$$\|U(t,s;m)\| \leq e^{c(t-s)}, \qquad t \geq s.$$

(3) U is strongly continuous in T_Δ.

(4) The following identities hold in $\mathcal{B}(\mathcal{H})$

$$U(t,s;m) = U(t,\tau;m)U(\tau,s;m), \qquad 0 \leq s \leq \tau \leq t \leq T; \tag{2.1a}$$

$$U(s,s;m) = I, \qquad s \in [0,T]. \tag{2.1b}$$

(5) On the domain D_o, U is strongly continuously differentiable relative to t and s. Furthermore U satisfies the equations of motion on T_Δ^o

$$i\hbar \frac{\partial}{\partial t} U(t,s;m)f = H(t,m)U(t,s;m)f, \qquad f \in D_o; \tag{2.1c}$$

$$-i\hbar \frac{\partial}{\partial s} U(t,s;m)f = U(t,s;m)H(s,m)f, \qquad f \in D_o. \tag{2.1d}$$

Some comments concerning the definition will be helpful. The domain conserving property (1) means that the product $H(t,m)U(t,s;m)f$, $f \in D_o$ occurring in equation (2.1c) is well defined. The equation of motion (2.1c) ensures that $\psi(t) = U(t,s;m)f$ satisfies (1.2a). The condition (2.1b) incorporates the Cauchy initial data condition (1.2b). For $m > 0$, $H(t,m)$ is most often self-adjoint and then $U(t,s;m)$ is a unitary operator on the Hilbert space $\mathcal{H}$, from which it follows that $\|U(t,s;m)\| = 1$. However when $\operatorname{Im} m > 0$, $H(t,m)$ is no longer self-adjoint and $\|U(t,s;m)\|$ has the t,s-growth bound given in property (2). Finally the semigroup property (2.1a) is a restricted version of the group property that holds for the static evolution problem described in (1.4). If one has backward $(t < s)$ evolution rather than forward evolution it is necessary to extend the mass into the lower half complex plane. In this section we prove that for a broad class of potentials a, ϕ, and V, the Schrödinger evolution operators $U(t, s; m)$ exist.

Fortunately there is a mathematical theory of linear differential equations in Banach space with unbounded operator coefficients that is readily applicable to the problem at hand. We briefly summarize this theory and show how to apply it to the complex mass evolution problem.

Suppose E is a Banach space and $A(t) : D(A(t)) \subset E \to E$ is an unbounded operator for all $t \in [0, T]$. It is always assumed that for all $t \in [0, T]$ the $A(t)$'s are closed and have a common domain $D(A(t)) = D(A)$ that is dense in E. The linear differential equation for evolution in the time interval $[0, T]$ is then

$$\frac{df}{dt} = A(t)f, \qquad 0 \leq t \leq T. \tag{2.2a}$$

As in the quantum evolution problem (1.2) we say $f(t)$ is a *solution of equation* (2.2a) on the interval $[s, T]$ ($0 \leq s \leq T$), if $f(t)$ has values in $D(A)$, possesses a strong derivative $\dot{f}(t)$ and satisfies (2.2a) on the segment $[s, T]$. The *Cauchy problem in the triangle* T_Δ is the problem of finding for each fixed $s \in [0, T]$ a solution $f(t, s)$ of (2.2a) on the segment $[s, T]$ that satisfies the initial data condition

$$f(s, s) = f_o \in D(A). \tag{2.2b}$$

The Cauchy problem in T_Δ is well understood if the operators $A(t)$ have certain standard properties. For a given operator $A : D(A) \subset E \to E$ the resolvent set $\rho(A) \subseteq \mathbb{C}$ is the set of points in $\mathbb{C}$ for which the resolvent operator is bounded, *i.e.*, for all $z \in \rho(A)$

$$R(A, z) = (A - z)^{-1} \in \mathcal{B}(E). \tag{2.3}$$

With this terminology it is useful to group together the properties of $A(t)$ that suffice to describe evolution operators for problem (2.2).

DEFINITION 2. The family of operators $\{A(t) : t \in [0, T]\}$ is said to be in the *evolution class* if

(1) the $A(t)$ are densely defined and closed with a t-invariant domain $D(A)$;

(2) the $A(t)$ are strongly continuously differentiable on the domain $D(A)$; and

(3) the half axis $[0, \infty) \subseteq \rho(A(t))$, $t \in [0, T]$ and

$$\|R(A(t), \lambda)\| \leq \frac{1}{1 + \lambda}, \qquad \lambda \geq 0. \tag{2.4}$$

The basic existence result for Banach space valued evolution problems is summarized by the following statement.

THEOREM 1. *Suppose the family of operators $\{A(t) : t \in [0,T]\}$ is in the evolution class. Then there exists a unique function (family of evolution operators) $\mathcal{U} : T_\Delta \to \mathcal{B}(E)$ such that*

(1) $\mathcal{U}(t,s) : D(A) \to D(A), \qquad (t,s) \in T_\Delta.$

(2) *The operator $\mathcal{U}(t,s)$ is uniformly bounded in T_Δ. Specifically*

$$\|\mathcal{U}(t,s)\| \leq 1.$$

(3) *The operator $\mathcal{U}(t,s)$is strongly continuous in T_Δ.*

(4) *The following operator identities hold in T_Δ:*

$$\mathcal{U}(t,s) = \mathcal{U}(t,\tau)\mathcal{U}(\tau,s), \qquad 0 \leq s \leq \tau \leq t \leq T, \tag{2.5a}$$

$$\mathcal{U}(s,s) = I, \qquad s \in [0,T]. \tag{2.5b}$$

(5) *On the domain $D(A)$, $\mathcal{U}$ is strongly continuously differentiable relative to t and s. Furthermore $\mathcal{U}$ satisfies (in T_Δ) the equations of motion for all $f \in D(A)$*

$$\frac{\partial}{\partial t}\mathcal{U}(t,s)f = A(t)\mathcal{U}(t,s)f, \tag{2.5c}$$

$$\frac{\partial}{\partial s}\mathcal{U}(t,s)f = -\mathcal{U}(t,s)A(s)f. \tag{2.5d}$$

These results are found in Chapter 3 of Krein's book [1]. The proof is based on the determination of $\mathcal{U}(t,s)$ via the limit of an approximating sequence of operators that have an explicit construction.

It is apparent that the evolution operators of Theorem 1 will provide the basis for constructing the Schrödinger evolution operators of Definition 1. The Banach space E appropriate for the quantum problem is clearly the Hilbert space $\mathcal{H} = L^2(\mathbf{R}^d)$. Similarly, one must define $A(t)$ in terms of the Hamiltonian operators $H(t,m)$. The simplest choice of setting $A(t) = (i\hbar)^{-1}H(t,m)$ leads to the difficulty that the resolvent estimate (2.4) is not generally obeyed. However if we set

$$A(t) = -\frac{i}{\hbar}H(t,m) - cI \tag{2.6}$$

where I is the identity operator on $\mathcal{H}$ and c is some suitable constant in $\mathbf{R}$, then (for $\operatorname{Im} m \geq 0$) we shall be able to verify that $\{A(t) : t \in [0,T]\}$ are in the evolution class for a wide class of potentials a, ϕ and V.

The divergence of a vector valued function in $\mathbf{R}^d$ will be indicated by $\nabla \cdot a$. The symbol ∇^β will denote the partial derivative with respect to $x \in \mathbf{R}^d$ specified by the multi-index $\beta = (\beta_1, \ldots, \beta_d)$ with length $|\beta| = \beta_1 + \cdots + \beta_d$, *i.e.*,

$$\nabla^\beta = \left(\frac{\partial}{\partial x_1}\right)^{\beta_1} \cdots \left(\frac{\partial}{\partial x_d}\right)^{\beta_d}.$$

The symbol ∂ will denote the partial derivative with respect to t. The norm of the Banach space $L^p(\mathbf{R}^d, \mathbf{C}^r)$, the space of r-dimensional vector valued functions on $\mathbf{R}^d$, will be $\|\cdot\|_p$, where $p \in [1,\infty]$. The values of r in our applications are either $r = 1$ or d and the context will determine the correct choice. The symbol $C^n(\Omega)$ denotes the space of n times continuously differentiable functions on the domain $\Omega \subset \mathbf{R}^d$. The subscript o, *i.e.*, $C_o^n(\Omega)$, will indicate the $C^n(\Omega)$ functions of compact support.

HYPOTHESIS (H1). The vector potential $a : \mathbf{R}^d \times [0,T] \to \mathbf{R}^d$ satisfies the following properties.

(1) $a \in C^1(\mathbf{R}^d \times [0,T])$;

(2) a, ∂a, and $\nabla^\beta a$ ($|\beta| = 1$) are uniformly bounded $L^\infty(\mathbf{R}^d, \mathbf{C}^d)$ functions on the interval $[0,T]$. For some postive $M_i < \infty$, $i = 1 \sim 3$,

$$\|a(\cdot,t)\|_\infty \leq M_1, \qquad \|\partial a(\cdot,t)\|_\infty \leq M_2, \qquad \|\nabla^\beta a(\cdot,t)\|_\infty \leq M_3;$$

(3) $a(\cdot,t) \in L^\infty(\mathbf{R}^d, \mathbf{C}^d)$ and $\nabla \cdot a(\cdot,t) \in L^\infty(\mathbf{R}^d, \mathbf{C})$ are continuously differentiable functions on $[0,T]$ in their respective L^∞ topologies with derivatives $\dot{a}(\cdot,t)$ and $\nabla \cdot \dot{a}(\cdot,t)$.

Let us consider the basic mathematical properties of the Hamiltonians $H(t,m)$. This will be done in two stages. First we discuss the behavior of the Hamiltonians with the differential form $(2m)^{-1}[i\hbar\nabla + a(x,t)] \cdot [i\hbar\nabla + a(x,t)]$. (Here $\cdot$ represents the Euclidean dot product in $\mathbf{R}^d$.) Hamiltonians of this type describe the mechanical kinetic energy of the system. They are always non-negative. Then the effect of adding scalar perturbations to the system will be discussed.

The behavior of 4-vector potentials in electromagnetism is a good guide to selecting a reasonable class of functions for $a(x,t)$. Recall that the electric and magnetic fields are recovered (for $N = 1$) from the 4-vector potential by

$$E = -\nabla\phi - \partial a, \qquad B = \nabla \times a.$$

In addition, the physical potentials (a,ϕ) are solutions of Maxwell's inhomogeneous wave equation. This requires that (a,ϕ) are continuously differentiable to second order in x and t. These standard features of electromagnetism suggest that the class of functions for $a(x,t)$ should involve smooth differentiable functions in both x and t.

Let $\tilde{H}_o(m)$ be the minimal operator with domain $C_o^\infty(\mathbf{R}^d)$ associated with the Laplacian in $\mathbf{R}^d$

$$\tilde{H}_o(m) = -\frac{\hbar^2}{2m}\Delta. \tag{2.7}$$

For $m > 0$, it is well known that $\tilde{H}_o(m)$ acting in the space $L^2(\mathbf{R}^d)$ is essentially self-adjoint with a self-adjoint closure we denote by $H_o(m)$ (Ref. 2, Chap.V). Furthermore the spectrum of $H_o(m)$, $\sigma(H_o(m))$, is the interval $[0,\infty)$. For complex m, by writing $\tilde{H}_o(m) = m^{-1}\tilde{H}_o(1)$, we see that $\tilde{H}_o(m)$ is closable with closure $H_o(m) = m^{-1}H_o(1)$. The domain of $H_o(m)$ for all $m \in \mathbf{C}_+$ is given by

$$D(H_o(m)) \equiv D_o = \{\psi \in \mathcal{H} : \alpha^2\hat{\psi}(\alpha) \in L^2(\mathbf{R}^d)\}$$

where $\hat{\psi}$ denotes the Fourier transform of ψ.

Next we consider the minimal Hamiltonian operator on $C_o^\infty(\mathbf{R}^d)$ associated with the partial differential operator

$$\tilde{H}_1(t,m) = \frac{1}{2m}\left(\frac{\hbar}{i}\nabla - a(\cdot,t)\right)^2. \tag{2.8}$$

Expanding out the square in (2.8) we can write $\tilde{H}_1(t,m)$ as

$$\tilde{H}_1(t,m) = \tilde{H}_o(m) + \tilde{W}(t,m), \tag{2.9}$$

where the perturbing operator $\tilde{W}(t,m)$, again having domain $C_o^\infty(\mathbf{R}^d)$, is given by

$$\tilde{W}(t,m) = \left\{\frac{i\hbar}{m}a(\cdot,t)\cdot\nabla + \frac{i\hbar}{2m}(\nabla\cdot a)(\cdot,t) + \frac{1}{2m}a(\cdot,t)^2\right\}. \tag{2.10}$$

Suppose temporarily that $m > 0$. An integration by parts shows that the H-inner product identity holds

$$\langle\psi, \tilde{W}(t,m)\varphi\rangle = \langle\tilde{W}(t,m)\psi, \varphi\rangle, \qquad \psi,\varphi \in C_o^\infty(\mathbf{R}^d).$$

Thus we have that $\tilde{W}(t,m)$ is symmetric and hence it is closable. We denote the closure of $\tilde{W}(t,1)$ by $W(t,1)$. Since $\tilde{W}(t,m) = m^{-1}\tilde{W}(t,1)$, we see that $\tilde{W}(t,m)$ is closable for all $m \in \mathbf{C}_+$ with the closure $W(t,m) = m^{-1}W(t,1)$.

LEMMA 1. *Let $a \in (H1)$. Then for each $(t,m) \in [0,T] \times \mathbf{C}_+$ the operator $\tilde{H}_1(t,m)$ has the closure $H_1(t,m)$ satisfying*

(1) *$D(H_1(t,m)) = D_o$;*

(2) *For all $\psi \in D_o$*

$$H_1(t,m)\psi = H_o(m)\psi + W(t,m)\psi; \tag{2.11}$$

(3) *For $m > 0$, $H_1(t,m)$ is self-adjoint and bounded from below by zero;*

(4) *$H_1(t,m)$ is strongly differentiable on the interval $[0,T]$ with respect to the domain D_o.*

Proof. We first show that $\tilde{W}(t,m)$ is $\tilde{H}_o(m)$-bounded. From the assumption $a \in (H1)$ it follows that both $(\nabla\cdot a)(x,t)$ and $a(x,t)^2$ give rise to bounded operators that are uniformly bounded in t. Thus we need only prove our assertion for $a(x,t)\cdot\nabla$. Let $\psi \in C_o^\infty(\mathbf{R}^d)$ and consider the following:

$$\|a(\cdot,t)\cdot\nabla\psi\|^2 = \int |a(x,t)\cdot(\nabla\psi)(x)|^2\,dx \le {M_1}^2 \int \alpha^2|\hat{\psi}(\alpha)|^2\,d\alpha.$$

Here $\hat{\psi}$ represents the $L^2(\mathbf{R}^d)$ Fourier transform of ψ. Likewise, the Fourier transform of $\nabla\psi$ has the pointwise values $i\alpha\hat{\psi}(\alpha)$. The Plancherel theorem, $\|\nabla\psi\|_2 = \|\alpha\hat{\psi}\|_2$, has been employed in obtaining the final inequality above. In the Fourier transform space the Laplacian is represented by multiplication by $-\alpha^2$. Then it is simple to verify that the gradient operator obeys the estimate

$$\|\nabla\psi\|^2 \leq \frac{1}{\delta^2}\|\psi\|^2 + \delta^2 \left|\frac{2m}{\hbar^2}\right|^2 \|\tilde{H}_o(m)\psi\|^2, \qquad \psi \in C_o^\infty(\mathbf{R}^d) \tag{2.12}$$

for each $\delta > 0$. Combining these two previous results gives

$$\|a(\cdot,t)\cdot\nabla\psi\| \leq M_1\{\frac{1}{\delta}\|\psi\| + \delta\left|\frac{2m}{\hbar^2}\right|\|\tilde{H}_o(m)\psi\|\}, \qquad \psi \in C_o^\infty(\mathbf{R}^d).$$

This shows that $\tilde{W}(t,m)$ is $\tilde{H}_o(m)$-bounded. Now the parameter δ can be made arbitrarily small, thus demonstrating that $\tilde{W}(t,m)$ has $\tilde{H}_o(m)$-bound zero. We also note that the estimate is time independent.

Let $m > 0$ for the moment. Then $\tilde{W}(t,m)$ is symmetric and by a theorem of Kato (Ref. 2, Chap.V, Theorem 4.4), $\tilde{H}_o(m) + \tilde{W}(t,m)$ is essentially self-adjoint with its closure $H_1(t,m) = H_o(m) + W(t,m)$. Furthermore $H_1(t,m)$ will be self-adjoint and have domain D_o. To extend the domain and closure properties to $m \in \mathbf{C}_+$, we simply note that $\tilde{H}_1(t,m) = m^{-1}\tilde{H}_1(t,1)$. Of course $H_1(t,m)$ will no longer be self-adjoint for a complex mass parameter. This establishes (1) and (2).

Next we show (3). Let $m > 0$. To show that $H_1(t,m) \geq 0$ we must verify

$$\langle\psi, H_1(t,m)\psi\rangle \geq 0, \qquad \psi \in D_o.$$

However from the closedness of $H_1(t,m)$ we need only prove this on a core of $H_1(t,m)$ and then extend this to all of D_o. One such core is $C_o^\infty(\mathbf{R}^d)$, upon which the Hamiltonian $H_1(t,m) = \tilde{H}_1(t,m)$. Using the definition of $\tilde{H}_1(t,m)$ and integrating by parts we get

$$\langle\psi, \tilde{H}_1(t,m)\psi\rangle = \frac{1}{2m}\left\|\left[\frac{\hbar}{i}\nabla - a(\cdot,t)\right]^2\psi\right\|^2 \geq 0.$$

The strong differentiability of $H_1(t,m)$ with respect to t is an immediate consequence of property (3) in $(H1)$. □

Next we consider the perturbations of $H_1(t,m)$ by a time dependent (possibly unbounded) operator $V(t)$. Recall that the general evolution theory summarized in Theorem 1 requires that the operators $A(t)$ which generate the evolution have time independent domains. In the approach adopted here the question of domain stability is controlled via the following theorem (cf. Ref. 2, Chap.IV Theorem 1.1).

THEOREM 2. *Let A and B be operators from Banach space $\mathcal{B}_1$ to $\mathcal{B}_2$ and let B be A-bounded with A-bound less than 1. Then $S = A + B$ is closable if and only if A is closable; in this case the closures of A and S have the same domain. In particular S is closed if and only if A is closed.*

We employ this theorem to prove

LEMMA 2. *Let $V(t) : D_o \to \mathcal{H}$ have a t-uniform $H_o(1)$-bound zero, namely*

$$\|V(t)\psi\| \leq \alpha_o\|\psi\| + \beta_o\|H_o(1)\psi\|, \qquad \psi \in D_o, \quad t \in [0,T] \tag{2.13}$$

where α_o and β_o are non-negative constants independent of t and β_o may be chosen arbitrarily small.

(1) *For each $(t,m) \in [0,T] \times \mathbf{C}_+$*

$$H(t,m) = H_1(t,m) + V(t) \tag{2.14}$$

is a closed operator with domain D_o.

(2) *Suppose further that $V(t) : D_o \to \mathcal{H}$ is strongly differentiable on $[0,T]$. Then for all $m \in \mathbf{C}_+$, $H(t,m)$ is strongly differentiable.*

(3) *If $V(t)$ is symmetric for $t \in [0,T]$ and $m > 0$, then $H(t,m)$ is self-adjoint and bounded from below by $-\gamma I$, where $\gamma \geq 0$ is some t-independent constant.*

Proof. Consider (1). For each $(t,m) \in [0,T] \times \mathbf{C}_+$, $H_1(t,m)$ is closed. The space $C_o^\infty(\mathbf{R}^d)$ is a core of the operator $H_o(1)$. Thus the inequality (2.12) may be extended to D_o. Adding the bounded operators associated with $(\nabla \cdot a)(x,t)$ and $a(x,t)^2$ gives one the inequality

$$\|W(t,1)\psi\| \leq \alpha_1\|\psi\| + \beta_1\|H_o(1)\psi\|, \qquad \psi \in D_o$$

with adjustable constants α_1, β_1. The values of $\beta_1 > 0$ can be as small as desired — *i.e.*, $W(t,1)$ has $H_o(1)$- bound zero. Employing (2.11) with $m = 1$ implies that

$$\|W(t,1)\psi\| \leq \frac{\alpha_1}{1-\beta_1}\|\psi\| + \frac{\beta_1}{1-\beta_1}\|H_1(t,1)\psi\|, \qquad \psi \in D_o. \tag{2.15}$$

Similarly using bound (2.13) together with (2.11) for $m = 1$ leads to

$$\|V(t)\psi\| \leq \alpha_o\|\psi\| + \beta_o\|H_1(t,1)\psi\| + \beta_o\|W(t,1)\psi\|.$$

Combining this inequality with (2.15) and noting that $H_1(t,1) = mH_1(t,m)$, one finds (for $\beta_1 < 1$)

$$\|V(t)\psi\| \leq A_1\|\psi\| + B_1\|H_1(t,m)\psi\|, \qquad \psi \in D_o, \tag{2.16}$$

where

$$A_1 = \alpha_o + \frac{\alpha_1}{1-\beta_1}, \qquad B_1 = \frac{\beta_o|m|}{1-\beta_1}.$$

Since β_o is arbitrarily small, (2.16) shows that $V(t)$ is $H_1(t,m)$-bounded with bound zero. Applying Theorem 2 establishes the domain stability statement (1).

Property (2) is an immediate consequence of (1) and the fact that $H_1(t,m)$ and $V(t)$ are both strongly differentiable for $t \in [0,T]$.

Finally consider (3). Here $H_1(t,m)$ is self-adjoint and $V(t)$ is symmetric. In inequality (2.16) a sufficiently small value of β_o means that B_1 will be less than 1. These circumstances allow the application of the Kato-Rellich Theorem (Ref. 2, Chap.V, Theorem 4.11) which asserts that $H(t,m)$ is self-adjoint and bounded from below. In particular the lower bound is $-\gamma I$ where $\gamma = A_1(1-B_1)^{-1}$. □

Let $A(t)$ be obtained from $H(t,m)$ by (2.6) for some $c \in \mathbf{R}$. If the perturbation $V(t)$ of $H_1(t,m)$ is such that the statements (1)–(3) of Lemma 2 hold, then the first two of the three required properties for $\{A(t) : t \in [0,T]\}$ to be in the evolution class are satisfied. It remains therefore to show that for some particular choice of c, the decay estimate (2.4) of the resolvent $R(A(t),\lambda)$ is obeyed. If $m > 0$, the operator $H(t,m)$ has a real spectrum contained in the interval $[-\gamma,\infty)$. Set $c = 1 + \gamma\hbar^{-1}$. The resolvent of $A(t)$ is given by

$$R(A(t),\lambda) = \frac{i\hbar}{H(t,m) - i\hbar(1+\gamma\hbar^{-1}+\lambda)I}.$$

It has the desired estimate

$$\|R(A(t),\lambda)\| \leq \frac{1}{1+\lambda},$$

valid for all $\lambda \geq 0$, $t \in [0,T]$ and $m > 0$. Actually $c = 1$ will suffice in order to obtain the above estimate, however if $c = 1+\gamma\hbar^{-1}$ then (2.6) may be used for both Theorems 3 and 4.

In these circumstances $\{A(t) : t \in [0,T]\}$ is in the evolution class and the results of Theorem 1 apply. The Schrödinger evolution operators are obtained from the operators $\mathcal{U}$ in Theorem 1 by the relationship

$$U(t,s;m) = e^{c(t-s)}\mathcal{U}(t,s). \tag{2.17}$$

A simple calculation shows that if $\mathcal{U}(t,s)$ satisfies the differential equation of motion (2.5c) of Theorem 1 then $U(t,s;m)$ obeys the abstract Schrödinger equation (2.1c). Summarizing these conclusions we have the following theorem.

THEOREM 3. *Let the vector potential $a \in (H1)$. If $V(t) : D_o \to \mathcal{H}$ is symmetric, strongly differentiable on $[0,T]$ and has a t-uniform $H_o(1)$-bound zero then for each $m > 0$ the family of Hamiltonian operators $\{H(t,m) : t \in [0,T]\}$ generates a Schrödinger evolution operator, $U(\cdot,\cdot;m) : T_\Delta \to \mathcal{B}(\mathcal{H})$.*

A variant of these results applies to the evolution problem if we allow m to have values in the upper half complex mass plane. Here the assumptions on the potential $V(t)$ are altered somewhat. We eliminate the requirement that $V(t)$ be symmetric but do assume that $V(t)$ is uniformly bounded. In greater detail we have

THEOREM 4. *Let the vector potential $a \in (H1)$. If $V(t) \in \mathcal{B}(\mathcal{H})$ is uniformly bounded and strongly differentiable on $[0,T]$ then for each $m \in \mathbf{C}_+$ the family* *of Hamiltonians* $\{H(t,m) : t \in [0,T]\}$ *generates a Schrödinger evolution operator,* $U(\cdot,\cdot;m) :\ T_\Delta \to\ \mathcal{B}(\mathcal{H})$.

The proof of this statement is similar to that above for Theorem 3 and the details may be found in Ref. 6 (Theorems 2 and 3). Theorem 4 is an example of the evolution defined throughout the $\mathbf{C}_+$ mass plane.

At this point it is useful to make an assessment of the result obtained. From the point of view of possible applications to quantum evolution it is important that the allowed class of potentials include all the known forces in atomic and sub-atomic physics. Let the index $i = 1 \sim N$ label particles whose coordinate positions are specified by $\vec{r}_i \in \mathbf{R}^3$, *i.e.,* $x = (\vec{r}_1, \dots, \vec{r}_N)$. The vectors $\vec{r}_{ij} = \vec{r}_i - \vec{r}_j$ will define the relative separation of particles i and j. The N-body Coulomb potential is given as the sum of all pairwise charged particle interactions plus a (possible) interaction with fixed charge Q at $\vec{r} = 0$. This potential is

$$V_c(x) = \sum_{i=1}^{N} \frac{Qq_i}{|\vec{r}_i|} + \sum_{i<j} \frac{q_i q_j}{|\vec{r}_{ij}|}.$$

Kato has proved [7] that the operator V_c is $H_o(1)$-bounded with $H_o(1)$-bound 0.

Similarly, if the pair interactions v_{ij} have conventional short range static forces such as the Yukawa potential

$$v_{ij}(\vec{r}_{ij}) = g_{ij} \frac{e^{-\beta_{ij}|\vec{r}_{ij}|}}{|\vec{r}_{ij}|},$$

(for real constants g_{ij} and $\beta_{ij} > 0$), then again the associated operators have $H_o(1)$-bound 0. These examples show we may include in $V(x,t)$ the important static forces in atomic physics.

The origin of time dependence in the scalar potential would normally come from the external electromagnetic potential $\phi(x,t)$. If $\phi(x,t)$ is real valued and belongs to $L^\infty(\mathbf{R}^d)$ for $t \in [0,T]$ and is continuously differentiable on $[0,T]$ with respect to the $L^\infty(\mathbf{R}^d)$ norm, then properties (2) and (3) of Lemma 2 are easily seen to be fullfilled. Of course the t- uniform $H_o(1)$-bound zero criteria of Lemma 2 is trivially satisfied. These comments show us that we have the capability of establishing the existence of the two-time Schrödinger evolution operators $U(t,s;m)$ for nearly all systems of physical interest.

One final comment concerns the inclusion of spin dependence. If for example the particles all have spin 1/2 and $\vec{S}_i$ denotes the spin operator for the ith particle [*i.e.,* $\vec{S} = (\hbar/2)\vec{\sigma}$, where $\vec{\sigma} = (\sigma_1, \sigma_2, \sigma_3)$ are the 2×2 Pauli spin matrices] then the differential form of the Hamiltonian becomes

$$H(x,p,t,m) = \{\frac{1}{2m}[p - qa(x,t)]^2 + q\phi(x,t)\}I + V(x,t) - \sum_{i=1}^{N} \mu_i \vec{S}_i \cdot \vec{B}(\vec{r}_i, t).$$

Here $\vec{B}(\vec{r}_i,t)$ is the induction field experienced by the ith particle at position $\vec{r}_i$ and μ_i is the magnetic moment of the ith particle. The Hamiltonian acts on a

$2N$ component spinor (vector) valued wave function and here I is the identity operator in this spinor space. The relevant Hilbert space for the state vector is $\mathcal{H} = \prod_{j=1}^{2N} L^2(\mathbf{R}^{3N})_j$. The general evolution theory readily extends to this non-Abelian quantum problem.

3. The complex mass propagator. This section investigates the nature of the propagator K of (1.5). We give a mathematical definition of the propagator that will suffice for all evolution problems discussed in the remainder of this paper. For a class of analytic potentials a, ϕ and V that are the Fourier images of complex valued bounded measures we will use the complex mass embedding method to construct a convergent series representation of the propagator $K(x,t;y,s;m)$. The propagator for free evolution is well known to be [8]

$$K_o(t-s;x-y;m) = \left[\frac{m}{2\pi i\hbar(t-s)}\right]^{d/2} \exp\left\{im\frac{|x-y|^2}{2\hbar(t-s)}\right\}. \tag{3.1}$$

Substitution of (3.1) into (1.6a) verifies that K_o is the solution of the Schrödinger equation with the Hamiltonian $H_o(m)$. In what sense is K_o an integral kernel of the free evolution operator $U_o(t,s;m)$? For $m > 0$, K_o has no decay (only oscillation) as $|x-y| \to \infty$ and for this reason K_o is not an $L^2(\mathbf{R}^d \times \mathbf{R}^d)$ kernel nor even a Carleman kernel. Recall that a kernel $K : \mathbf{R}^d \times \mathbf{R}^d \to \mathbf{C}$, acting on the space of $L^2(\mathbf{R}^d)$ functions, is Carelman [9] if for *a.a.* $x \in \mathbf{R}^d$, $K(x,\cdot)$ is $L^2(\mathbf{R}^d, dy)$. With these caveats in mind we seek a definition for the propagator which should be broad enough to include all cases of physical interest. The notation $L^p_o(\mathbf{R}^d)$ denotes the $L^p(\mathbf{R}^d)$ functions having arbitrary compact support.

DEFINITION 3. Fix $m \in \mathbf{C}_+$. A two parameter family (in T^o_Δ) of complex valued functions $K(\cdot,t;\cdot,s;m)$, that are measurable and locally integrable on $\mathbf{R}^d \times \mathbf{R}^d$, is called the ***propagator*** for Schrödinger evolution $\{U(t,s;m) : (t,s) \in T^o_\Delta\}$ if for all $f \in L^\infty_o(\mathbf{R}^d)$

$$[U(t,s;m)f](x) = \int K(x,t;y,s;m)f(y)\,dy, \qquad a.a.\ x. \tag{3.2}$$

Observe that since $L^\infty_o(\mathbf{R}^d)$ is dense in $\mathcal{H}$ and $U(t,s;m)$ is bounded, statement (3.2) leads to the identity

$$U(t,s;m)f = \ell.i.m. \int K(\cdot,t;y,s;m)f(y)\,dy,$$

which holds for all $f \in \mathcal{H} = L^2(\mathbf{R}^d)$. Definition 3 also implies that for each $(t,s) \in T^o_\Delta$ the propagator $K(x,t;y,s;m)$ is uniquely defined almost everywhere on $\mathbf{R}^d \times \mathbf{R}^d$. It is evident that the free kernel satisfies Definition 3. Note that the point $t = s$ is excluded from the specification of the propagator since K_o is not defined there. Definition 3 is essentially the same as the one Simon has used in his study of Schrödinger semigroups [10].

Let us consider the specific construction of K for a class of analytic potentials. The purpose of this discussion is to give an overview of the methods used to implement this construction. The detailed proofs (which are often of substantial length) of most of the conclusions cited here can be found in Refs. [6] and [11].

The Hamiltonian operator for the full quantum system is defined in Lemma 2 by (2.14) to be $H(t,m)$. Using (2.11) we can write $H(t,m)$ as a perturbed $H_o(m)$ operator,

$$H(t,m) = H_o(m) + \mathcal{V}(t,m), \tag{3.3}$$

$$\mathcal{V}(t,m) = W(t,m) + Y(t). \tag{3.4}$$

Here $Y(t)$ is the bounded operator defined by the total scalar potential

$$v(x,t) = q\phi(x,t) + V(x,t). \tag{3.5}$$

Here and in the sequel it is convenient to set $q = 1$.

Using formal arguments it is evident that the integral equation equivalent to the abstract Schrödinger equation of motion (2.1c) is

$$U(t,s;m)f = U_o(t,s;m)f - \left(\frac{i}{\hbar}\right)\int_s^t d\tau\, U_o(t,\tau;m)\mathcal{V}(\tau,m)U(\tau,s;m)f \tag{3.6}$$

where $U_o(t,s;m)$ is the free evolution operator generated by $H_o(m)$. Iterating (3.6) leads to the formula

$$D_n(t,s;m)f = \left(-\frac{i}{\hbar}\right)^n \int_< d\mathbf{t}_n\, U_o(t,t_n;m)\mathcal{V}(t_n,m)U_o(t_n,t_{n-1};m) \times \cdots \times \mathcal{V}(t_1,m)U_o(t_1,s;m)f \tag{3.7}$$

where $\mathbf{t}_n = (t_1,\ldots,t_n)$ and $<$ is a short-hand notation for the n-dimensional time ordered domain $\Delta_n(t,s) = \{\mathbf{t}_n \in \mathbf{R}^n : s \le t_1 \le t_2 \le \cdots \le t_n \le t\}$.

The series obtained by summing $D_n(t,s;m)f$ over $n = 0 \sim \infty$ is called the Dyson series [3]. In the circumstances where $\mathcal{V}(\cdot,m)$ is a uniformly bounded operator on the interval $[0,T]$ then it is well known (Ref. 1, Chap.II) that the sum of terms (3.7) converges strongly to $U(t,s;m)f$. However, for the problem at hand (with $a \neq 0$) the term $(i\hbar/m)a(\cdot,t)\cdot\nabla$ is unbounded no matter how nicely $a(x,t)$ and $v(x,t)$ behave. As a general method of obtaining $U(t,s;m)$ the Dyson series has many drawbacks. Because $\mathcal{V}(t,m)$ is unbounded, one has the problem of showing that the product of unbounded operators on the right hand side of (3.7) have the necessary domain and range consistency, that they define (operator valued) integrable functions, and that the resulting integrals form a convergent series.

In spite of these difficulties we will use the Dyson series to construct both $U(t,s;m)$ and its kernel $K(x,t;y,s;m)$. The study of the operators $D_n(t,s;m)$ and their associated kernels $d_n(x,t;y,s;m)$ is much simplified if $\operatorname{Im} m > 0$. For example

$d_n(x,t;y,s;m)$ and $K(x,t;y,s;m)$ turn out to be Carleman kernels if $\operatorname{Im} m > 0$. This improved behavior can be traced to the fact that if $\operatorname{Im} m > 0$ and $t > s$ then $K_o(t-s; x-y; m)$ has Gaussian decay as $|x-y| \to \infty$ and thus defines (since it is clearly a function of only $x-y$) an $L^1(\mathbf{R}^d)$ convolution kernel. In the final stage of our approach it is shown that both $U(t,s;m)$ and $K(x,t;y,s;m)$ are continuous (in some appropriate norms) as $\operatorname{Im} m > 0 \to 0^+$. In this way the physical propagator for real mass is obtained. This continuation in complex mass to determine the propagator for real mass is similar to Nelson's method [12] of using the analytic continuation in mass to define the Feynman path integral in terms of kernels constructed from Wiener measures.

In order to construct explicit formulae for $d_n(x,t;y,s;m)$ we introduce a family of potentials that are the Fourier transforms of complex bounded measures on $\mathbf{R}^d$. The vector and combined scalar potentials are assumed to have the form

$$a(x,t) = \int e^{i\alpha \cdot x}\, d\gamma(t), \tag{3.8a}$$

$$v(x,t) = \int e^{i\alpha \cdot x}\, d\nu(t). \tag{3.8b}$$

In these integrals the measures $\gamma(t)$ and $\nu(t)$ are time dependent, while the variable of integration $\alpha \in \mathbf{R}^d$ (the wave vector) is *not* displayed in the measure symbol $d\gamma(t)$ or $d\nu(t)$.

Our measures ($\gamma(t)$ and $\nu(t)$) will be chosen from the Banach spaces $\mathcal{M}(\mathbf{R}^d, \mathbf{C}^r)$, ($r = d$ or 1) of bounded $\mathbf{C}^r$ valued Borel measures γ on $\mathbf{R}^d$, which have complex valued Fourier images (3.8). The space $\mathcal{M}^*(\mathbf{R}^d, \mathbf{C}^r)$ is the subspace of $\mathcal{M}(\mathbf{R}^d, \mathbf{C}^r)$ whose images are real valued. The norm $\|\cdot\|$ for $\mathcal{M}(\mathbf{R}^d, \mathbf{C}^r)$ and $\mathcal{M}^*(\mathbf{R}^d, \mathbf{C}^r)$ is defined using the total variation measure $|\gamma|$, via

$$\|\gamma\| = |\gamma|(\mathbf{R}^d) < \infty.$$

The same symbol $\|\cdot\|$ is used as the norm for a variety of different spaces. The context will determine its correct meaning. An additional restriction on the measures is the requirement that they have compact support. Let $S_k \subset \mathbf{R}^d$ be the closed ball of radius k and center at 0. Then $\mathcal{M}^*(S_k, \mathbf{C}^r)$ will denote the Banach subspace of measures in $\mathcal{M}^*(\mathbf{R}^d, \mathbf{C}^r)$ that have their support contained by S_k.

A time dependent measure is defined by the map

$$\gamma(\cdot) : [0,T] \to \mathcal{M}^*(\mathbf{R}^d, \mathbf{C}^r).$$

From this point of view, $\gamma(t)$ is a Banach space valued function of t. In the space $\mathcal{M}^*(\mathbf{R}^d, \mathbf{C}^r)$ one has the conventional definitions of continuity and differentiability with respect to $\|\cdot\|$. The symbol $\dot{\gamma}(t)$ denotes the derivative of $\gamma(\cdot)$ at t. With this terminology in place we may state the hypothesis on the potentials that is required for the remainder of this paper.

DEFINITION 4. *Potential class (A)*: The potentials a and v are said to be in class (A) if a and v are the Fourier images, Eqs.(3.8), of time dependent measures $\gamma(\cdot)$ and $\nu(\cdot)$ satisfying

(1) $\gamma(t) \in \mathcal{M}^*(S_{k/2}, \mathbf{C}^d), \qquad t \in [0,T]$,
(2) $\nu(t) \in \mathcal{M}^*(S_k, \mathbf{C}), \qquad t \in [0,T]$,
(3) both $\gamma(\cdot)$ and $\nu(\cdot)$ are continuously differentiable on $[0,T]$.

The functions $a(\cdot,t)$ and $v(\cdot,t)$ in class (A) are respectively $\mathbf{R}^d$ and $\mathbf{R}$ valued analytic functions. The requirement that $k < \infty$ means that the electric and magnetic fields have a space frequency cutoff k. Potentials in class (A) are suitable for the N-body problem and the description of the electromagnetic fields since they do not require any decay as $|x| \to \infty$. This class of potentials is advantageous in that it allows us to obtain explicit formulae, in the form of parametric integrals, for the kernels $d_n(x,t;y,s;m)$. The limitations of this class are also apparent. Unbounded potentials like those that occur in the Coulomb or Yukawa interactions are prohibited. The use of this class of potentials consisting of Fourier images of complex bounded measures was initiated by Ito [13] in the study of the Feynman path integrals and played a central role in the results Albeverio and Høegh-Krohn [14] obtained for the path integral. In the situation where there is no electomagnetic field present and $v(x,t)$ is static, Osborn and Fujiwara [15] constructed the propagator K for the analytic semigroup $\{e^{-zH} : z \in \mathbf{C}, \operatorname{Re} z \geq 0\}$. Some useful constants related to a and v that often appear in the subsequent estimates are

$$\nu_T = \sup \|\nu(t)\|, \qquad \gamma_T = \sup \|\gamma(t)\|, \qquad \dot{\gamma}_T = \sup \|\dot{\gamma}(t)\|,$$

where each supremum is taken over $t \in [0,T]$. Furthermore note that the part of the Hamiltonian free of derivatives can also be represented as the Fourier transform of a measure, *i.e.*,

$$\frac{1}{2m} a(x,t)^2 + v(x,t) = \int e^{i\alpha\cdot x}\, d\mu(t),$$

where

$$\mu(t) = \frac{1}{2m}\gamma(t) * \gamma(t) + \nu(t).$$

Here the $*$ denotes the scalar valued convolution of two vector valued measures. For potentials in class (A), $\mu(t) \in \mathcal{M}^*(S_k, \mathbf{C})$ and it has the t-independent norm estimate,

$$\|\mu(t)\| \leq \mu_T \equiv \frac{1}{2|m|}\gamma_T^2 + \nu_T.$$

The first consequence of the assumption a and v being in class (A) is that the vector potential a is easily seen to satisfy $(H1)$ and that $v(x,t)$ defines an operator $Y(t) \in \mathcal{B}(\mathcal{H})$ which is uniformly bounded and strongly continuously differentiable on $[0,T]$. Thus the conclusions of Theorem 4 apply and $\{H(t,m) : t \in [0,T]\}$ generates a Schrödinger evolution $U(\cdot,\cdot;m) : T_\Delta \to \mathcal{B}(\mathcal{H})$ for all $m \in \mathbf{C}_+$.

The continuity in (complex) mass of the evolution operators is summarized by

PROPOSITION 1. *Assume a and v are in class (A) and let $U\colon T_\Delta \times \mathbb{C}_+ \to \mathcal{B}(\mathcal{H})$ be the evolution operator of Theorem 4. For each fixed $(t,s) \in T_\Delta$, $U(t,s;m)$ is strongly continuous in $\mathbb{C}_+$.*

Proof. In the notation for $A(t)$ of (2.6) and the associated evolution $\mathcal{U}$, we make explicit the m- dependence and write $A(t,m)$ and $\mathcal{U}(t,s;m)$. Let $f \in D_o$ and $m, m' \in \mathbb{C}_+$ with $m \neq m'$. The formula (2.17) shows that $U(t,s;m)$ and $\mathcal{U}(t,s;m)$ have the same continuity properties in m. Thus it suffices to consider the m-continuity of $\mathcal{U}$.

The function $\mathcal{U}(t,\tau;m')\mathcal{U}(\tau,s;m)f$ is strongly differentiable in τ with the result

$$\frac{\partial}{\partial \tau}[\mathcal{U}(t,\tau;m')\mathcal{U}(\tau,s;m)]f = \mathcal{U}(t,\tau;m')[A(\tau,m) - A(\tau,m')]\mathcal{U}(\tau,s;m)f. \tag{3.9}$$

The equality is a consequence of (2.5c-d). For fixed t, s, m and m' the right side of (3.9) is strongly continuous for $\tau \in [s,t]$ so we may integrate (3.9) (in the strong Riemann sense) to obtain

$$[\mathcal{U}(t,s;m) - \mathcal{U}(t,s;m')]f = \int_s^t d\tau\, \mathcal{U}(t,\tau;m')[A(\tau,m) - A(\tau,m')]\mathcal{U}(\tau,s;m)f.$$

In the evaluation of the left hand side, the initial condition (2.5b) has been utilized. The bound $\|\mathcal{U}(t,\tau;m')\| \leq 1$ gives

$$\|[\mathcal{U}(t,s;m) - \mathcal{U}(t,s;m')]f\| \leq \int_s^t d\tau\, \|[A(\tau,m) - A(\tau,m')]\mathcal{U}(\tau,s;m)f\|. \tag{3.10}$$

Set $\delta = (1 - m/m')$ and note that

$$A(\tau,m) - A(\tau,m') = \delta\{A(\tau,m) + i\hbar^{-1}Y(\tau) + (1 + \gamma\hbar^{-1})I\}.$$

Recall that $A(\tau,m)\mathcal{U}(\tau,s;m)f$ is strongly continuous for $\tau \in [s,t]$ and that $Y(\tau)$ is uniformly bounded. This is sufficient to verify that the right hand side of (3.10) vanishes as $\delta \to 0$. This proves $\mathcal{U}(t,s;m)$ is strongly continuous in m on D_o. Because $\mathcal{U}(t,s;m)$ is uniformly bounded with respect to m and the domain D_o is dense in $\mathcal{H}$ we can extend the above result to all $f \in \mathcal{H}$. This demonstrates that $\mathcal{U}(t,s;m)$ is strongly continuous in $\mathbb{C}_+$. □

One could also prove that $U(t,s;m)$ is a $\mathcal{B}(\mathcal{H})$ valued analytic function on $\mathbb{C}_>$, but our analysis will only require continuity on $\mathbb{C}_+$.

Now consider the Dyson series for potentials which are in class (A). Let $\mathcal{S}$ denote the Schwartz space of $C^\infty(\mathbf{R}^d)$ functions of rapid decrease. One begins by investigating the iterations of the map $U_o(t_2,t_1;m)\mathcal{V}(t_1,m)$, $(s \leq t_1 \leq t_2)$, acting on an element of $\mathcal{S}$. Suppose $f \in \mathcal{S}$ and let $\hat{f} \in \mathcal{S}$ be its Fourier transform. The operator $\mathcal{V}(t_1,m)$ on $\mathcal{S}$ has the differential form

$$\frac{i\hbar}{m}a(x,t_1)\cdot\nabla + \frac{i\hbar}{2m}(\nabla\cdot a)(x,t_1) + \frac{1}{2m}a(x,t_1)^2 + v(x,t_1).$$

Using the integral relation (3.8a) for $a(x,t_1)$ leads to

$$(3.11)\quad \frac{i\hbar}{m}a(x,t_1)\cdot(\nabla f)(x) = -\frac{1}{(2\pi)^{d/2}}\frac{\hbar}{m}\int d\alpha_o\,\hat{f}(\alpha_o)\int e^{ix\cdot(\alpha_o+\alpha_1)}\alpha_o\cdot d\gamma(\alpha_1,t_1);$$

(3.12)

$$\frac{i\hbar}{2m}(\nabla\cdot a)(x,t_1)f(x) = -\frac{1}{(2\pi)^{d/2}}\frac{\hbar}{2m}\int d\alpha_o\,\hat{f}(\alpha_o)\int e^{ix\cdot(\alpha_o+\alpha_1)}\alpha_1\cdot d\gamma(\alpha_1,t_1);$$

with a similar formula applying to $(2m)^{-1}a(x,t_1)^2 + v(x,t_1)$. In this way explicit integral formulae are obtained for $U_o(t_2,t_1;m)\mathcal{V}(t_1,m)f$. Note in particular that the effect of $U_o(t_2,t_1;m)$ acting on the function in either (3.11) or (3.12) is simply described by the replacement

$$\exp\{ix\cdot(\alpha_o+\alpha_1)\} \to \exp\{-\frac{i\hbar}{2m}(t_2-t_1)(\alpha_o+\alpha_1)^2\}\exp\{ix\cdot(\alpha_o+\alpha_1)\}$$

in the corresponding integrals above.

It is convenient to define the operator quantity which enters the nth Dyson iterate. For $m\in\mathbf{C}_+$, $(t,s)\in T_\Delta$ and $\mathbf{t}_n\in\Delta_n(t,s)$, let

$$I_n(t,s;\mathbf{t}_n;m) \equiv \left(\frac{-i}{\hbar}\right)^n U_o(t,t_n;m)\mathcal{V}(t_n,m)U_o(t_n,t_{n-1};m)\times\cdots\times\mathcal{V}(t_1,m)U_o(t_1,s;m).$$

Put $\Delta_n(T) = \{(t,s,\mathbf{t}_n) : (t,s)\in T_\Delta, \mathbf{t}_n\in\Delta_n(t,s)\}$ and set

$$g(x-y;t,s,m) \equiv \left(\frac{|m|}{2\pi\hbar(t-s)}\right)^{d/2}\exp\{-\frac{1}{2\hbar}\frac{\operatorname{Im} m}{t-s}|x-y|^2 + c_1|x-y| + c_2\}$$

where

$$c_1 = \frac{|m|}{k(t-s)},\qquad c_2 = \frac{|m|}{2k^2(t-s)} + \frac{|m|\mu_T}{k\gamma_T}.$$

The basic properties of the nth Dyson iterate are:

Proposition 2. *Assume a and v are in class (A) and $m\in\mathbf{C}_+$*

(1) *If $\tau_1,\tau_2\geq 0$, then the operator $\exp[-i\tau_2 H_o(m)/\hbar]\mathcal{V}(\tau_1,m)$ maps $\mathcal{S}\to\mathcal{S}$.*

(2) *Fix $n\geq 1$, $(t,s)\in T_\Delta$ and $f\in\mathcal{S}$. The $\mathcal{S}$-valued function $I(\cdot,\cdot;\cdot;m)f$ is a continuous mapping from $\Delta_n(T)\to\mathcal{H}$ with respect to the $\|\cdot\|_2$ norm.*

(3) *Fix $n\geq 1$, $(t,s)\in T_\Delta$ and $f\in\mathcal{S}$. The nth Dyson iterate $D_n(\cdot,\cdot;m)f : T_\Delta\to\mathcal{H}$ is defined by the n-dimensional strong Riemann integral*

$$(3.13)\qquad D_n(t,s;m)f = \int_{\Delta_n(t,s)} d\mathbf{t}_n\, I_n(t,s;\mathbf{t}_n;m)f.$$

(4) *For each $n\geq 1$, there exists a family (in $(t,s)\in T^o_\Delta$) of continuous functions $d_n(\cdot,t;\cdot,s;\cdot) : \mathbf{R}^d\times\mathbf{R}^d\times\mathbf{C}_+\to\mathbf{C}$, such that*

$$(3.14)\qquad (D_n(t,s;m)f)(x) = \int d_n(x,t;y,s;m)f(y)\,dy,\qquad f\in C_0^\infty(\mathbf{R}^d).$$

The function d_n has the pointwise estimate

$$(3.15)\qquad |d_n(x,t;y,s;m)| \leq \Theta^n g(x-y;t,s,m),$$

where $\Theta\equiv 2ek(t-s)|m|^{-1}\gamma_T$. If $\operatorname{Im} m > 0$, $D_n(\cdot,\cdot;m) : T_\Delta\to\mathcal{B}(\mathcal{H})$ and d_n is a Carleman kernel.

Proof. These results (and others) are established in Lemmas 4, 5 and 9 of Ref. 6. In fact, in Ref. 6 an explicit parametric integral formula is obtained for $d_n(x,t;y,s;m)$, cf. Eq.(6.20), which we do not quote because of its length. □

Having characterized the nth Dyson iterate let us turn our attention to the problem of summing these terms over $n = o \sim \infty$. The first observation is that the sum over n of d_n is always convergent if $\Theta < 1$.

LEMMA 3. *Assume a and v are in class (A). Let $m \in \mathbf{C}_+$ and $(t,s) \in T^o_\Delta$. If $\Theta < 1$, the sum over $d_n(x,t;y,s;m)$ is absolutely convergent and gives an x,y jointly continuous function*

$$K(x,t;y,s;m) = \sum_{n=0}^{\infty} d_n(x,t;y,s;m). \tag{3.16}$$

The function K has the pointwise estimate

$$|K(x,t;y,s;m)| \leq \frac{1}{1-\Theta} g(x-y;t,s,m). \tag{3.17}$$

Proof. The convergence and estimate are an immmediate consequence of the bound (3.15). □

The estimate (3.17) means that if $\operatorname{Im} m > 0$ then $K(x,t;y,s;m)$ may be interpreted as the integral kernel of a bounded operator acting on $L^2(\mathbf{R}^d)$. The integral,

$$h(x) = \int K(x,t;y,s;m) f(y)\, dy, \tag{3.18}$$

is well defined for f in a large class of functions. Let $f \in L^2(\mathbf{R}^d)$. The estimate

$$|h(x)| \leq \frac{1}{1-\Theta} \int g(x-y;t,s,m)|f(y)|\, dy \tag{3.19}$$

is in the form of a convolution. Because $g(\cdot;t,s,m) \in L^1(\mathbf{R}^d)$ we can apply Young's inequality (Ref. 16, p.29) to show that $h \in L^2(\mathbf{R}^d)$. Thus (3.18) defines a family of bounded operators, $K(t,s;m) : \mathcal{H} \to \mathcal{H}$ for each $(t,s,m) \in T^o_\Delta \times \mathbf{C}_>$. Specifically

$$h = K(t,s;m) f \tag{3.20}$$

and h has the Young inequality bound

$$\|h\|_2 \leq \frac{1}{1-\Theta} \|g(\cdot;t,s,m)\|_1 \|f\|_2. \tag{3.21}$$

We are now ready for the following proposition.

PROPOSITION 3. *Assume a and v are in class (A). Let $(t,s,m) \in T^o_\Delta \times \mathbf{C}_>$ and assume $\Theta < 1$.*

(1) $\|K(t,s;m)\| \leq (1-\Theta)^{-1}\|g(\cdot;t,s,m)\|_1$.

(2) *The operator valued series defined by the sum over $n = 0 \sim \infty$ of $D_n(t,s;m)$ is convergent in the operator norm topology to $K(t,s;m)$, i.e.,*

$$K(t,s;m) = \sum_{n=0}^{\infty} D_n(t,s;m). \tag{3.22}$$

(3) *Let $U(t,s;m)$ be the Schrödinger evolution given in Theorem 4. Then*

$$K(t,s;m) = U(t,s;m). \tag{3.23}$$

Proof. (1) is a direct consequence of (3.21). For (2) an application of the Young convolution inequality argument gives

$$\|D_n(t,s;m)\| \leq \Theta^n \|g(\cdot;t,s,m)\|_1. \tag{3.24}$$

This estimate implies statement (2).

Consider (3). The details of the proof are found in Ref. 6. The basic idea is the following. For functions $f \in D_o$, whose Fourier transforms have compact support, the $D_n(t,s;m)f$'s can be shown to satisfy a recurrence relation. Summing this relation over n implies that $K(t,s;m)f$ is a solution of the evolution problem in $\mathcal{H}$. The uniqueness of the solutions to the abstract evolution problem then require that

$$K(t,s;m)f = U(t,s;m)f, \qquad f \in \mathcal{H}, \quad \operatorname{Im} m > 0. \tag{3.25}$$

This establishes (3). □

The results of Propositions 2 and 3 show that $K(\cdot,t;\cdot,s;m)$ is the propagator for the Schrödinger evolution $U(t,s;m)$ for complex mass. Namely for *a.a.* x

$$[U(t,s;m)f](x) = \int K(x,t;y,s;m)f(y)\,dy. \tag{3.26}$$

Provided it is possible to continue the identity (3.26) down to the real mass axis, K will be the propagator for the evolution in the physical quantum problem. Continuity in mass of $U(t,s;\cdot)$ is controlled by Proposition 1. For (x,y,m) in a compact subset of $\mathbf{R}^d \times \mathbf{R}^d \times \mathbf{C}_+$, $K(x,t;y,s;m)$ is jointly continuous. This is enough to give us

THEOREM 5. *Assume a and v are in class (A). Let M_o be a compact subset of $\mathbf{C}_+$ whose smallest absolute mass value is m_o. If $(t,s) \in T^o_\Delta$ and $t-s < (2ek\gamma_T)^{-1}m_o$*

then for $m \in M_o$ $K(x,t;y,s;m)$ is the propagator (in the sense of Definition 3) of the Schrödinger evolution operator $U(t,s;m)$.

4. Singular perturbation expansions. In this section we give an account of the gauge invariant large mass expansion of the propagator K.

The values of the electric and magnetic fields as well as expectation values of observables in quantum mechanics are independent of the choice of gauge. Thus it is desirable that any approximations for K preserve this $U(1)$-gauge group invariance. Suppose $\lambda : \mathbf{R}^d \times [0,T] \to \mathbf{R}$ is a $C^1(\mathbf{R}^d \times [0,T])$ gauge potential. Then the transformed potentials (a, ϕ) are related to their original values by

$$a(x,t;\lambda) = a(x,t) + \nabla\lambda(x,t), \tag{4.1a}$$

$$v(x,t;\lambda) = v(x,t) - \partial\lambda(x,t). \tag{4.1b}$$

The propagator K in Sec. 3 was constructed from potentials in class (A) that were the Fourier transforms of bounded measures. It is reasonable to inquire if the potential class (A) is stable under gauge transformations. Specifically, is there a natural class of gauge transformations that leave class (A) invariant? An affirmative answer is provided by the following construction.

DEFINITION 5. *Gauge class ($\mathcal{G}$)*: Let $k < \infty$. The gauge potential λ is said to be in the class $\mathcal{G}(k)$ if λ is the Fourier transform,

$$\lambda(x,t) = \int e^{i\alpha\cdot x}\, d\zeta(t),$$

of a time dependent measure, $\zeta(t)$, satisfying

(1) $\zeta(t) \in \mathcal{M}^*(S_{k/2}, \mathrm{C})$, $\quad t \in [0,T]$,

(2) $\zeta(\cdot)$ is twice continuously differentiable on $[0,T]$.

Note that $\lambda(x,t)$ defines a t-dependent family of bounded operators on $\mathcal{H}$. Namely let $\Lambda(t) \in \mathcal{B}(\mathcal{H})$ be specified, for each $t \in [0,T]$, by the identity

$$[\Lambda(t)f](x) = \lambda(x,t)f(x), \qquad f \in \mathcal{H}.$$

The operator norm of $\Lambda(t)$ obeys $\|\Lambda(t)\| \leq \|\zeta(t)\|$. Similarily, the strong t-derivative of $\Lambda(t)$ is also a uniformly bounded operator on $[0,T]$ and is given by

$$[\dot{\Lambda}(t)f](x) = \dot{\lambda}(x,t)f(x), \qquad f \in \mathcal{H}.$$

The Abelian $U(1)$ gauge group is conventionally taken to be the family of unitary operators, $\{\exp[q(i\hbar)^{-1}\Lambda(t)] : q \in \mathbf{R}\}$. However in (3.5) and thereafter the charge coupling constant q was incorporated into the definition of a and ϕ. To be notationally consistent, here, it is necessary to keep $q = 1$ and write the unitary gauge operator as $\exp[(i\hbar)^{-1}\Lambda(t)]$.

It is easy to see that the measure images of the two gauge transformation equations (4.1a-b) are

$$\gamma(t;\Lambda) = \gamma(t) + \nabla\zeta(t), \tag{4.2a}$$

$$\nu(t;\Lambda) = \nu(t) - \dot{\zeta}(t), \tag{4.2b}$$

for all $t \in [0, T]$. Here $\nabla\zeta \in \mathcal{M}^*(S_{k/2}, \mathbf{C})$ is defined by

$$\nabla\zeta(t)(e) = \int_e i\alpha\, d\zeta(t), \qquad e \in B,$$

where B denotes the Borel subsets of $\mathbf{R}^d$. The definition of $\mathcal{G}(k)$ ensures that the right-sides of (4.2a-b) are, respectively, in $\mathcal{M}^*(S_{k/2}, \mathbf{C}^d)$ and $\mathcal{M}^*(S_k, \mathbf{C})$, and satisfy the t-differentiability conditions required in (A).

Time evolution, whether described in terms of $U(t,s;m)$ or its kernel form K, has a well known [17] gauge dependence. Let $H(t,m;\Lambda)$ be the Hamiltonian operator determined by (1.1) with $a(x,t;\Lambda)$ and $v(x,t;\Lambda)$ substituting for $a(x,t)$ and $v(x,t)$. Further, let $U(t,s;m;\Lambda)$ be the family of complex mass Schrödinger evolution operators (described in Theorem 4) generated by $\{H(t,m;\Lambda) : t \in [0,T]\}$. It follows without difficulty from (2.1c) that

$$U(t,s;m;\Lambda) = e^{i\Lambda(t)/\hbar} U(t,s;m) e^{-i\Lambda(s)/\hbar}. \tag{4.3a}$$

This is the operator-valued form of the $U(1)$ gauge dependence of the evolution process. In an obvious notation its kernel analog reads

$$K(x,t;y,s;m;\Lambda) = e^{i\lambda(x,t)/\hbar} K(x,t;y,s;m) e^{-i\lambda(y,s)/\hbar}. \tag{4.3b}$$

Let us begin the investigation of the large mass asymptotic expansion of K. It is convenient to introduce the inverse mass variable $u = m^{-1}$ as well as the notation $Q \equiv (x,t;y,s)$ to characterize the pair of space-time points that enter as arguments of the propagator. Since u is the natural small parameter in the asymptotic expansions we obtain, we will replace m with u as the argument of K_o, K, d_n, etc. First we note that the free propagator K_o has an essential singularity at $u = 0$. Such singularities are common in singular perturbation problems. Nevertheless we attempt an expansion of the exact K about the point $u = 0$. This is possible if we know precisely the form of the singularity at $u = 0$. A good clue as to what is going on comes from examining the formulae defining the kernel of the nth Dyson iterate, $d_n(Q;u)$. These formulae show that for every $n \geq 1$ they factor into a product of the type

$$d_n(Q;u) = K_o(t-s; x-y; u)\tilde{d}_n(Q;u) \tag{4.4}$$

where $\tilde{d}_n(Q;u)$ is given by an explicit parametric integral [Ref. 11, Eqs.(3.4) and (3.5b)] that is non-singular in the neighborhood of $u = 0$. In fact for each $Q \in T_\Delta \times \mathbf{R}^d \times \mathbf{R}^d$, $\tilde{d}_n(Q;u)$ is an entire function of u.

Recall that the sum over n of d_n is convergent and constructs the kernel $K(Q;u)$. The identity (4.4) suggests that

$$K(Q;u) = K_o(t-s;x-y;u)F(Q;u) \tag{4.5}$$

where F is the sum over n of $\tilde{d}_n$. If F is well behaved (smooth and bounded) in the neighborhood of $u=0$ then (4.5) gives us a description of the singular behavior of K near the point $u=0$. The convergence properties of the series for F is similar (but simpler) than the series involving d_n. Recall that $\mathbf{C}_<$ ($\mathbf{C}_\leq$) represents the lower half plane $\operatorname{Im} u < 0$ ($\operatorname{Im} u \leq 0$). Furthermore denote the open semi-disk of radius u_o by $\mathcal{D}_<(u_o) = \{u \in \mathbf{C}_< : |u| < u_o\}$ and its closure by $\mathcal{D}_\leq(u_o)$. The behavior of F is given by

PROPOSITION 4. *Assume the potentials a and v are in class (A). Let $u_o < (2ekT\gamma_T)^{-1}$.*

(1) *For each $(Q;u) \in T_\Delta \times \mathbf{R}^d \times \mathbf{R}^d \times \mathcal{D}_\leq(u_o)$ the sum over n of $\tilde{d}_n(Q;u)$ is absolutely convergent and provides a pointwise definition of the function F, i.e.,*

$$F(Q;u) = \sum_{n=0}^{\infty} \tilde{d}_n(Q;u). \tag{4.6}$$

(2) *The function F has partial derivatives to arbitrary order in x, y which are (jointly) continuous on the domain $T_\Delta \times \mathbf{R}^d \times \mathbf{R}^d \times \mathcal{D}_\leq(u_o)$.*

(3) *F has first order partial derivatives with respect to t and s which are continuous functions on $T_\Delta^o \times \mathbf{R}^d \times \mathbf{R}^d \times \mathcal{D}_\leq(u_o)$.*

(4) *For each $Q \in T_\Delta \times \mathbf{R}^d \times \mathbf{R}^d$, $F(Q;\cdot)$ is holomorphic in $\mathcal{D}_<(u_o)$ and continuous in $\mathcal{D}_\leq(u_o)$.*

(5) *Let K be the propagator defined in Theorem 5. K admits the factorization (4.5) on the domain $T_\Delta^o \times \mathbf{R}^d \times \mathbf{R}^d \times \mathcal{D}_\leq(u_o)$.*

The proof of Proposition 4 may be found in Ref. 11. It is worth noting how one establishes (2) and (3). The explicit expressions for $\tilde{d}_n$ show that it has the partial derivatives with respect to x and y, to arbitrary order, which are continuous on the domain $T_\Delta \times \mathbf{R}^d \times \mathbf{R}^d \times \mathbf{C}$. Furthermore if $\nabla^\gamma_{x,y}$ is the partial derivative with respect to the variable set $(x_1, \ldots, x_d, y_1, \ldots, y_d)$ and specified by the multi-index $\gamma = (\gamma_1, \ldots, \gamma_{2d})$, then one can find bounds on $\nabla^\gamma_{x,y}\tilde{d}_n(Q;u)$ which are sufficient to prove

$$\nabla^\gamma_{x,y} \sum_{n=0}^{\infty} \tilde{d}_n(Q;u) = \sum_{n=0}^{\infty} \nabla^\gamma_{x,y}\tilde{d}_n(Q;u), \tag{4.7}$$

i.e., the series for F may be differentiated term-by-term. Each term in the sum on the right is uniformly continuous in the compact subsets of $T_\Delta \times \mathbf{R}^d \times \mathbf{R}^d \times \mathcal{D}_\leq(u_o)$. A similar analysis applies to the statement

$$\frac{\partial}{\partial t} \sum_{n=0}^{\infty} \tilde{d}_n(Q;u) = \sum_{n=0}^{\infty} \frac{\partial}{\partial t}\tilde{d}_n(Q;u). \tag{4.8}$$

The continuously differentiable properties of F (and thereby K) in x and t let us establish that the propagator K is a $C^1(\mathbf{R}^d \times (s,T))$ solution of the time dependent Schrödinger equation (1.6a), interpreted as a classical partial differential equation in the open time region T^o_Δ. If the boundary points $t = s$ are added to T^o_Δ, the propagator is the fundamental solution of (1.6a).

PROPOSITION 5. *Assume potentials a and v are in class (A). Let $u_o < (2ekT\gamma_T)^{-1}$ and $u \in \mathcal{D}_{\leq}(u_o)\backslash\{0\}$. The propagator K of Theorem 5 satisfies the Schrödinger partial differential equation*

$$i\hbar\frac{\partial}{\partial t}K(x,t;y,s;u) = H(x,-i\hbar\nabla,t,u)K(x,t;y,s;u) \tag{4.9a}$$

identically for all $Q \in T^o_\Delta \times \mathbf{R}^d \times \mathbf{R}^d$. Furthermore, for all $\varphi \in C^d_o(\mathbf{R}^d)$

$$\lim_{t\to s+}\int K(x,t;y,s;u)\varphi(y)\,dy = \varphi(x), \qquad x \in \mathbf{R}^d. \tag{4.9b}$$

Proof. (sketch) The idea of the proof is to use the fact that $U(t,s;u)f$ is a solution (for $f \in D_o$) of the abstract evolution equation to show that $K(Q;u)$ is a solution of the classical partial differential equation.

If $\varphi \in C^d_o(\mathbf{R}^d)$ then $\varphi \in D_o \cap L^2_o(\mathbf{R}^d)$. Let the function $\Phi : \mathbf{R}^d \times (s,T) \to \mathbf{C}$ be the integral

$$\Phi(x,t) = \int K(x,t;y,s;u)\varphi(y)\,dy.$$

From Theorems 4 and 5 one has that $\Phi(\cdot,t) \in D_o$ and

$$U(t,s;u)\varphi = \Phi(\cdot,t), \qquad t \in [s,T].$$

The boundedness and smoothness properties of F allow one to show

$$[H(t,u)\Phi(\cdot,t)](x) = \int H(x,-i\hbar\nabla,t,u)K(x,t;y,s;u)\varphi(y)\,dy, \tag{4.10a}$$

$$[i\hbar\frac{d}{dt}\Phi(\cdot,t)](x) = \int i\hbar\frac{\partial}{\partial t}K(x,t;y,s;u)\varphi(y)\,dy. \tag{4.10b}$$

Since $\Phi(\cdot,t) \in D_o$, the left side of (4.10a) is well defined. In (4.10b) the time derivative of $\Phi(\cdot,t)$ is the strong derivative of a D_o-valued function. The $\mathcal{H}$-valued equation of motion (2.1c) requires that the left hand side of (4.10a) and (4.10b) are equal almost everywhere, so

$$\int\left\{i\hbar\frac{\partial}{\partial t}K(x,t;y,s;u) - H(x,-i\hbar\nabla,t,u)K(x,t;y,s;u)\right\}\varphi(y)\,dy = 0. \tag{4.11}$$

For fixed x, t, s and u the function in the curly bracket is a continuous function of y and thus it is in $L^2_{\mathcal{K}}$ for any compact set $\mathcal{K} \subset \mathbf{R}^d$. Interpreting (4.11) as an inner product on $L^2_{\mathcal{K}}$ shows that (4.11) holds almost everywhere in $\mathbf{R}^d_x \times \mathbf{R}^d_y$. The x,y continuity of all the functions in (4.11) shows it holds everywhere in $T^o_\Delta \times \mathbf{R}^d \times \mathbf{R}^d$.

Statement (4.9b) is obtained by using the multi-dimensional stationary phase formula [18]. For suitably smooth $h : \mathbf{R}^d \times \mathbf{C}_{\leq} \to \mathbf{C}$

$$\lim_{\lambda \to 0} \frac{1}{(\pi i \lambda)^{d/2}} \int_{\mathbf{R}^d} e^{iy^2/\lambda} h(y, \lambda)\, dy = h(0,0), \qquad \lambda \in \mathbf{C}_{\leq}. \ \square$$

Let us resume our study of the small u asymptotic expansion of K. Since $F(Q;u)$ is a smooth function of $u \in \mathcal{D}_{\leq}(u_o)$, we can expect to find that F admits a small u asymptotic expansion. Because each $\tilde{d}_n$ is an entire function of u, it can be expanded as an $(M-1)^{th}$ order polynomial in u with a remainder that is $O(|u|^M)$. Summing over all these expansions of $\tilde{d}_n$ gives one the following which is established in Ref. 11.

THEOREM 6. *Let $u_o < (2ekT\gamma_T)^{-1}$. For all integers $M \geq 1$ and each $Q \in T_\Delta \times \mathbf{R}^d \times \mathbf{R}^d$, $F(Q;u)$ has the small u asymptotic expansion in $\mathcal{D}_{\leq}(u_o)$*

$$F(Q;u) = \sum_{j=0}^{M-1} u^j P_j(Q) + u^M E_M(Q;u). \tag{4.12}$$

The complex valued, u-independent, coefficient functions $P_j : T_\Delta \times \mathbf{R}^d \times \mathbf{R}^d \to \mathbf{C}$, $j = 0 \sim M-1$, and the error function $E_M : T_\Delta \times \mathbf{R}^d \times \mathbf{R}^d \times \mathcal{D}_{\leq}(u_o) \to \mathbf{C}$ possess continuous partial derivatives up to arbitrary order in (x,y), throughout their domains of definition. On the more restricted domains $T_\Delta^o \times \mathbf{R}^d \times \mathbf{R}^d$ and $T_\Delta^o \times \mathbf{R}^d \times \mathbf{R}^d \times \mathcal{D}_{\leq}(u_o)$, P_j and E_M have continuous first order partial derivatives with respect to t and s.

Furthermore for all compact subsets Ω of $\mathbf{R}^d \times \mathbf{R}^d$, P_j and E_M obey the estimates

$$|P_j(Q)| \leq \left(\frac{t-s}{u_o T}\right)^j C_j(\Omega), \qquad Q \in T_\Delta \times \Omega, \tag{4.13}$$

$$|E_M(Q;u)| \leq \left(\frac{t-s}{u_o T}\right)^M C_M(\Omega), \qquad (Q,u) \in T_\Delta \times \Omega \times \mathcal{D}_{\leq}(u_o) \tag{4.14}$$

for some finite constants $C_j(\Omega)$, $j = 0 \sim M$.

The expansion (4.12) may be differentiated to arbitrary order with respect to x and y, and to first order in t and s. For all $|\gamma| \geq 1$, the space derivatives $\nabla_{x,y}^\gamma E_M$ are $O((t-s)^M)$, uniformly with respect to (x,y,u) taking values in compact subsets of $\mathbf{R}^d \times \mathbf{R}^d \times \mathcal{D}_{\leq}(u_o)$. Moreover the time derivatives $\partial_t E_M$ and $\partial_s E_M$ are $O((t-s)^{M-1})$ for (Q,u) taking values in compact subsets of $T_\Delta^o \times \mathbf{R}^d \times \mathbf{R}^d \times \mathcal{D}_{\leq}(u_o)$.

Theorem 6 achieves an asymptotic expansion of $F(Q;u)$ for small u and via the identity (4.5) we have an expansion of the propagator K. Explicit forms of the estimating constants C_j are found in Ref. 11. However expansion (4.12) is not in a gauge invariant form. The modification of the propagator under a gauge transformation generated by $\lambda \in \mathcal{G}(k)$ is given in (4.3). Since the free propagator, K_o, is gauge independent, representation (4.5) tells us that $F(Q;u)$ is a gauge dependent function. A second difficulty with Theorem 6 resides in our knowledge of the coefficients $P_j(Q)$. The constructive method, based on explicit sums like that of (4.6), suffices to establish the existence of u as a valid small expansion parameter for $F(Q;u)$. But it is so complicated in detail that one cannot generally obtain explicit expressions for the coefficient functions $P_j(Q)$.

Both of these drawbacks can be resolved. First let us calculate the coefficient $P_o(Q)$. This is the exception to the rule above on the difficulty of constructing an explicit functional form for $P_j(Q)$. The asymptotic expansion (4.12) remains applicable if we set $u = 0$, so

$$P_o(Q) = F(Q;0) = \sum_{n=0}^{\infty} \tilde{d}_n(Q;0). \tag{4.15}$$

In the case where $u = 0$ the reduced (*i.e.*, d_n with the factor K_o removed) Dyson kernels, $\tilde{d}_n$, greatly simplify and take the form

$$\tilde{d}_n(Q;0) = \frac{1}{n!}[(i\hbar)^{-1}J(Q)]^n \tag{4.16}$$

for all $n \geq 0$, where the function $J(Q)$ is

$$J(Q) = \int_0^1 d\xi \, \{(t-s)v\big(w(\xi;Q)\big) - (x-y)\cdot a\big(w(\xi;Q)\big)\}. \tag{4.17}$$

In this expression w is the linear path in $\mathbf{R}^d \times [0,T]$ connecting the intial space-time point (y,s) to the final point (x,t);

$$w(\xi) \equiv w(\xi;Q) = \big(y + \xi(x-y), s + \xi(t-s)\big), \qquad \xi \in [0,1]. \tag{4.18}$$

Clearly J is real valued and independent of both m and $\hbar$. Putting together (4.15), (4.16), and (4.17) gives

$$P_o(Q) = e^{(i\hbar)^{-1}J(Q)}. \tag{4.19}$$

In the calculation above note that if we try to set $u \equiv m^{-1} = 0$ in $d_n(Q;u)$ we obtain an ill defined result since each $d_n(Q;u)$ has an essential singularity at $u = 0$.

These considerations suggest we introduce the representation

$$F(Q;u) = T(Q;u)e^{(i\hbar)^{-1}J(Q)}. \tag{4.20}$$

The function $\exp[(i\hbar)^{-1}J(Q)]$ is unimodular and so (4.20) may be interpreted as a definition of $T(Q;u)$ in terms of $F(Q;u)$ and $J(Q)$. The function $J(Q)$ is u-independent so the small u expansion of $F(Q;u)$ implies the existence of a small u expansion $T(Q;u)$. The decomposition (4.20) is of interest because $T(Q;u)$ will turn out to be gauge invariant. To see this note the following. Let $\lambda \in \mathcal{G}(k)$ generate a gauge transformation Λ in which $a(x,t)$ and $v(x,t)$ are replaced by their transformed potentials $a(x,t;\Lambda)$ and $v(x,t;\Lambda)$. A simple calculation shows

$$J_\Lambda(Q) = J(Q) - \lambda(x,t) + \lambda(y,s). \tag{4.21}$$

Upon writing the gauge transformation identity (4.3b) for propagators in terms of J, T and J_Λ, T_Λ, it is evident that (4.21) implies

$$T_\Lambda(Q;u) = T(Q;u), \qquad \lambda \in \mathcal{G}(k). \tag{4.22}$$

i.e., that $T(Q;u)$ is the same function for all gauges in $\mathcal{G}(k)$. Clearly $J(Q)$ carries all the gauge dependence of the propagator K. The functions T and F both suffice to completely determine the propagator K and both have well defined small u expansions. However the large mass asymptotic expansion of T is of greater interest for physical applications since T and the associated expansion coefficients are gauge invariant. The next task is to recast the small u expansion of $F(Q;u)$ into an expansion for $T(Q;u)$. At the same time an efficient procedure for determining the coefficient functions of the expansion is introduced. In order to prepare for the next theorem, recall the following formulas from electromagnetism. The vector $f : \mathbf{R}^d \times [0,T] \to \mathbf{R}^d$ will represent the electric force on the system plus the contribution from V,

$$f_i(x,t) = -(\nabla^i v)(x,t) - \partial a_i(x,t). \tag{4.23a}$$

The magnetic part of the electromagnetic field tensor is given by

$$A_{ij}(x,t) = (\nabla^i a_j)(x,t) - (\nabla^j a_i)(x,t). \tag{4.23b}$$

In (4.23) the indices i and j denote the Cartesian components of vectors and rank two tensors in $\mathbf{R}^d$. The symbol ∇^i is the partial derivative with respect to the ith component of the vector argument of $x \in \mathbf{R}^d$. On the other hand the notation ∇_1 (and Δ_1) describes the gradient (Laplacian) with respect to the first vector argument of a function - e.g. $(\nabla_1 J)(Q)$ is the x-gradient of J. Similarly ∂ is a partial derivative with respect to a scalar argument; in (4.23a) it is the time argument.

The functions f_i and A_{ij} are well known to be gauge invariant quantities. We define the gauge invariant function $\bar{f} : T_\Delta \times \mathbf{R}^d \times \mathbf{R}^d \to \mathbf{R}^d$, in terms of (4.23) and the path $w(\xi)$ of (4.18), by

$$\bar{f}_i(Q) = \int_0^1 d\xi\, \xi \left[(t-s) f_i(w(\xi)) + \sum_{j=1}^{d} (x-y)_j A_{ij}(w(\xi))\right]. \tag{4.24}$$

Note how similar in form (4.24) is to integral (4.17) defining J. The function $(t-s)^{-1}\bar{f}$ has the interpretation of a ξ-weighted average Lorentz force experienced by a system of classical particles moving with constant velocity $(x-y)/(t-s)$ from y to x. Because f_i and A_{ij} are gauge invariant so is $\bar{f}_i$. It is helpful to recall a basic identity linking $\bar{f}$ to J and a, namely

$$\bar{f}(Q) = -(\nabla_1 J)(Q) - a(x,t). \tag{4.25}$$

Equation (4.25) is verified by using definition (4.17) for J and then forming the partial derivative $(\nabla_1^i J)(Q)$. An integration by parts with respect to ξ leads to (4.25). For more details see the discussion prior to Eq. (4.11) in Ref. 19. Given that a and v are in class (A), it is evident that $\bar{f}_i$ is differentiable to arbitrary order in x, y and differentiable to first order in t and s on the domain $T_\Delta \times \mathbf{R}^d \times \mathbf{R}^d$.

The final statement of the large mass asymptotic expansion takes the form

THEOREM 7. *Assume the potentials a and v are in class (A) and that $u_o < (2ekT\gamma_T)^{-1}$. For all integers $M \geq 1$ the gauge invariant function $T : T_\Delta \times \mathbf{R}^d \times \mathbf{R}^d \times \mathcal{D}_\leq(u_o) \to \mathbf{C}$ has the small u asymptotic expansion in $\mathcal{D}_\leq(u_o)$*

$$T(Q;u) = \sum_{j=0}^{M-1} u^j T_j(Q) + u^M \tilde{E}_M(Q;u). \tag{4.26}$$

The coefficients T_j are defined in terms of P_j by

$$T_j(Q) = e^{-(i\hbar)^{-1}J(Q)} P_j(Q), \qquad j = 0, \dots, M-1 \tag{4.27}$$

and the error function $\tilde{E}_M$ in terms of E_M by

$$\tilde{E}_M(Q;u) = e^{-(i\hbar)^{-1}J(Q)} E_M(Q;u). \tag{4.28}$$

The functions T_j and $\tilde{E}_M$ are the same functions for all gauges $\lambda \in \mathcal{G}(k)$ and have the same differentiability properties as P_j and E_M. The error term satisfies $|\tilde{E}_M| = |E_M|$ and $\tilde{E}_M(Q;u)$ is bounded in $\mathcal{D}_\leq(u_o)$ by the u-independent estimate (4.14).

For all $Q \in T_\Delta \times \mathbf{R}^d \times \mathbf{R}^d$, $T_o(Q) = 1$ and the higher order coefficient functions $T_j(Q)$ are uniquely determined by the manifestly gauge invariant transport recurrence relation,

$$\begin{aligned} T_j(Q) = \frac{(t-s)}{2} \int_0^1 d\xi \, \{ & i\hbar \Delta_1 T_{j-1}(w(\xi;Q);y,s) \\ & - 2\bar{f}(w(\xi;Q);y,s) \cdot \nabla_1 T_{j-1}(w(\xi;Q);y,s) \\ + [(i\hbar)^{-1} \bar{f}(w(\xi;Q);y,s)^2 & - \nabla_1 \cdot \bar{f}(w(\xi;Q);y,s)] T_{j-1}(w(\xi;Q);y,s) \}. \end{aligned} \tag{4.29}$$

Proof. Expansion (4.26) is obtained from (4.12) by multiplication with the function $\exp\{-(i\hbar)^{-1}J(Q)\}$. Since a and v are in class (A) and J is defined by (4.17) and (4.18) it follows for each $(t,s) \in T_\Delta$ that $\exp\{-(i\hbar)^{-1}J(Q)\}$ is a $C^\infty(\mathbf{R}^d \times \mathbf{R}^d)$ function with partial derivatives in x and y that are (jointly) continuous functions on $T_\Delta \times \mathbf{R}^d \times \mathbf{R}^d$. Furthermore, the phase function J has first order derivatives with respect to t and s that are continuous on the domain $T_\Delta \times \mathbf{R}^d \times \mathbf{R}^d$. Thus T, T_j and $\tilde{E}_M$ have the same differentiability properties as F, P_j and E_M cited in Proposition 4 and Theorem 6. The coefficients T_j and error term $\tilde{E}_M$ are gauge invariant because both T and the expansion parameter u are gauge invariant.

Consider the recurrence relation (4.29). For $(Q,u) \in T_\Delta^o \times \mathbf{R}^d \times \mathbf{R}^d \times \mathcal{D}_\leq(u_o)$ the representation

$$K(Q;u) = K_o(t-s;x-y;u) e^{(i\hbar)^{-1}J(Q)} T(Q;u) \tag{4.30}$$

follows from Proposition 4(5) and (4.20). In addition, Proposition 2 states that $K(Q;u)$ is the pointwise solution of (4.9a). Inserting (4.30) into (4.9a) determines the partial differential equation satisfied by T;

(4.31)

$$\begin{aligned} [\partial_1 + (i\hbar)^{-1}(\partial_1 J)]T = & \frac{u}{2}\Big\{ (i\hbar)\Delta_1 T - 2\bar{f} \cdot (\nabla_1 T) + [(i\hbar)^{-1}\bar{f}^2 - (\nabla \cdot \bar{f})]T \Big\} \\ & - \Big\{ \frac{(x-y)}{(t-s)} \cdot \nabla_1 T - (i\hbar)^{-1} [\frac{(x-y)}{(t-s)} \cdot \bar{f} + v] T \Big\}. \end{aligned}$$

In obtaining the right side of (4.31) we have employed the identity (4.25). In (4.31) the omitted arguments of the various functions are $T = T(Q;u)$, $v = v(x,t)$, $\bar{f} = \bar{f}(Q)$ and $J = J(Q)$.

The next step is to put expansion (4.26) into (4.31) and then to equate the common coefficients of the differing powers of the independent functions u^j, $j = 0 \sim M-1$. In carrying out this substitution we rely on the facts that $T_j(Q)$ and $\tilde{E}_M(Q;u)$ have x derivatives of order two and t derivatives of order one which are continuous throughout the domain $T^o_\Delta \times \mathbf{R}^d \times \mathbf{R}^d \times \mathcal{D}_\leq(u_o)$. Furthermore, we use the fact that the x and t derivatives of $\tilde{E}_M(Q;u)$ are uniformly bounded in $\mathcal{D}_\leq(u_o)$ for each $Q \in T^o_\Delta \times \mathbf{R}^d \times \mathbf{R}^d$. The coefficient proportional to $(u)^0$ cancels identically by virtue of the definition of J. This mass independent term is exceptional in that it is the only part of (4.31) that is gauge dependent.

The coefficient function $T_j(Q)$ is related to $T_{j-1}(Q)$ by the family [in the parameters $(y,s) \in \mathbf{R}^d \times (0,T)$] of linear partial differential equations

$$[(t-s)\partial_1 + (x-y)\cdot\nabla_1]T_j(Q) = (t-s)g_{j-1}(Q) \tag{4.32a}$$

where

$$g_{j-1}(Q) = \frac{1}{2}\Big\{i\hbar\Delta_1 - 2\bar{f}\cdot\nabla_1 - [\nabla_1\cdot\bar{f} - (i\hbar)^{-1}\bar{f}^2]\Big\}T_{j-1}(Q). \tag{4.32b}$$

The continuity of $\nabla_1^\gamma T_{j-1}$, for $0 \leq |\gamma| \leq 3$, on $T_\Delta \times \mathbf{R}^d \times \mathbf{R}^d$ and the continuity of $\nabla_1^\gamma \bar{f}$, for $0 \leq |\gamma| \leq 1$, on $\mathbf{R}^d \times [0,T]$, means that g_{j-1} is a continuously differentiable function of $(x,t) \in \mathbf{R}^d \times (s,T)$.

If we make the substitution $(x,t) \to w(\xi;Q)$ in the partial differential equation (4.32), it is transformed into the ordinary differential equation

$$\frac{d}{d\xi}T_j\big(w(\xi;Q);y,s\big) = (t-s)g_{j-1}\big(w(\xi;Q);y,s\big). \tag{4.33}$$

Integrate (4.33) with respect to $\xi \in [0,1]$. Noting that $w(1;Q) = (x,t)$ and $w(0;Q) = (y,s)$ and using inequality (4.13), which implies that $T_j(y,s;y,s) = 0$ if $j \geq 1$, we establish recurrence relation (4.29). Expressions for T_1 and T_2 are found in Ref. 11. □

5. Conclusions. We have demonstrated that for each space time coordinate $Q = (x,t;y,s)$, the propagator K admits the multiple factorization (4.30). The $u = m^{-1}$ singular behavior of K as well as its gauge dependence are completely characterized by representation (4.30). Within the closed semi-disk $\mathcal{D}_\leq(u_o)$ and for each value of Q with $t > s$, the propagator $K(Q;u)$ has only one singular point — namely an essential singularity at $u = 0$. This singularity is entirely carried by the free propagator $K_o(t-s;x-y;u)$. The mass-independent unimodular phase factor $\exp\{(i\hbar)^{-1}J(Q)\}$ carries all the gauge dependence of $K(Q;u)$. As a result, the function $T(Q;u)$ is gauge independent and sufficiently smooth in $\mathcal{D}_\leq(u_o)$ to be described by the asymptotic expansion of Theorem 7. Since u is the multiplier of the highest order differential operator (the Laplacian) in the differential equation (4.9a) obeyed by $K(Q;u)$, the asymptotic expansion (4.26) together with (4.30) provide a detailed characterization of the singular perturbation behavior as $u \to 0$ of the fundamental solution of the time-dependent Schrödinger equation (4.9a).

The asymptotic expansion (4.26) of $T(Q;u)$ has a number of attractive features. It is a nonperturbative approximation for the propagator that represents K in terms of the first and higher order derivatives of the fields a, ϕ and V. The expansion is valid for all $Q \in T_\Delta \times \mathbf{R}^d \times \mathbf{R}^d$ to an arbitrary order M. The expansion has uniquely determined gauge invariant coefficients and error term, both of which are uniformly bounded for Q taking values in compact subsets of $T_\Delta \times \mathbf{R}^d \times \mathbf{R}^d$. The expansion is robust (stable) in the sense that identity (4.26) may be differentiated to first order in t or s and to arbitrary order in x and y. The resultant identities are also valid asymptotic expansions with an error term whose order estimate remains $O(|u|^M)$. This stability feature of the expansion means that one may determine from (4.26) the time evolving expectation values of all operators that are representable as sums of partial derivatives with locally integrable coefficients. This class of operators includes most observables (self-adjoint operators) of interest in quantum mechanics.

Consider in detail the physical meaning of the phase J. The construction of the propagator in Theorem 5 requires only that the potentials a and v be in class (A). However the circumstances of greatest interest in physics occur when the external fields arise as solutions of Maxwell's equations. Let the index $i = 1 \sim N$ label the particle coordinates in $x = (\vec{r}_1, \dots, \vec{r}_N)$. Each particle i interacts with a 4-potential $\{\vec{a}_i(\vec{r}_i,t), \phi_i(\vec{r}_i,t)\}$. All N 4-potentials determine the total fields appearing in Hamiltonian (1.1) by the relations

$$a(x,t) = \big(\vec{a}_1(\vec{r}_1,t), \dots, \vec{a}_N(\vec{r}_N,t)\big)$$
$$\phi(x,t) = \sum_{i=1}^{N} \phi_i(\vec{r}_i,t).$$

Now assume that $\{\vec{a}_i, \phi_i\}_{i=1}^N$ are solutions of Maxwell's equations. Structurally J is similar to the well-known Dirac magnetic phase factor which plays the central role in the Aharonov-Bohm effect. If $\Gamma(\vec{r}\,', \vec{r})$ is a smooth directed path in $\mathbf{R}^3$ from initial point $\vec{r}\,'$ to final point $\vec{r}$ and $\vec{a}(\vec{r}\,'')$ is a static vector potential then the one-body Dirac phase factor is

$$\exp\Big\{-\frac{q}{i\hbar}\int_{\Gamma(\vec{r}\,',\vec{r})} \vec{a}(\vec{r}\,'')\cdot d\vec{r}\,''\Big\}. \tag{5.1}$$

In comparing J to the Dirac phase it is helpful to split J into two parts. Let J_V denote the contribution to J that is proportional to V, and write

$$J(Q) = J_L(Q) + J_V(Q).$$

Then J_L is defined solely in terms of the electromagnetic potentials. Let $\{\vec{w}_i(\xi), \tau_i(\xi)\}$ be the projection of path $w(\xi;Q)$ onto the space-time coordinates of the ith particle.

If $y = (\vec{r}_1{}', \ldots, \vec{r}_N{}')$ then $\vec{w}_i(\xi) = \vec{r}_i{}' + \xi(\vec{r}_i - \vec{r}_i{}')$ and $\tau_i = s + \xi(t-s)$. Then J_L is given by the formula

$$(5.2) \quad J_L(Q) = q \sum_{i=1}^{N} \int_0^1 \{(t-s)\phi(\vec{w}_i(\xi), \tau_i(\xi)) - (\vec{r}_i - \vec{r}_i{}') \cdot \vec{a}_i(\vec{w}_i(\xi), \tau_i(\xi))\}\, d\xi.$$

For each value of $\xi \in [0,1]$ the integrand above is a Lorentz scalar formed by the product of two Lorentz 4-vectors. Inspecting (5.1) and (5.2) shows that J_L is an extended version of the Dirac phase in which time dependent potentials are allowed and in which the scalar fields $\{\phi_i\}$ are adjoined in such a way that the phase J_L is a Lorentz scalar. The static Dirac path $\Gamma(\vec{r}\,', \vec{r})$ is extended to the linear space-time path $\{\vec{w}_i(\xi), \tau_i(\xi)\}$.

It may appear unexpected that Lorentz invariant features appear in a problem whose particle dynamics are strictly non-relativistic. But of course the phase J_L is an average of the electromagnetic potentials $\{\vec{a}_i, \phi_i\}$ with respect to the path $\{\vec{w}_i(\xi), \tau_i(\xi)\}$. The path integrals of the form (5.2) would define a Lorentz invariant for any smooth space-time path connecting (y,s) to (x,t). The residual effect of our constructive solution is that it selects the particular path in J_L to be $w(\xi; Q)$.

As is well known, quantum systems exhibit semiclassical behavior if the particle mass is large. The semiclassical aspect of the representation (5.2) and its companion asymptotic expansion (4.26) is reflected in the geometrical character of the transport averages over the linear path $w(\xi; Q)$ that enter the phase factor $J(Q)$ and the expansion coefficients $T_j(Q)$. Let $\tau = s + \xi(t-s)$ be the running time variable. Then (4.18) leads to

$$x(\tau) = y + \frac{\tau - s}{t - s}(x - y).$$

The path $x(\tau)$ is the geodesic for the free evolution problem having initial point (y,s) and end point (x,t). Such straight line geodesics are characteristic of the eikonal [4] approximation. These semi-classical properties have emerged directly from the exact Dyson series description of quantum evolution without the need to resort to any semi-classical ansatz about the analytic character of K near $u = 0$.

It is interesting to contrast the large mass expansion of Theorem 7 with available results for the WKB approximation for the same dynamical system. A major difference between the $\hbar \to 0$ asymptotics of the propagator and the $u \equiv m^{-1} \to 0$ asymptotics is that $\hbar$ appears only in the quantum evolution problem (1.2) whereas the variable mass parameter m enters both the quantum evolution and the companion classical evolution problems. In Ref. 19 a detailed comparison of the $\hbar \to 0$ and the $u \to 0$ limits was used to formally obtain expansion (4.26) from the higher-order WKB representation of K. Such an approach is instructive in how the formula (4.26) emerges from the classical trajectories having two fixed end points that enter the WKB approximation but it suffers from the drawback that it is difficult to make rigorous. Here we have not had to make any ansatz concerning single valuedness of the action, the absence of caustics or the type of singular behavior the propagator has in the limit $\hbar \to 0$. We note finally that expansion (4.26) is relatively easy to use in the calculation of observables since the phase factor J and expansion coefficients T_i are explicitly given expressions of the fields whereas the analogous calculation in the WKB approximation requires one to first solve the difficult two point boundary value problems for the Hamiltonian dynamical equations (*i.e.,* solve the Hamiltonian-Jacobi equation) in order to obtain the action function and expansion coefficients.

REFERENCES

[1] S. G. KREIN, *Linear Differential Equations in Banach Space*, American Mathematical Society, Providence, 1971.

[2] T. KATO, *Perturbation Theory for Linear Operators, 2nd. ed.*, Springer, Berlin, 1984.

[3] F. J. DYSON, *The S matrix in quantum electrodynamics*, Phys. Rev., 75 (1949), pp. 1736–1754.

[4] V. GUILLEMIN AND S. STEINBERG, *Geometrical Asymptotics*, American Mathematical Society, Providence, 1977.

[5] J. D. JACKSON, *Classical Electrodynamics, 2nd. ed.*, Wiley, New York, 1975. Section 6.5

[6] T. A. OSBORN, L.PAPIEZ AND R. CORNS, *Constructive representations of propagators for quantum systems with electromagnetic fields*, J. Math. Phys., 28 (1987), pp. 103–123.

[7] T. KATO, *Fundamental properties of Hamiltonian operators of the Schrödinger type*, Trans. Am. Math. Soc., 70 (1951), pp. 195–211.

[8] R. P. FEYNMAN AND H. R. HIBBS, *Quantum Mechanical Path Integrals*, McGraw-Hill, New York, 1965, p. 42.

[9] P. R. HALMOS AND V. S. SUNDER, *Bounded Integral Operators on L^2 Spaces*, Springer, Berlin, 1978.

[10] B. SIMON, *Schrödinger semigroups*, Bull. Am. Math. Soc., 7 (1982), pp. 447–526.

[11] L. PAPIEZ, T. A. OSBORN AND F. H. MOLZAHN, *Quantum systems with external electromagnetic fields: the large mass asymptotics*, Univ. of Manitoba preprint, (1987).

[12] E. NELSON, *Feynman integrals and the Schrödinger equation*, J. Math. Phys., 5 (1964), pp. 332-343.

[13] K. ITO, *Weiner integral and Feynman integral*, in *Proceedings of the Fourth Berkeley Symposium on Mathematical Statistics and Probability, Vol. II*, Univ. of Calif. Press, Berkeley, 1961, pp. 227–238.

[14] S. A. ALBEVERIO AND R. J. HØEGH-KROHN, *Mathematical Theory of Feynman Path Integrals*, Springer, Berlin, 1974.

[15] T. A. OSBORN AND Y. FUJIWARA, *Time evolution kernels: uniform asymptotic expansions*, J. Math. Phys., 24 (1983), pp. 1093–1103.

[16] M. REED AND B. SIMON, *Methods of Modern Mathematical Physics II: Fourier Analysis, Self-Adjointness*, Academic, New York, 1975.

[17] K. MORIYASU, *An Elementary Primer for Gauge Theory*, World Scientific, Singapore, 1983.

[18] M. V. FEDORIUK, *The stationary phase method and pseudo-differential operators*, Russian Math. Surveys, 26:3 (1971), pp. 65–115.

[19] T. A. OSBORN AND F. H. MOLZAHN, *Structural connections between two semiclassical approximations: The WKB and the Wigner-Kirkwood approximations*, Phys. Rev. A, 34 (1986), pp. 1696–1707.

NEW TIME-DEPENDENT AND TIME-INDEPENDENT COMPUTATIONAL METHODS FOR MOLECULAR COLLISIONS

DONALD J. KOURI†, YAN SUN†*, RICHARD C. MOWREY†, JOHN Z. H. ZHANG†*, DONALD G. TRUHLAR‡, KENNETH HAUG‡, AND DAVID W. SCHWENKE‡

Abstract. This paper presents two approaches to gas-phase molecular collisions. The collision process is described by quantum mechanics, and an introduction to scattering-type solutions of the Schrödinger equation is presented. Then gas-phase reactive scattering is treated by means of expansions using quadratically integrable basis functions of auxiliary "amplitude density" functions. This converts the scattering problem into one of linear algebra, highly suited to solution utilizing vector processing supercomputers. An alternative approach based on the time-dependent Schrödinger equation is also presented. This method utilizes the powerful Fast Fourier Transform method to facilitate the time evolution of an appropriate wavepacket from some initial time to a post-collision time. The final packet may then be analyzed appropriately for the scattering information.

Key words. Collision theory, wave packets, amplitude density, Fast Fourier Transform, integral equations, chemical reactions, rearrangement scattering, basis-set expansions

1. Introduction. We are interested in the solution of the Schrödinger equation describing collisions of molecular systems. The description of the collision may be made using the time-dependent Schrödinger equation of the form

$$i\hbar\frac{\partial}{\partial t}|\Psi\rangle = H|\Psi\rangle, \tag{1.1}$$

or the time-independent Schrödinger equation

$$H|\chi\rangle = E|\chi\rangle. \tag{1.2}$$

We shall employ the bra-ket notation of Dirac [1]. In (1.1), $\hbar$ is Planck's constant divided by 2π, t denotes time, $|\Psi\rangle$ is the (time-dependent) quantum mechanical state vector of the system, and H is the Hamiltonian operator for the system. In (1.2), E is the total energy of the system, $|\chi\rangle$ now denotes the time-independent state vector of the system, related to $|\Psi\rangle$ by

$$|\Psi\rangle = |\chi\rangle \exp(-iEt/\hbar), \tag{1.3}$$

---

†Department of Chemistry and Department of Physics, University of Houston, 4800 Calhoun Road, Houston, TX 77004, USA. Supported in part by the National Science Foundation under grant CHE86–00363 and in part by the R. A. Welch Foundation under grant E–608. Partial support by the Minnesota Supercomputer Institute is also acknowledged.

*Visiting Scholar, Minnesota Supercomputer Institute.

‡Department of Chemistry and Supercomputer Institute, University of Minnesota, Minneapolis, MN 55455, USA. Supported in part by the National Science Foundation under grants CHE83–79144 and CHE86–17063 and by the Minnesota Supercomputer Institute.

and H is the same as in (1.1). Thus, $|\chi\rangle$ is a stationary state of the system.

In the following, we shall consider a prototype system of a structureless atom colliding with a diatom, which we assume can both vibrate and rotate, and which may or may not react. Examples of nonreactive collision systems are the collision of a He atom with H_2, CO, or I_2. Examples of reactive collisions are D + H_2 → HD + H, O + H_2 → OH + H, H + HBr → H_2 + Br, and F + HD → HF + D or DF + H. The objective is to calculate the probability that a collision with a particular initial state leads to a particular final state. The collision is described using so-called relative coordinates $\underset{\sim}{R}$ and $\underset{\sim}{r}$, where $\underset{\sim}{R}$ is the vector from the diatomic center of mass to the atomic projectile and $\underset{\sim}{r}$ is the internuclear vector of the diatom. If we label the atoms in the system A, B, and C, then for later considerations, it is convenient to introduce so-called "arrangement channel" labels. There are then three possible sets of coordinates which can be used to describe the system: $\underset{\sim}{R}_1$, $\underset{\sim}{r}_1$,; $\underset{\sim}{R}_2$, $\underset{\sim}{r}_2$; and $\underset{\sim}{R}_3$, $\underset{\sim}{r}_3$. The subscript 1 implies that atom A is the projectile and BC the molecule, 2 implies that atom B is the projectile and AC the molecule, and 3 implies that atom C is the projectile and AB the molecule. The Hamiltonian operator H may be written as

$$H = H_1 + V_1 = H_2 + V_2 = H_3 + V_3 \tag{1.4}$$

where

$$H_\alpha = \lim_{R_\alpha \to \infty} H, \tag{1.5}$$

$$V_\alpha = H - H_\alpha, \quad \alpha = 1, 2, 3. \tag{1.6}$$

Thus, H_α is the unperturbed Hamiltonian describing the motion of atom α (1 for A, 2 for B, and 3 for C) and the molecule in arrangement α, where the atom and molecule are infinitely separated. Clearly, V_α is then the interaction between atom α and the diatom in arrangement α, which is responsible for the scattering in the collision. The Hamiltonian H_α consists of a kinetic energy operator for the motion of atom α relative to the center of mass of the α-diatom and a kinetic energy operator and binding potential for the relative motion of the two atoms comprising the α-arrangement diatom. Thus,

$$H_\alpha = T_\alpha + V^\alpha, \tag{1.7}$$

where T_α is the α-arrangement kinetic energy operator and V^α is the α-diatom binding potential. More explicitly,

$$T_\alpha = -\frac{\hbar^2}{2\mu}\frac{1}{R_\alpha}\frac{\partial^2}{\partial R_\alpha^2}R_\alpha + \frac{\hbar^2}{2\mu R_\alpha^2}\underset{\sim}{\ell}_\alpha^2 - \frac{\hbar^2}{2\mu}\frac{1}{r_\alpha}\frac{\partial^2}{\partial r_\alpha^2}r_\alpha + \frac{\hbar^2}{2\mu r_\alpha^2}\underset{\sim}{j}_\alpha^2 \tag{1.8}$$

where the generalized reduced mass μ is defined by [2]

$$\mu = \left(\frac{m_A m_B m_C}{m_A + m_B + m_C}\right)^{1/2}, \tag{1.9}$$

$\hbar^2 \underset{\sim}{\ell}_\alpha^2$ is the quantum mechanical operator for the square of the orbital angular momentum of atom α with respect to the α-diatom center of mass, and $\hbar^2 \underset{\sim}{j}_\alpha^2$ is the square of the rotational angular momentum of the two atoms in the α-diatom measured relative to their center of mass. In spherical polar coordinates,

$$\underset{\sim}{R}_\alpha = (R_\alpha, \theta_\alpha, \phi_\alpha) = (R_\alpha, \hat{R}_\alpha) \tag{1.10}$$

$$\underset{\sim}{r}_\alpha = (r_\alpha, \gamma_\alpha, \xi_\alpha) = (r_\alpha, \hat{r}_\alpha) \tag{1.11}$$

where $\theta_\alpha(\gamma_\alpha)$ is the polar angle of $\underset{\sim}{R}_\alpha(\underset{\sim}{r}_\alpha)$, $\phi_\alpha(\xi_\alpha)$ is the azimuthal angle of $\underset{\sim}{R}_\alpha(\underset{\sim}{r}_\alpha)$, and $\hat{R}_\alpha$ and $\hat{r}_\alpha$ are unit vectors along $\underset{\sim}{R}_\alpha$ and $\underset{\sim}{r}_\alpha$, respectively. Finally, there is a unitary transformation connecting the coordinates of any two arrangements α and α' [2],

$$\begin{pmatrix} \underset{\sim}{r}_{\alpha'} \\ \underset{\sim}{R}_{\alpha'} \end{pmatrix} = \begin{pmatrix} A_{\alpha'\alpha} & B_{\alpha'\alpha} \\ -B_{\alpha'\alpha} & A_{\alpha'\alpha} \end{pmatrix} \begin{pmatrix} \underset{\sim}{r}_{\alpha} \\ \underset{\sim}{R}_{\alpha} \end{pmatrix}, \tag{1.12}$$

where

$$A_{\alpha'\alpha} = -\left[\frac{m_\alpha m_{\alpha'}}{(m_\alpha + m_{\alpha'})(m_{\alpha'} + m_{\alpha''})}\right]^{1/2}, \tag{1.13a}$$

$$B_{\alpha'\alpha} = -\left[\frac{m_{\alpha''}(m_\alpha + m_{\alpha'} + m_{\alpha''})}{(m_\alpha + m_{\alpha''})(m_{\alpha'} + m_{\alpha''})}\right]^{1/2}. \tag{1.13b}$$

Here $m_\alpha(m_{\alpha'}, m_{\alpha''})$ is the mass of the projectile (atom) in arrangement $\alpha(\alpha', \alpha'')$. Of course, it is obvious that any pair of (non-parallel) vectors in the plane of the 3 particles constitutes a basis for any vector in the plane.

We assume the Born-Oppenheimer electronic adiabatic approximation throughout this chapter, and the total Born-Oppenheimer potential energy surface for the ABC system is the sum of V^α and V_α, which sum is independent of α.

1.1. Multichannel Lippmann-Schwinger equation. We now can discuss scattering solutions of the time-independent Schrödinger equation (1.2). We write (1.2) as

$$(E - H_\alpha)|\chi\rangle = V_\alpha|\chi\rangle, \tag{1.14}$$

and apply the inverse of $E - H_\alpha$ to both sides of this equation:

$$|\chi\rangle = (E - H_\alpha)^{-1} V_\alpha |\chi\rangle. \tag{1.15}$$

However, in the limit that the atom and diatom in arrangement α do not interact, $V_\alpha \equiv 0$ and therefore $|\chi\rangle \equiv 0$ which is a trivial solution. Physically, it corresponds to no particles being present. In order to avoid this, we treat $V_\alpha|\chi\rangle$ as an inhomogeneity in (1.14), and solve the equation using (1.15) as a *particular solution* and add to it a solution of the homogeneous equation

$$(E - H_\alpha)|\Phi\rangle = 0. \tag{1.16}$$

Then the solution is

$$|\chi\rangle = |\Phi\rangle + (E - H_\alpha)^{-1} V_\alpha |\chi\rangle. \tag{1.17}$$

However, this is still not satisfactory since by (1.16), $V_\alpha|\chi\rangle$ in general will contain a nonvanishing projection along $|\Phi\rangle$, which belongs to the null space of $E - H_\alpha$, so that $(E - H_\alpha)^{-1}$ on $V_\alpha|\chi\rangle$ is undefined. To avoid this, we seek a solution of the form

$$|\chi\rangle = |\phi\rangle + (E - H_\alpha + i\epsilon)^{-1} V_\alpha |\chi\rangle, \quad \epsilon > 0, \tag{1.18}$$

and at the proper stage, take the limit as $\epsilon \to 0$ through positive values. Equation (1.18) is the Lippmann-Schwinger equation [3]. In the coordinate representation, H_α contains differential operators, so that (1.18) is an integral equation in that representation.

The final feature of the state function which should be specified in order to completely characterize the collision is to denote the various quantum numbers associated with the system. In order to do this, we expect at least one quantum number for each degree of freedom in the system (a total of 6 using the relative coordinates $\underset{\sim}{R}_\alpha, \underset{\sim}{r}_\alpha$). These can be the three components of the linear momentum vector $\hbar \underset{\sim}{k}^0_\alpha$, for the motion of atom α relative to the center of mass of diatom α, and the vibration-rotational state quantum numbers of the diatom v^0_α, j^0_α, m^0_α, where v^0_α, j^0_α determine the initial energy, $E_{v^0_\alpha j^0_\alpha}$, of molecule α, v^0_α specifies its vibrational state, $\hbar\sqrt{j^0_\alpha(j^0_\alpha + 1)}$ is the magnitude of its initial rotational angular momentum, and $\hbar m^0_\alpha$ is its angular momentum along some laboratory-fixed z-axis. Thus,

$$\begin{aligned} &|\chi(\underset{\sim}{k}^0_\alpha, v^0_\alpha, j^0_\alpha, m^0_\alpha)\rangle \\ &= |\Phi(\underset{\sim}{k}^0_\alpha, v^0_\alpha, j^0_\alpha, m^0_\alpha)\rangle + (E - H_\alpha + i\epsilon)^{-1} V_\alpha |\chi(\underset{\sim}{k}^0_\alpha, v^0_\alpha, j^0_\alpha, m^0_\alpha)\rangle. \end{aligned} \tag{1.19}$$

Alternatively, we can utilize states of well defined initial total angular momentum magnitude $\hbar\sqrt{J(J+1)}$, z-component $\hbar M$, and radial momentum $\hbar k^0_\alpha$, in place of states of well defined initial relative linear momentum. This is advantageous because in the absence of external torques on the atom-diatom system, J and M are conserved; *i.e.*, they are good quantum numbers. Indeed, states characterized by one set of quantum numbers can be constructed from linear superpositions of states characterized by the other set. Thus, one may show that [4]

$$|\chi(\underset{\sim}{k}^0_\alpha, v^0_\alpha, j^0_\alpha, m^0_\alpha)\rangle = \sum_{\substack{J v_\alpha j_\alpha \\ \ell_\alpha \ell^0_\alpha}} |Z^J_{v_\alpha j_\alpha \ell_\alpha \ell^0_\alpha}\rangle |\chi^J_{k^0_\alpha}(v_\alpha, j_\alpha, \ell_\alpha | v^0_\alpha, j^0_\alpha, \ell^0_\alpha)\rangle \tag{1.20a}$$

where

$$|Z^J_{v_\alpha j_\alpha \ell_\alpha \ell^0_\alpha}\rangle = \sqrt{4\pi} i^{\ell^0_\alpha} \sqrt{2\ell^0_\alpha + 1} \langle \ell^0_\alpha 0 j^0_\alpha m^0_\alpha | J j^0_\alpha \rangle |\Delta_{v_\alpha j_\alpha}\rangle |\Lambda^{Jm^0_\alpha}_{j_\alpha \ell_\alpha}\rangle, \quad (1.20b)$$

and where $|\Lambda^{Jm^0_\alpha}_{j_\alpha \ell_\alpha}\rangle$ is an angular momentum eigenstate vector characterized by

$$\underset{\sim}{J}^2_{op} |\Lambda^{Jm^0_\alpha}_{j_\alpha \ell_\alpha}\rangle = J(J+1) |\Lambda^{Jm^0_\alpha}_{j_\alpha \ell_\alpha}\rangle, \quad (1.21)$$

$$J_z |\Lambda^{Jm^0_\alpha}_{j_\alpha \ell_\alpha}\rangle = m^0_\alpha |\Lambda^{Jm^0_\alpha}_{j_\alpha \ell_\alpha}\rangle, \quad (1.22)$$

$$\underset{\sim}{j}^2_\alpha |\Lambda^{Jm^0_\alpha}_{j_\alpha \ell_\alpha}\rangle = j_\alpha(j_\alpha + 1) |\Lambda^{Jm^0_\alpha}_{j_\alpha \ell_\alpha}\rangle, \quad (1.23)$$

and

$$\underset{\sim}{\ell}^2_\alpha |\Lambda^{Jm^0_\alpha}_{j_\alpha \ell_\alpha}\rangle = \ell_\alpha(\ell_\alpha + 1) |\Lambda^{Jm^0_\alpha}_{j_\alpha \ell_\alpha}\rangle. \quad (1.24)$$

The $|\Lambda^{Jm^0_\alpha}_{j_\alpha \ell_\alpha}\rangle$ are the so-called total angular momentum eigenstates for the composite system and are constructed by vector coupling the angular momentum eigenstate vectors $|Y_{j_\alpha m_\alpha}\rangle$ and $|Y_{\ell_\alpha \mu_\alpha}\rangle$ of the diatomic rotor and orbital motion of atom α relative to the center of mass of the α-arrangement diatom:

$$|\Lambda^{Jm^0_\alpha}_{j_\alpha \ell_\alpha}\rangle = \sum_{m_\alpha \mu_\alpha} \langle \ell_\alpha \mu_\alpha j_\alpha m_\alpha | J m^0_\alpha \rangle |Y_{\ell_\alpha \mu_\alpha}\rangle |Y_{j_\alpha m_\alpha}\rangle, \quad (1.25)$$

where $\langle \hat{R}_\alpha | Y_{\ell_\alpha \mu_\alpha}\rangle$ and $\langle \hat{r}_\alpha | Y_{j_\alpha m_\alpha}\rangle$ are spherical harmonics, and the $\langle \ell_\alpha \mu_\alpha j_\alpha m_\alpha | J m^0_\alpha \rangle$ are the usual (real) Clebsch-Gordan vector coupling coefficients (see, *e.g.*, [5]). In particular, they vanish unless

$$\mu_\alpha + m_\alpha = m^0_\alpha. \quad (1.26)$$

The $|\Delta_{v_\alpha j_\alpha}\rangle$ are the α-diatom vibrational eigenstate vectors with vibrational energy $E_{v_\alpha j_\alpha}$.

It is important to realize that in general, the vibrational spectrum will include both a discrete *and* continuous portion. Thus, the sum over v_α in Eq. (1.20) really includes a discrete summation and an integral over the continuum. The role of the continuum can be of particular importance when reactive collisions are considered, since, *e.g.*, breaking the BC bond to form a new molecule AB implies that the formerly bound relative motion of the BC bond becomes unbound as the new product molecule BA and atom C separate. In fact, if one tries to solve the Schrödinger equation for a reactive system, and does not explicitly couple together all arrangements which are connected by reaction, the continuum vibrational manifold becomes absolutely essential in order to achieve convergence. In this paper, we shall avoid this problem by explicitly coupling all arrangements to one another. This ameliorates the problem of the continuum states, unless one is close to or above the breakup threshold for producing three free particles.

There is also another reason for explicitly including different arrangements in the wave function. A general motivation for transforming from differential equations to integral equations in quantum mechanical collision theory is that integral equations build in the boundary conditions. For rearrangement collisions, *i.e.*, collisions in which the particles may be found asymptotically in more than one arrangement, we must enforce boundary conditions in more than one asymptotic region, *e.g.*, not only when R_α is large but also when $R_{\alpha'}$ is large. Notice, however, that in the coordinate representation, $|\chi^J_{k^0_\alpha}(v_\alpha j_\alpha \ell_\alpha | v^0_\alpha j^0_\alpha \ell^0_\alpha)\rangle$ needs to depend only on R_α. Thus we now will see below that eq. (1.20) may be interpreted as an expansion in vibrational-rotational-orbital internal basis functions with R_α-dependent coefficients. Furthermore the internal basis functions are eigenfunctions of H_α [except that, since these basis functions do not depend on R_α, they are not eigenfunctions of the first term of Eq. (1.8)]. An expansion of the wave function in terms of eigenfunctions of H_α with coefficients that depend on the radial translational coordinate is called—for historical reasons and because it assumes the states retained when the expansion is necessarily truncated are the "closely coupled" ones—the close coupling (CC) method [6, 7]. With this form of wave function it is easy to enforce the boundary conditions at large R_α but complicated to enforce them for large $R_{\alpha'}$ or $R_{\alpha''}$. For electron-atom or electron-molecule scattering, these limits would correspond to exchanging which electron is at infinity, and a more transparent treatment would involve using wave functions with the correct permutational symmetry. This is sometimes called close coupling with exchange (CCE), and it involves using multiconfiguration Hartree-Fock (MCHF) wave functions with one electron in a continuum orbital in at least some of the configurations [6, 8–29]. This method was extended to positron-atom collisions, where the channels are not identical by symmetry, by Smith [30]. The analog for atom-molecule collisions is to couple wave functions corresponding to different atoms reaching infinity ($R_\alpha = \infty$ with r_α finite, or $R_{\alpha'} = \infty$ with $r_{\alpha'}$ finite). There are several ways to do this [31], and the way corresponding most closely to the MCHF method was introduced by Micha [32] and Miller [33] and is considered in Section 2.2. In the nuclear physics literature similar physical ideas are sometimes studied under the aegis of cluster expansions (see, *e.g.*, [34], and references therein), but the formalisms developed in this context are not usually useful in chemistry since they are developed for the case where the potential is a sum of pairwise interactions, which is a poor approximation for chemical reactions. We note, however, that a set of equations equivalent to Miller's was introduced in the nuclear physics literature by Hahn [35, 36]. In the remainder of this section we present some further discussion of the integral equation that follows from the single-arrangement expansion of eq. (1.20). This is most useful for nonreactive collisions, such as the He + H_2 collisions mentioned above, where the goal of the calculation is to calculate the distribution of the products in scattering angle and quantum states, but only one chemical arrangement of the products is energetically accessible. We return to multi-arrangement expansions and chemical reactions in Sect. 2.3.

It will be convenient in future equations to use n_α to denote the collection $(v_\alpha, j_\alpha, \ell_\alpha)$ and n^0_α to denote $(v^0_\alpha, j^0_\alpha, \ell^0_\alpha)$, etc.

It may be shown that the abstract state vectors $|\chi^J_{k^0_\alpha}(n_\alpha|n^0_\alpha)\rangle$ satisfy:

$$|\chi^J_{k^0_\alpha}(n_\alpha|n^0_\alpha)\rangle = \delta_{v_\alpha v^0_\alpha}\delta_{j_\alpha j^0_\alpha}\delta_{\ell_\alpha \ell^0_\alpha}|j(\ell^0_\alpha, k^0_\alpha)\rangle + \sum_{n'_\alpha}(E - h_{n_\alpha} + i\epsilon)^{-1}V^J(n_\alpha|n'_\alpha)|\chi^J_{k^0_\alpha}(n'_\alpha|n^0_\alpha)\rangle. \tag{1.27}$$

where $|j(\ell^0_\alpha, k^0_\alpha)\rangle$ is specified below Eqs. (1.34) and (1.35). The operator h_{n_α} describes the relative translational motion of the α-arrangement atom and diatom when the internal energy of the diatom is $E_{v_\alpha j_\alpha}$ and the relative orbital angular momentum is $\hbar\sqrt{l_\alpha(l_\alpha+1)}$. As such, h_{n_α} can be written as the sum of the rotor energy $E_{v_\alpha j_\alpha}$ and the radial kinetic energy operator (Hamiltonian) h_{l_α}. The operator $V^J(n_\alpha|n'_\alpha)$ is responsible for the fact the the α atom and diatom interact, and therefore energy and angular momentum can be exchanged between the internal diatomic vib-rotation and the translation of the atom relative to the diatom center of mass. The indices correspond to a transition $n_\alpha \to n'_\alpha$ for $V^J(n_\alpha|n'_\alpha)$ or $n^0_\alpha \to n_\alpha$ for $\chi^J_{k^0_\alpha}(n_\alpha|n^0_\alpha)$, while a single total angular momentum label J indicates that, in the absence of external torques, the total angular momentum cannot change (it is "conserved"). The absence of the projection label m^0_α indicates that the collision is *independent* of the orientation of the total angular momentum vector J. This is an example of the Wigner-Eckart theorem [5]. We may express Eq. (1.27) in more familiar form by putting the abstract vectors into a convenient basis (representation). It is convenient computationally to use the coordinate representation. This is achieved by judicious use of the resolution of the identity in the coordinate representation

$$1 = \int_0^\infty dR_\alpha R^2_\alpha |R_\alpha\rangle\langle R_\alpha|, \tag{1.28}$$

where R_α is an eigenvalue of the radial position operator in arrangement α. The vectors $|R_\alpha\rangle$ are normalized according

$$\langle R_\alpha|R'_\alpha\rangle = \frac{1}{R^2_\alpha}\delta(R_\alpha - R'_\alpha), \tag{1.29}$$

as is usual for a radial-type variable. Then we insert (1.28) into (1.27) and operate with $\langle R_\alpha|$ to obtain

$$\begin{aligned}\langle R_\alpha|\chi^J_{k^0_\alpha}(n_\alpha|n^0_\alpha)\rangle &= \delta_{v_\alpha v^0_\alpha}\delta_{j_\alpha j^0_\alpha}\delta_{\ell_\alpha \ell^0_\alpha}\langle R_\alpha|j(\ell^0_\alpha, k^0_\alpha)\rangle \\ &+ \sum_{n'_\alpha}\int_0^\infty dR'_\alpha R'^2_\alpha \langle R_\alpha|(E - E_{v_\alpha j_\alpha} - h_{\ell_\alpha} + i\epsilon)^{-1}|R'_\alpha\rangle \\ &\times \langle R'_\alpha|V^J(n_\alpha|n'_\alpha)|\chi^J_{k^0_\alpha}(n'_\alpha|n^0_\alpha)\rangle.\end{aligned} \tag{1.30}$$

Now we assume that the potential operator $V^J(n_\alpha|n'_\alpha)$ is *diagonal* in the coordinate representation, so that

$$V^J(n_\alpha|n'_\alpha)|R'_\alpha\rangle = V^J(n_\alpha|n'_\alpha|R'_\alpha)|R'_\alpha\rangle, \tag{1.31}$$

where its eigenvalue, $V^J(n_\alpha|n'_\alpha|R'_\alpha)$, is in fact the potential matrix element connecting vib-rotational states n'_α and n_α. Since the potential is a Hermitian (self-adjoint) operator it has real eigenvalues and one has that

$$\langle R_\alpha|V^J(n_\alpha|n'_\alpha) = V^J(n_\alpha|n'_\alpha|R_\alpha)\langle R_\alpha|. \tag{1.32}$$

We define

$$\chi^J(n_\alpha|k^0_\alpha, n^0_\alpha|R_\alpha) \equiv \langle R_\alpha|\chi^J_{k^0_\alpha}(n_\alpha|n^0_\alpha)\rangle, \tag{1.33}$$

and

$$j_{\ell^0_\alpha}(k^0_\alpha R_\alpha) \equiv \langle R_\alpha|j(\ell^0_\alpha, k^0_\alpha)\rangle, \tag{1.34}$$

so that (1.30) becomes

$$\begin{aligned} \chi^J(n_\alpha|k^0_\alpha, n^0_\alpha|R_\alpha) &= \delta_{n_\alpha n^0_\alpha}\delta_{j_\alpha j^0_\alpha}\delta_{\ell_\alpha \ell^0_\alpha} j_{\ell^0_\alpha}(k^0_\alpha R_\alpha) \\ &+ \sum_{n'_\alpha}\int_0^\infty dR'_\alpha R'^2_\alpha \langle R_\alpha|(E - E_{v_\alpha j_\alpha} - h_{\ell_\alpha} + i\epsilon)^{-1}|R'_\alpha\rangle \\ &\times V^J(n_\alpha|n'_\alpha|R'_\alpha)\chi^J(n'_\alpha|k^0_\alpha, n^0_\alpha|R'_\alpha). \end{aligned} \tag{1.35}$$

In fact, $j_{l^0_\alpha}(k^0_\alpha R_\alpha)$ can be shown to be a spherical Bessel function of order ℓ^0_α [4]. In this form, it is clear that the collision problem has now been expressed as coupled integral equations for the radial components $\chi^j(n_\alpha|k^0_\alpha n^0_\alpha|R_\alpha)$ (projection along $|R_\alpha\rangle$) of the state vector $|\chi^j_{k^0_\alpha}(n_\alpha|n^0_\alpha)\rangle$. The final step in arriving at a form of the equations which can provide the basis of actual computations is to obtain a detailed expression for the radial coordinate representation matrix element $\langle R_\alpha|(E - E_{v_\alpha j_\alpha} - h_{l_\alpha} + i\epsilon)^{-1}|R'_\alpha\rangle$ of the operator $(E - E_{v_\alpha j_\alpha} - h_{l_\alpha} + i\epsilon)^{-1}$. To do this, it is convenient to introduce the eigenstates of h_{l_α} given by

$$h_{l_\alpha}|j(k, \ell_\alpha)\rangle = \frac{\hbar^2 k^2}{2\mu}|j(k, \ell_\alpha)\rangle. \tag{1.36}$$

In fact, the radial coordinate representative of this state, $\langle R_\alpha|j(k, l_\alpha)\rangle$ is given by

$$\langle R_\alpha|j(k, l_\alpha)\rangle = j_{l_\alpha}(kR_\alpha), \tag{1.37}$$

where again, $j_{l_\alpha}(kR_\alpha)$ is the usual spherical Bessel function of order l_α [29]. Then the identity can be resolved as

$$1 = \frac{2}{\pi}\int_0^\infty dk k^2 |j(k, \ell_\alpha)\rangle\langle j(k, \ell_\alpha)|, \tag{1.38}$$

with the normalization

$$\langle j(k, \ell_\alpha)|j(k', \ell_\alpha)\rangle = \frac{\pi}{2}\frac{\delta(k - k')}{k^2}. \tag{1.39}$$

Then using (1.36)–(1.38), we write

$$(140) \quad \langle R_\alpha|(E - E_{v_\alpha j_\alpha} - h_{\ell_\alpha} + i\epsilon)^{-1}|R'_\alpha\rangle = \frac{2}{\pi}\int_0^\infty dk \frac{k^2 j_{\ell_\alpha}(kR_\alpha) j_{\ell_\alpha}(kR'_\alpha)}{(E - E_{v_\alpha j_\alpha} - \frac{\hbar^2 k^2}{2\mu} + i\epsilon)},$$

where we have made use of the fact that

$$(1.41) \qquad f(h_{\ell_\alpha})|j(k,\ell_\alpha)\rangle = f(k)|j(k,\ell_\alpha)\rangle,$$

for well behaved functions $f(h_{l_\alpha})$. The presence of the $+i\epsilon$ in (1.40) ensures the validity of (1.41) for

$$(1.42) \qquad f(h_{\ell_\alpha}) = (E - E_{v_\alpha j_\alpha} - h_{\ell_\alpha} + i\epsilon)^{-1}.$$

The right hand side of (1.42)) is immediately recognized as a standard expression for a Green's function. The integral over k can be carried out by well known contour integration using the Cauchy residue theorem. The result is

(1.43)
$$\frac{2}{\pi}\int_0^\infty dk \frac{k^2 j_{\ell_\alpha}(kR_\alpha) j_{\ell_\alpha}(kR'_\alpha)}{(E - E_{v_\alpha j_\alpha} - \frac{\hbar^2 k^2}{2\mu} + i\epsilon)} = -\frac{2\mu}{\hbar^2} k_{v_\alpha j_\alpha} j_{\ell_\alpha}(k_{v_\alpha j_\alpha} R_\alpha^{<}) h^+_{\ell_\alpha}(k_{v_\alpha j_\alpha} R_\alpha^{>}),$$

where

$$(1.44) \qquad \frac{\hbar^2}{2\mu} k^2_{v_\alpha j_\alpha} = E - E_{v_\alpha j_\alpha},$$

and $h^+_{l_\alpha}(k_{v_\alpha j_\alpha} R_\alpha)$ is an outgoing spherical Hankel function of order l_α, satisfying

$$(1.45) \qquad \lim_{R_\alpha \to \infty} h^+_{\ell_\alpha}(k_{v_\alpha j_\alpha} R_\alpha) = \exp[i(k_{v_\alpha j_\alpha} R_\alpha - \ell_\alpha \pi/2)]/k_{v_\alpha j_\alpha} R_\alpha.$$

The $R_\alpha^{>}$ and $R_\alpha^{<}$ are respectively the greater and lesser of R_α, R'_α. (We also note that by construction, $\hbar^2 {k_\alpha^0}^2/2\mu$ equals $E - E_{v_\alpha^0 j_\alpha^0}$.) Thus, (1.35) becomes

$$(1.46) \qquad \underset{\sim}{\chi}^J(R_\alpha) = \underset{\sim}{j}(R_\alpha) - \frac{2\mu}{\hbar^2}\underset{\sim}{k}\int_0^\infty dR'_\alpha {R'_\alpha}^2 \underset{\sim}{j}(R_\alpha^{<}) \underset{\sim}{h}^+(R_\alpha^{>}) \underset{\sim}{V}^J(R'_\alpha) \underset{\sim}{\chi}^J(R'_\alpha),$$

where it has been convenient to introduce a compact matrix notation such that

$$(1.47) \qquad [\underset{\sim}{\chi}^J(R_\alpha)]_{n_\alpha n_\alpha^0} = \chi^J(n_\alpha|k_\alpha^0 n_\alpha^0|R_\alpha),$$

$$(1.48) \qquad [\underset{\sim}{k}]_{n_\alpha n_\alpha^0} = \delta_{v_\alpha v_\alpha^0}\delta_{j_\alpha j_\alpha^0}\delta_{\ell_\alpha \ell_\alpha^0} k_{v_\alpha^0 j_\alpha^0},$$

$$(1.49) \qquad [\underset{\sim}{j}(R_\alpha^{<})]_{n_\alpha n_\alpha^0} = \delta_{v_\alpha v_\alpha^0}\delta_{j_\alpha j_\alpha^0}\delta_{\ell_\alpha \ell_\alpha^0} j_{\ell_\alpha^0}(k_{v_\alpha^0 j_\alpha^0} R_\alpha^{<}),$$

$$(1.50) \qquad [\underset{\sim}{h}^+(R_\alpha^{>})]_{n_\alpha n_\alpha^0} = \delta_{v_\alpha v_\alpha^0}\delta_{j_\alpha j_\alpha^0}\delta_{\ell_\alpha \ell_\alpha^0} h^+_{\ell_\alpha^0}(k_{v_\alpha^0 j_\alpha^0} R_\alpha^{>}),$$

and

$$(1.51) \qquad [\underset{\sim}{V}^J(R_\alpha)]_{n_\alpha n'_\alpha} = V^J(n_\alpha|n'_\alpha|R_\alpha).$$

The scattering information is contained in the wavefunction for the region where one places the detectors (to identify the resulting particles scattered in a given direction by the collision process). Since one wants to be sure the collision is entirely over, the detector is placed a large distance from the region where the collision occurs. The quantity of interest is the number of projectiles per second scattered into the solid angle of acceptance subtended by the detector. If one knows the flux of scattered particles (*i.e.*, the number of scattered particles per second per unit area). then multiplying this by the area of the detector will give the number of scattered particles per second. This number of scattered particles per second should be proportional to the incident flux of projectiles times the area of the target molecule which is effective in deflecting the projectile into the detector. If the detector can distinguish the final states of the molecule and projectile (including the vib-rotor state and relative momentum of the final atom), then the area of the target causing that resulting final state will, in general, depend on the final state measured. It is this target area which is the quantity of interest; it is the so-called state-to-state scattering cross section. It is determined by the behavior of the scattered wavefunction at large distances, $R_\alpha \to \infty$. We can ask for the behavior of $\chi^J(v_\alpha j_\alpha l_\alpha|k^0_\alpha v^0_\alpha j^0_\alpha l^0_\alpha|R_\alpha)$ in this limit by using (1.46):

$$(1.52) \qquad \chi^J(n_\alpha|k^0_\alpha, n^0_\alpha|R_\alpha) \underset{R_\alpha\to\infty}{\sim} \delta_{v_\alpha v^0_\alpha}\delta_{j_\alpha j^0_\alpha}\delta_{\ell_\alpha \ell^0_\alpha} j_{\ell^0_\alpha}(k^0_\alpha R_\alpha)$$
$$- \frac{2\mu}{\hbar^2} k_{v_\alpha j_\alpha} \sum_{n'_\alpha} \int_0^\infty dR'_\alpha {R'_\alpha}^2 h^+_{\ell_\alpha}(k_{v_\alpha j_\alpha} R^>_\alpha) j_{\ell_\alpha}(k_{v_\alpha j_\alpha} R^<_\alpha)$$
$$\times V^J(n_\alpha|n'_\alpha|R'_\alpha)\chi^J(n'_\alpha|k^0_\alpha n^0_\alpha|R'_\alpha).$$

But the potential $V^J(n_\alpha|n'_\alpha|R'_\alpha)$ must tend to zero as R'_α gets large in order for the integral over R'_α from 0 to ∞ to be meaningful (since χ^J does *not* go to zero as $R'_\alpha \to \infty$ because the particles *can* separate from one another in a collision). This means that when R_α becomes large, for a given accuracy required for the integral over R'_α, R_α can be made larger than any R'_α contributing to the integral. Then $R^>_\alpha \equiv R_\alpha$, $R^<_\alpha \equiv R'_\alpha$ and we obtain

$$(1.53) \qquad \chi^J(n_\alpha|k^0_\alpha, n^0_\alpha|R_\alpha) \underset{R_\alpha\to\infty}{\sim} \delta_{v_\alpha v^0_\alpha}\delta_{j_\alpha j^0_\alpha}\delta_{\ell_\alpha \ell^0_\alpha} j_{\ell^0_\alpha}(k^0_\alpha R_\alpha)$$
$$+ \frac{i}{2}\frac{1}{\sqrt{k^0_\alpha k_{v_\alpha j_\alpha}}R_\alpha} \exp[i(k_{v_\alpha j_\alpha}R_\alpha - \ell_\alpha \frac{\pi}{2})]T^J(n_\alpha|n^0_\alpha),$$

where

$$(1.54) \qquad T^J(n_\alpha|n^0_\alpha) \equiv \frac{2\mu i}{\hbar^2}\sqrt{k^0_\alpha k_{v_\alpha j_\alpha}} \sum_{n'_\alpha} \int_0^\infty dR'_\alpha {R'_\alpha}^2 j_{\ell^0_\alpha}(k^0_\alpha R_\alpha)$$
$$\times V^J(n_\alpha|n'_\alpha|R'_\alpha)\chi^J(n'_\alpha|k^0_\alpha, n^0_\alpha|R'_\alpha).$$

The quantity $T^J(n_\alpha|n_{\alpha_0})$ is termed the transition amplitude matrix. For the case where the molecule is rigid (*i.e.*, nonvibrating) it reduces to the transition amplitude of Arthurs and Dalgarno [4] and is analogous to that introduced by Percival and Seaton [10]. In terms of this amplitude, the physically relevant scattering amplitude, $f(v_\alpha j_\alpha m_\alpha | v_\alpha^0 j_\alpha^0 m_\alpha^0 | \hat{R}_\alpha)$, is given by

$$
(1.55) \qquad f(v_\alpha, j_\alpha, m_\alpha | v_\alpha^0, j_\alpha^0, m_\alpha^0 | \hat{R}_\alpha) = \sqrt{\pi / k_\alpha^0 k_{v_\alpha j_\alpha}} \sum_{J \ell_\alpha \ell_\alpha^0} i^{\ell_\alpha^0 - \ell_\alpha + 1}
\times \sqrt{2\ell_\alpha^0 + 1} Y_{\ell_\alpha m_\alpha^0 - m_\alpha}(\hat{R}_\alpha) < \ell_\alpha m_\alpha^0 - m_\alpha j_\alpha m_\alpha | J m_\alpha^0 >
\times \langle \ell_\alpha^0 0 j_\alpha^0 m_\alpha^0 | J m_\alpha^0 \rangle T^J(n_\alpha | n_\alpha^0).
$$

The physical differential cross section for scattering the projectile in the direction $\hat{R}_\alpha$ with the quantum state changing from relative momentum $\hbar \underset{\sim}{k}_\alpha^0$, molecular internal state $v_\alpha^0, j_\alpha^0, m_\alpha^0$ to final relative momentum $(k_{v_\alpha j_\alpha} \hat{R}_\alpha)$, molecular internal state $v_\alpha, j_\alpha, m_\alpha$ is given by

$$
(1.56) \qquad \frac{d\sigma}{d\hat{R}_\alpha}(v_\alpha, j_\alpha, m_\alpha | v_\alpha^0, j_\alpha^0, m_\alpha^0) = \left(\frac{k_{v_\alpha j_\alpha}}{k_\alpha^0} \right) |f(v_\alpha, j_\alpha, m_\alpha | v_\alpha^0, j_\alpha^0, m_\alpha^0 | \hat{R}_\alpha)|^2.
$$

Integrating this over all final directions yields the state-to-state integral cross section or effective target size for this transition. Thus, once the $T^J(n_\alpha | n_\alpha^0)$ are determined for all J, ℓ_α, and ℓ_α^0 contributing to the sums in (1.55), the physical scattering cross sections can be calculated. Determining the T^J-elements requires solution of the χ^J integral equations from $R_\alpha = 0$ out to the asymptotic region where the potential elements V^J are essentially zero.

In Section 2, we will discuss two methods for solving the $\underset{\sim}{\chi}^J(R_\alpha)$ equations.

1.2. Close coupling wave packet method. We also wish to present an introduction to the time-dependent approach to scattering. Our discussion will for the most part follow that of Sun, Mowrey, and Kouri [37]. Most of the necessary apparatus is presented above. Returning to (1.1)–(1.3), it is clear that knowledge of the $|\chi^J\rangle$ determines the corresponding time-dependent wavefunction $|\Psi^J\rangle$ at all times t. However (1.1) has more general solutions than the stationary ones which satisfy (1.3). In general, solutions of (1.1) may be constructed as superpositions of stationary states $|\chi^J\rangle$, with such superpositions being called wavepackets. Furthermore, unlike the $|\chi^J\rangle$, which are non-normalizable, the wavepacket solutions $|\Psi^J\rangle$ of (1.1) may be constructed to be normalizable. A fundamental feature of (1.1) is that it is a *first order* partial differential equation in time; *i.e.*, it constitutes an initial value problem. Thus specification of the packet at *any* time t uniquely specifies its behavior for all other times. If we were to construct a wavepacket describing our atom-diatom collision system in the distant past, prior to the collision, it would correspond to a configuration in which the particles did not interact. This packet would then be a superposition of noninteracting states (eigenstates of H_α). If the packet were then evolved forward under the action of the *full* Hamiltonian H (according to (1.1)), it would produce a final wavepacket which would also be a superposition of the $|\chi^J\rangle$ stationary states with the *same* coefficients. This is the basis of the time-dependent approach. We will emphasize methods that take advantage of the broadness of the wavepacket in energy.

As in our discussion of stationary states, we again make a close coupling expansion of the wave function, *i.e.*, we write

(1.57a)
$$|\Psi\rangle = \sum_{Jv_\alpha \ell_\alpha^0} |Z^J_{v_\alpha \ell_\alpha^0}\rangle |\Psi^J(n_\alpha|n_\alpha^0|t)\rangle$$

and

(1.57b)
$$\Psi^J(n_\alpha|n^0_{\alpha_0}|R_\alpha, t) = \langle R_\alpha|\Psi^J(n_\alpha|n^0_{\alpha_0}|t)\rangle.$$

This is called the close coupling wave packet (CCWP) method, and it was originally introduced in different contexts by two of the authors [38, 39] and by Jackson and Metiu [40]. For the present purposes, it is convenient to employ wavepackets corresponding to well defined total angular momenta J. Thus we consider a wavepacket given initially ($t = 0$) by

(1.58a)
$$\Psi^J(n_\alpha|n_\alpha^0|R_\alpha, 0) = \delta_{v_\alpha v_\alpha^0}\delta_{j_\alpha j_\alpha^0}\delta_{\ell_\alpha \ell_\alpha^0}\int_0^\infty dk_\alpha^0 {k_\alpha^0}^2 A_{\ell_\alpha^0}(k_\alpha^0) j_{\ell_\alpha^0}(k_\alpha^0 R_\alpha).$$

This implies that at time $t = 0$, the packet corresponds to well defined quantum numbers J and n_α^0 but with a superposition of relative translational kinetic energies $\hbar^2 {k_\alpha^0}^2/2\mu$. The detailed shape of the packet is determined by the weighting function $A_{\ell_\alpha^0}(k_\alpha^0)$. Specification of the spatial extent of the initial packet, $\Phi(R_\alpha, R_\alpha^0)$, uniquely determines $A_{\ell_\alpha^0}(k_\alpha^0)$. Here, $\Phi(R_\alpha, R_\alpha^0)$ is a quadratically integrable function of R_α centered about R_α^0. Then

(1.58b)
$$A_{\ell_\alpha^0}(k_\alpha^0) \equiv \frac{2}{\pi}\int_0^\infty dR_\alpha R_\alpha^2 j_{\ell_\alpha^0}(k_\alpha^0 R_\alpha)\Phi(R_\alpha, R_\alpha^0).$$

We note that (1.1) may be formally integrated to yield

(1.59)
$$|\Psi(t)\rangle = \exp(-iHt/\hbar)|\Psi(t=0)\rangle.$$

If the initial packet $|\Psi(t=0)\rangle$ is a superposition of stationary states $|\chi\rangle$,

(1.60)
$$|\Psi(t=0)\rangle = \int_0^\infty dE\mathcal{A}(E)|\chi_E\rangle,$$

then one immediately obtains

(1.61)
$$|\Psi(t)\rangle = \int_0^\infty dE\mathcal{A}(E)|\chi_E\rangle \exp(-iEt/\hbar).$$

If the coefficients $\mathcal{A}(E)$ could be related to the $A_{\ell_\alpha^0}(k_\alpha^0)$ in an initial packet of the form (1.58), and such a packet were *numerically* evolved forward in time according to (1.1), then the numerical packet would have to be equal to the formal expression (1.59). Now the most important feature of the time-dependent method is, in fact, that numerically evolving a packet defined initially by (1.57)–(1.58) forward in time using (1.1) *automatically* generates a wavepacket at t given formally by [37] (see also, *e.g.*, [41]):

(1.62)
$$\Psi^J(n_\alpha|n_\alpha^0|R_\alpha,t) = \int_0^\infty dk_\alpha^0 {k_\alpha^0}^2 A_{\ell_\alpha^0}(k_\alpha^0)\chi^J(n_\alpha|k_\alpha^0,n_\alpha^0|R_\alpha)\exp[-i(E_{n_\alpha^0 j_\alpha^0}+\hbar^2{k_\alpha^0}^2/2\mu)t/\hbar],$$

where the $\chi^J(n_\alpha|k_\alpha^0,n_\alpha^0|R_\alpha)$ satisfy the integral equation given in (1.46). This provides the means of deriving how one extracts from a numerically propagated wavepacket the scattering information, namely, the transition amplitude $T^J(n_\alpha|n_\alpha^0)$. To do this, we re-write the integral equation for the $\chi^J(n_\alpha|k_\alpha^0,n_\alpha^0|R_\alpha)$ using the expression (1.40) for the Green's function rather than the contour integrated form (1.43). Thus,

(1.63)
$$\begin{aligned}\chi^J(n_\alpha|k_\alpha^0,n_\alpha^0|R_\alpha) &= \delta_{v_\alpha v_\alpha^0}\delta_{j_\alpha j_\alpha^0}\delta_{\ell_\alpha \ell_\alpha^0} j_{\ell_\alpha^0}(k_\alpha^0 R_\alpha)\\ &+\frac{2}{\pi}\sum_{n'_\alpha}\int_0^\infty dR'_\alpha R'^2_\alpha\int_0^\infty \frac{dk'_\alpha {k'_\alpha}^2 j_{\ell_\alpha}(k'_\alpha R_\alpha) j_{\ell_\alpha}(k'_\alpha R'_\alpha)}{\left(\frac{\hbar^2{k_\alpha^0}^2}{2\mu}+E_{v_\alpha^0 j_\alpha^0}-\frac{\hbar^2{k'_\alpha}^2}{2\mu}+i\epsilon\right)}\\ &\times V^J(n_\alpha|n'_\alpha|R'_\alpha)\chi^J(n'_\alpha|k_\alpha^0,n_\alpha^0|R'_\alpha).\end{aligned}$$

We notice that the sum over n'_α and integral over R'_α are similar to those which defined the transition amplitude $T^J(n_\alpha|n_\alpha^0)$ except that certain factors $\sqrt{k_{n_\alpha j_\alpha}k_{v_\alpha^0 j_\alpha^0}}$ are not present, and more importantly, the energy $\frac{\hbar^2{k_\alpha^0}^2}{2\mu}+E_{v_\alpha^0 j_\alpha^0}$ *cannot* equal $\frac{\hbar^2{k'_\alpha}^2}{2\mu}+E_{v_\alpha j_\alpha}$ in general since k'_α is being integrated for each value of k_α^0. However, we introduce a modified amplitude

(1.64)
$$\tilde{T}^J(n_\alpha|n_\alpha^0) \equiv \frac{4\mu i}{\hbar^2}\sum_{n'_\alpha}\int_0^\infty dR'_\alpha {R'_\alpha}^2 j_{\ell_\alpha}(k'_\alpha R'_\alpha)V^J(n_\alpha|n'_\alpha|R'_\alpha)\chi^J(n'_\alpha|k_\alpha^0,n_\alpha^0|R'_\alpha).$$

Because the initial and final total energies are not equal, $\tilde{T}^J(n_\alpha|n_\alpha^0)$ is related to a so called "half-off-energy-shell T-matrix" element. In terms of it, (1.63) becomes

(1.65)
$$\begin{aligned}\chi^J(n_\alpha|k_\alpha^0,n_\alpha^0|R_\alpha) &= \delta_{v_\alpha v_\alpha^0}\delta_{j_\alpha j_\alpha^0}\delta_{\ell_\alpha \ell_\alpha^0} j_{\ell_\alpha^0}(k_\alpha^0 R_\alpha)\\ &+\frac{\hbar^2}{2\pi i\mu}\int_0^\infty \frac{dk'_\alpha {k'_\alpha}^2 j_{\ell_\alpha}(k'_\alpha R_\alpha)\tilde{T}(n_\alpha|n_\alpha^0)}{\left(\frac{\hbar^2{k_\alpha^0}^2}{2\mu}+E_{v_\alpha^0 j_\alpha^0}-\frac{\hbar^2{k'_\alpha}^2}{2\mu}-E_{v_\alpha j_\alpha}+i\epsilon\right)}.\end{aligned}$$

Then our wave packet expression (1.62) can be written as

(1.66)
$$\begin{aligned}\Psi^J(n_\alpha|n_\alpha^0|R_\alpha,t) &= \int_0^\infty dk_\alpha^0 {k_\alpha^0}^2 A_{\ell_\alpha^0}(k_\alpha^0)\\ &\times \exp[-i(E_{v_\alpha^0 j_\alpha^0}+\hbar^2{k_\alpha^0}^2/2\mu)t/\hbar]\Bigg\{\delta_{v_\alpha v_\alpha^0}\delta_{j_\alpha j_\alpha^0}\delta_{\ell_\alpha \ell_\alpha^0} j_{\ell_\alpha^0}(k_\alpha^0 R_\alpha)\\ &+\frac{\hbar^2}{2\pi i\mu}\int_0^\infty \frac{dk'_\alpha {k'_\alpha}^2 j_{\ell_\alpha}(k'_\alpha R_\alpha)\tilde{T}(n_\alpha|n_\alpha^0)}{\left(\frac{\hbar^2{k_\alpha^0}^2}{2\mu}+E_{v_\alpha^0 j_\alpha^0}-\frac{\hbar^2{k'_\alpha}^2}{2\mu}-E_{v_\alpha j_\alpha}+i\epsilon\right)}\Bigg\}.\end{aligned}$$

The physical interpretation of the appearance of half-off-energy-shell amplitudes is that at finite R_α, the total energy is still uncertain. We then form the time-dependent integral $I^J(k_\alpha, v_\alpha, j_\alpha, \ell_\alpha | v_\alpha^0, j_\alpha^0, \ell_\alpha^0 | t)$ given by

$$(1.67) \qquad I^J(k_\alpha, n_\alpha | n_\alpha^0 | t) = \int_0^\infty dR_\alpha R_\alpha^2 j_{\ell_\alpha}(k_\alpha R_\alpha) \Psi^J(n_\alpha | n_\alpha^0 | R_\alpha, t),$$

which is evaluated *numerically* at some time t such that the numerically propagated wavepacket [constructed initially via (1.57)] has passed through the region of interaction and back out into the large R_α region where the detector can be located. The final step required to express $T^J(n_\alpha | n_\alpha^0)$ in terms of the $I^J(k_\alpha, n_\alpha | n_\alpha^0)$ is to evaluate (1.67) analytically using (1.66). This is done by using (1.39), expressed as

$$(1.68) \qquad \int_0^\infty dR_\alpha R_\alpha^2 j_{\ell_\alpha}(k_\alpha R_\alpha) j_{\ell_\alpha}(k'_\alpha R_\alpha) = \frac{\pi}{2} \frac{\delta(k_\alpha - k'_\alpha)}{k_\alpha^2},$$

along with the Kronecker deltas, to obtain

(1.69)

$$I^J(k_\alpha, n_\alpha | n_\alpha^0) = \int_0^\infty dk_\alpha^0 {k_\alpha^0}^2 A_{\ell_\alpha^0}(k_\alpha^0) \exp[-i(E_{v_\alpha^0 j_\alpha^0} + \hbar^2 {k_\alpha^0}^2/2\mu)t/\hbar]$$
$$\times \left\{ \delta_{v_\alpha v_\alpha^0} \delta_{j_\alpha j_\alpha^0} \delta_{\ell_\alpha \ell_\alpha^0} \frac{\pi}{2} \frac{\delta(k_\alpha - k_\alpha^0)}{{k_\alpha^0}^2} + \frac{\hbar^2}{4\mu i} \tilde{T}^J(n_\alpha | n_\alpha^0) \right.$$
$$\left. \left[\frac{\hbar^2 {k_\alpha^0}^2}{2\mu} + E_{v_\alpha^0 j_\alpha^0} - \frac{\hbar^2 {k'_\alpha}^2}{2\mu} - E_{v_\alpha j_\alpha} + i\epsilon \right]^{-1} \right\}.$$

The first term containing $\delta(k_\alpha - k_\alpha^0)$ is trivial to evaluate, leading to the expression

$$(1.70) \qquad I^J(k_\alpha, n_\alpha | n_\alpha^0) = \delta_{v_\alpha v_\alpha^0} \delta_{j_\alpha j_\alpha^0} \delta_{\ell_\alpha \ell_\alpha^0} \frac{\pi}{2} A_{\ell_\alpha^0}(k_\alpha) \exp[-i \left(\frac{\hbar^2 k_\alpha^2}{2\mu} + E_{v_\alpha j_\alpha} \right) t/\hbar]$$
$$+ \frac{\hbar^2}{4\mu i} \int_0^\infty dk_\alpha^0 {k_\alpha^0}^2 A_{\ell_\alpha^0}(k_\alpha^0) \exp\left[-i \Big(E_{v_\alpha^0 j_\alpha^0} + \hbar^2 {k_\alpha^0}^2/2\mu \Big) t/\hbar \right]$$
$$\times \tilde{T}^J(n_\alpha | n_\alpha^0) / \left[\frac{\hbar^2 {k_\alpha^0}^2}{2\mu} + E_{v_\alpha^0 j_\alpha^0} - \frac{\hbar^2 k_\alpha^2}{2\mu} - E_{v_\alpha j_\alpha} + i\epsilon \right].$$

The integral over k_α^0 in the second term on the right-hand side of (1.70) is done using the Cauchy residue theorem. It is convenient to carry out the integral using a kinetic energy variable ϵ_α^0, defined by

$$(1.71) \qquad \epsilon_\alpha^0 = \hbar^2 {k_\alpha^0}^2/2\mu,$$

such that

$$(1.72) \qquad I^J(k_\alpha, n_\alpha | n_\alpha^0) =$$
$$\delta_{v_\alpha v_\alpha^0} \delta_{j_\alpha j_\alpha^0} \delta_{\ell_\alpha \ell_\alpha^0} \frac{\pi}{2} A_{\ell_\alpha^0}(k_\alpha) \exp[-i \left(E_{v_\alpha^0 j_\alpha^0} + \frac{\hbar^2 {k_\alpha^0}^2}{2\mu} \right) t/\hbar]$$
$$+ \frac{1}{4i\hbar} \int_0^\infty d\epsilon_\alpha^0 \sqrt{2\mu\epsilon_\alpha^0} A_{\ell_\alpha^0}[k_\alpha^0(\epsilon_\alpha^0)] \exp[-i(\epsilon_\alpha^0 + E_{v_\alpha^0 j_\alpha^0})t/\hbar]$$
$$\times \tilde{T}^J(n_\alpha | n_\alpha^0) / \left(\epsilon_\alpha^0 + E_{v_\alpha^0 j_\alpha^0} - \epsilon_\alpha - E_{v_\alpha j_\alpha} + i\epsilon \right)$$

where

$$\epsilon_\alpha \equiv \hbar^2 \frac{k_\alpha^2}{2\mu}. \tag{1.73}$$

The denominator of the second term on the RHS of Eq. (1.72) vanishes where

$$\epsilon_\alpha^0 = \epsilon_\alpha + E_{v_\alpha j_\alpha} - E_{v_\alpha^0 j_\alpha^0} - i\epsilon, \tag{1.74}$$

so that it produces a pole in the fourth quadrant of the complex ϵ_α^0 plane. The quantity $A_{\ell_\alpha^0}[k_\alpha^0(\epsilon_\alpha^0)]$ does not have singularities in the region of interest of the ϵ_α^0 plane, nor does $\exp(-i\epsilon_\alpha^0 t/\hbar)$. The quantity $\tilde{T}^J(n_\alpha|n_\alpha^0)$ has poles on the negative real axis of ϵ_α^0 (corresponding to the true bound states of the full Hamiltonian H) and in the fourth quadrant of the complex ϵ_α^0 plane (corresponding to resonance complexes associated with the full 3-particle system) [42]. The resonance poles of $\tilde{T}^J$ thus occur at

$$\epsilon_\alpha^0(m) = \epsilon_{\alpha,r}^0(m) - i\Gamma_\alpha^0(m)/2, \quad m = 1, 2, ...; \tag{1.75}$$

m labels the resonance, and $\epsilon_{\alpha,r}^0(m)$ and $\Gamma_\alpha^0(m)/2(>0)$ are the real and imaginary parts of $\epsilon_\alpha^0(m)$. It turns out that the dominant factor determining whether one can close the contour in the lower complex ϵ_α^0 plane is the factor $\exp(-i\epsilon_\alpha^0 t/\hbar)$. In particular, one must close the contour so as to enclose the pole given in (1.74), which is $-\epsilon$ below the real axis, and exclude the poles due to $\tilde{T}^J(n_\alpha|n_\alpha^0)$, which occur at $-\frac{1}{2}\Gamma_\alpha^0(m)$ below the real ϵ_α^0 axis. Clearly, it is necessary then to take ϵ to be less than the minimum of the $\frac{1}{2}\Gamma_\alpha^0(m)$ (but $\epsilon > 0$). Then the contour can pass between the pole due to (1.74) and those due to (1.75). The smaller the minimum $\frac{1}{2}\Gamma_\alpha^0(m)$, the smaller ϵ must be made. However, the smaller is ϵ, the larger must be the time t at which the integral $I^J(k_\alpha n_\alpha|n_\alpha^0)$ is calculated. This is because the vanishing of the added contour required to close the path in the integral over ϵ_α^0 is determined by the magnitude of $\exp(-\epsilon_{\alpha,i}^0 t/\hbar)$, where $-\epsilon_{\alpha,i}^0$ is the imaginary part of ϵ_α^0, and $\epsilon_{\alpha,i}^0$ *must* be greater then ϵ but *smaller* than $\min[\frac{1}{2}\Gamma_\alpha^0(m)]$. Physically, we note that the lifetimes of the resonances produced by the Hamiltonian H are determined essentially by an uncertainty-principle-type relation,

$$\tau \sim 2\hbar/\Gamma_\alpha^0(m); \tag{1.76}$$

the smaller $\Gamma_\alpha^0(m)$, the longer the resonance complex lasts. Thus, t *must* be longer than the lifetime of the longest lived resonance complex. The result of the contour integration is

$$I^J(k_\alpha, n_\alpha|n_\alpha^0|t) = \frac{\pi}{2} A_{\ell_\alpha^0}(k_\alpha^0)\exp[-i(E_{v_\alpha^0 j_\alpha^0} + \frac{\hbar^2 k_\alpha^{0\,2}}{2\mu})t/\hbar][\delta_{v_\alpha v_\alpha^0}\delta_{j_\alpha j_\alpha^0}\delta_{\ell_\alpha \ell_\alpha^0} - k_{v_\alpha^0 j_\alpha^0}\tilde{T}^J(n_\alpha|n_\alpha^0)], \tag{1.77}$$

where

$$(1.78) \qquad \frac{\hbar^2 k_\alpha^2}{2\mu} + E_{v_\alpha j_\alpha} \equiv \frac{\hbar^2 {k_\alpha^0}^2}{2\mu} + E_{v_\alpha^0 j_\alpha^0}.$$

Using(1.54), (1.64), and (1.78) along with the Kronecker deltas in the right-hand side of (1.77), we may write this as

$$(1.79) \qquad I^J(k_\alpha, n_\alpha|n_\alpha^0|t) = \frac{\pi}{2}\sqrt{\frac{k_\alpha^0}{k_\alpha}} A_{\ell_\alpha^0}(k_\alpha^0)$$
$$\times \exp[-i(E_{v_\alpha^0 j_\alpha^0} + \frac{\hbar^2 {k_\alpha^0}^2}{2\mu})t/\hbar][\delta_{v_\alpha v_\alpha^0}\delta_{j_\alpha j_\alpha^0}\delta_{\ell_\alpha \ell_\alpha^0} - T^J(n_\alpha|n_\alpha^0)].$$

This is trivially inverted to yield the following expression for the transition amplitude

$$(1.80) \qquad T^J(n_\alpha|n_\alpha^0) = \delta_{v_\alpha v_\alpha^0}\delta_{j_\alpha j_\alpha^0}\delta_{\ell_\alpha \ell_\alpha^0} - \frac{2}{\pi}\sqrt{\frac{k_\alpha}{k_\alpha^0}}\left[A_{\ell_\alpha^0(k_\alpha^0)}\right]^{-1}$$
$$\times \exp\left[i\left(E_{v_\alpha^0 j_\alpha^0} + \frac{\hbar^2 {k_\alpha^0}^2}{2\mu}\right)t/\hbar\right]\int_0^\infty dR_\alpha R_\alpha^2 j_{\ell_\alpha}(k_\alpha R_\alpha)\Psi^J(n_\alpha|n_\alpha^0|R_\alpha, t)$$

This result is well known [43–46], but the derivation [37] just presented may be more satisfactory than those available earlier. This completes the introductory remarks. We now turn to consider how one solves the relevant time-independent integral equations or time-dependent differential equations of scattering.

2. Time-independent integral equation methods. We shall begin in Sections 2.1 and 2.2 by discussing two noniterative methods for solving the single-arrangement integral equations (1.47) for the $\chi^J(n_\alpha|n_\alpha^0|R_\alpha)$. Then in Section 2.3, we consider a set of multi-arrangement integral equations and their solution by a generalization of the method of Section 2.2.

2.1. Propagation method. First we consider a procedure involving replacing the original integral equation, which is an inhomogeneous Fredholm equation of the second kind, by an inhomogeneous Volterra equation of the second kind [42, 47–50]. To do this, we eliminate the $R_\alpha^<, R_\alpha^>$ variables by splitting the integral over R'_α into one from 0 to R_α and another from R_α to ∞:

$$(2.1) \qquad \underset{\sim}{\chi}^J(R_\alpha) = \underset{\sim}{j}(R_\alpha) - \frac{2\mu}{\hbar^2}\underset{\sim\sim}{k h}^+(R_\alpha)\int_0^{R_\alpha} dr'_\alpha {R'_\alpha}^2 \underset{\sim}{j}(R'_\alpha)\underset{\sim}{V}^J(R'_\alpha)\underset{\sim}{\chi}^J(R'_\alpha)$$
$$-\frac{2\mu}{\hbar^2}\underset{\sim\sim}{k j}(R_\alpha)\int_{R_\alpha}^\infty dR'_\alpha {R'_\alpha}^2 \underset{\sim}{h}^+(R'_\alpha)\underset{\sim}{V}^J(R'_\alpha)\underset{\sim}{\chi}^J(R'_\alpha).$$

Now we add and subtract the quantity

$$(2.2) \qquad \frac{2\mu}{\hbar^2}\underset{\sim\sim}{k j}(R_\alpha)\int_0^{R_\alpha} dR'_\alpha {R'_\alpha}^2 \underset{\sim}{h}^+(R'_\alpha)\underset{\sim}{V}^J(R'_\alpha)\underset{\sim}{\chi}^J(R'_\alpha)$$

and combine appropriate terms to write

$$\underset{\sim}{\chi}^J(R_\alpha) = \underset{\sim}{j}(R_\alpha)[\underset{\sim}{1} - \frac{2\mu}{\hbar^2}\underset{\sim}{k}\int_0^\infty dR'_\alpha R'^2_\alpha \underset{\sim}{h}^+(R'_\alpha)\underset{\sim}{V}^J(R'_\alpha)\underset{\sim}{\chi}^J(R'_\alpha)]$$
$$+\frac{2\mu}{\hbar^2}\underset{\sim}{k}\int_0^{R_\alpha} dR'_\alpha R'^2_\alpha [\underset{\sim}{j}(R_\alpha)\underset{\sim}{h}^+(R'_\alpha) - \underset{\sim}{h}^+(R_\alpha)\underset{\sim}{j}(R'_\alpha)]\underset{\sim}{V}^J(R'_\alpha)\underset{\sim}{\chi}^J(R'_\alpha). \tag{2.3}$$

Here we have used the fact that diagonal matrices commute. Now although we do not know its value, the matrix $\int_0^\infty dR'_\alpha R'^2_\alpha \underset{\sim}{h}^+(R'_\alpha)\underset{\sim}{V}^J(R'_\alpha)\underset{\sim}{\chi}^J(R'_\alpha)$ is a constant matrix. We introduce a new matrix function $\underset{\sim}{U}^J(R_\alpha)$, defined by

$$\underset{\sim}{\chi}^J(R_\alpha) = \underset{\sim}{U}^J(R_\alpha)\underset{\sim}{C}^J \tag{2.4}$$

where

$$\underset{\sim}{C}^J = \underset{\sim}{1} - \frac{2\mu}{\hbar^2}\underset{\sim}{k}\int_0^\infty dR'_\alpha R'^2_\alpha \underset{\sim}{h}^+(R'_\alpha)\underset{\sim}{V}^J(R'_\alpha)\underset{\sim}{\chi}^J(R'_\alpha). \tag{2.5}$$

This matrix $\underset{\sim}{C}^J$ can be shown to possess an inverse for all (real) scattering energies, and in fact is related to a quantity called the Jost matrix (see, *e.g.*, [42]). As a result, one finds that $\underset{\sim}{U}^J(R_\alpha)$ satisfies

$$\underset{\sim}{U}^J(R_\alpha) = \underset{\sim}{j}(R_\alpha) + \frac{2\mu}{\hbar^2}\underset{\sim}{k}\int_0^{R_\alpha} dR'_\alpha R'^2_\alpha [\underset{\sim}{j}(R_\alpha)\underset{\sim}{h}^+(R'_\alpha)$$
$$- \underset{\sim}{h}^+(R_\alpha)\underset{\sim}{j}(R'_\alpha)]\underset{\sim}{V}^J(R'_\alpha)\underset{\sim}{\chi}^J(R'_\alpha) \tag{2.6}$$

which is the desired inhomogeneous Volterra equation of the second kind. By using (2.4), we rewrite (2.5) as

$$\underset{\sim}{C}^J = \underset{\sim}{1} - \frac{2\mu}{\hbar^2}\underset{\sim}{k}\int_0^\infty dR'_\alpha R'^2_\alpha \underset{\sim}{h}^+(R'_\alpha)\underset{\sim}{V}^J(R'_\alpha)\underset{\sim}{U}^J(R'_\alpha)\underset{\sim}{C}^J, \tag{2.7}$$

so that

$$\underset{\sim}{C}^J = [\underset{\sim}{1} + \frac{2\mu}{\hbar^2}\underset{\sim}{k}\int_0^\infty dR'_\alpha R'^2_\alpha \underset{\sim}{h}^+(R'_\alpha)\underset{\sim}{V}^J(R'_\alpha)\underset{\sim}{U}^J(R'_\alpha)]^{-1}. \tag{2.8}$$

Thus once $\underset{\sim}{U}^J(R_\alpha)$ is known so that $\int_0^\infty dR'_\alpha R'^2_\alpha \underset{\sim}{h}^+(R'_\alpha)\underset{\sim}{V}^J(R'_\alpha)\underset{\sim}{U}^J(R'_\alpha)$ can be calculated, one can determine $\underset{\sim}{C}^J$. To obtain the transition amplitude matrix $\underset{\sim}{T}^J$, defined by

$$[\underset{\sim}{T}^J]_{n_\alpha, n^0_\alpha} = T^J(n_\alpha|n^0_\alpha), \tag{2.9}$$

we note that it can also be written as

$$\underset{\sim}{T}^J = \frac{4\mu i}{\hbar^2}\underset{\sim}{k}^{1/2}\int_0^\infty dR'_\alpha R'^2_\alpha \underset{\sim}{j}(R'_\alpha)\underset{\sim}{V}^J(R'_\alpha)\underset{\sim}{\chi}^J(R'_\alpha)\underset{\sim}{k}^{1/2}. \tag{2.10}$$

By (2.4), this is

$$T^J_{\sim} = \frac{4\mu i}{\hbar^2} k^{1/2}_{\sim} \int_0^\infty dR'_\alpha {R'_\alpha}^2 j(R'_\alpha) V^J_{\sim}(R'_\alpha) U^J_{\sim}(R'_\alpha) C^J_{\sim} k^{1/2}_{\sim}; \tag{2.11}$$

and using (2.8), we obtain

$$\begin{aligned} T^J_{\sim} = \frac{4\mu i}{\hbar^2} k^{1/2}_{\sim} \int_0^\infty & dR'_\alpha {R'_\alpha}^2 j_{\sim}(R'_\alpha) V^J_{\sim}(R'_\alpha) U^J_{\sim}(R'_\alpha) \\ \times [1_{\sim} + \frac{2\mu}{\hbar^2} k_{\sim} \int_0^\infty & dR'_\alpha {R'_\alpha}^2 h^+_{\sim}(R'_\alpha) V^J_{\sim}(R'_\alpha) U^J_{\sim}(R'_\alpha)]^{-1} k^{1/2}_{\sim}. \end{aligned} \tag{2.12}$$

This then completely expresses $T^J_{\sim}$ in terms of integrals over the matrix function $U^J_{\sim}(R_\alpha)$, which satisfies (2.6).

It is easy to devise an efficient algorithm to solve (2.6) for $U^J_{\sim}(R_\alpha)$ noniteratively. We introduce a Newton-Cotes quadrature for the integral over R'_α, with grids points at $R_1, R_2, \ldots, R_N$ so that for the point $R_\alpha = R_N$,

$$\begin{aligned} U^J_{\sim}(R_N) = j_{\sim}(R_N) + \frac{2\mu}{\hbar^2} k_{\sim} \sum_{\gamma=1}^{N} W_\gamma R_\gamma^2 [j(R_N) h^+_{\sim}(R_\gamma) \\ - h^+_{\sim}(R_N) j_{\sim}(R_\gamma)] V^J_{\sim}(R_\gamma) U^J_{\sim}(R_\gamma), \end{aligned} \tag{2.13}$$

where W_γ is the integration weight at grid point γ, and with the starting condition being that $U^J_{\sim}(0) = j_{\sim}(0)$. In practice, one does not start the solution at $R_\alpha = 0$ but rather at a point where the diagonal potential matrix elements are large compared to the total energy E. Now consider the term $\gamma = N$ in the sum:

$$S_N = W_N R_N^2 [j_{\sim}(R_N) h^+_{\sim}(R_N) - h^+_{\sim}(R_N) j_{\sim}(R_N)] V^J_{\sim}(R_N) U^J_{\sim}(R_N). \tag{2.14}$$

But the diagonal matrices $j_{\sim}$ and $h^+_{\sim}$ commute, and therefore S_N vanishes identically. Thus, (2.13) becomes

$$\begin{aligned} U^J_{\sim}(R_N) = j_{\sim}(R_N) + \frac{2\mu}{\hbar^2} k_{\sim} \sum_{\gamma=1}^{N-1} W_\gamma R_\gamma^2 \\ \times [j_{\sim}(R_N) h^+_{\sim}(R_\gamma) - h^+_{\sim}(R_N) j_{\sim}(R_\gamma)] V^J_{\sim}(R_\gamma) U^J_{\sim}(R_\gamma), \end{aligned} \tag{2.15}$$

which is a simple recursion relation for $U^J_{\sim}(R_N)$ in terms of its preceding values. For convenience in calculations, we define two partial sums

$$\Sigma_{\sim}(1|N-1) = \frac{2\mu}{\hbar^2} k_{\sim} \sum_{\gamma=1}^{N-1} W_\gamma R_\gamma^2 j_{\sim}(R_\gamma) V^J_{\sim}(R_\gamma) U^J_{\sim}(R_\gamma), \tag{2.16a}$$

$$\Sigma_{\sim}(2|N-1) = 1_{\sim} + \frac{2\mu}{\hbar^2} k_{\sim} \sum_{\gamma=1}^{N-1} W_\gamma R_\gamma^2 h^+_{\sim}(R_\gamma) V^J_{\sim}(R_\gamma) U^J_{\sim}(R_\gamma). \tag{2.16b}$$

Then (2.14) is written as

$$U^J(R_N) = j(R_N)\Sigma(2|N-1) - h^+(R_N)\Sigma(1|N-1). \tag{2.17}$$

Finally, we note that when R_N is large enough that

$$V^J(R_N) \sim 0, \tag{2.18}$$

the sums will have converged. Further, these converged sums, which we denote as $\sum(j|\infty)$, $j = 1, 2$, are precisely the integrals occuring in the expression (2.12) for the T^J. Thus, we have that

$$T^J = \frac{4\mu i}{\hbar^2} k^{1/2}\Sigma(1|\infty)[\Sigma(2|\infty)]^{-1} k^{1/2}, \tag{2.19}$$

so that when we have recurred out to the region where the potential is zero, all the information is at hand to calculate the transition amplitude matrix. It is clear that this algorithm will scale as the cube of the number of quantum states (channels) n_α, since one must carry out matrix multiplications and calculate the inverse in (2.19).

An important practical note is that one must be careful that the columns of the solution matrix $U^J(R_N)$ remain linearly independent in order to compute the scattering information. This necessitates periodically carrying out so-called stabilization transformations (stabilization methods are also required and were first developed for coupled differential equations involving closed channels as in [51, 52]). In practice, every 5-10 steps or so in the recursion, the solution or one of the partial sum matrices is transformed into upper triangular form. Suppose the transformations that accomplish this are denoted by $B(k)$, for the kth transformation. Then at the end of the calculation, in place of the $\Sigma(1|\infty)$, and $\Sigma(2|\infty)$, one has transformed matrices:

$$\tilde{\sum}(1|\infty) = \sum(1|\infty)B(1)B(2)\ldots B(N), \tag{2.20}$$

$$\tilde{\sum}(2|\infty) = \sum(2|\infty)B(1)B(2)\ldots B(N). \tag{2.21}$$

It is trivial to show that

$$\tilde{\Sigma}(1|\infty)[\tilde{\Sigma}(2|\infty)]^{-1} \equiv \Sigma(1|\infty)[\Sigma(2|\infty)]^{-1}, \tag{2.22}$$

so that the correct T^J matrix results even though the stabilization transformations have been applied to the solution. Finally, we note that one may replace the complex equations for the $U^J(R_\alpha)$ by real equations. We simply note that the Hankel functions may be written as

$$h^+_{\ell_\alpha}(k_\alpha R_\alpha) = n_{\ell_\alpha}(k_\alpha R_\alpha) + i j_{\ell_\alpha}(k_\alpha R_\alpha), \tag{2.23}$$

where $n_{\ell_\alpha}(k_\alpha R_\alpha)$ is the spherical Neumann function of order ℓ_α with a sign convention such that the asymptotic condition is

$$\lim_{R_\alpha \to \infty} n_{\ell_\alpha}(k_\alpha R_\alpha) = \cos(k_\alpha R_\alpha - \ell_\alpha \pi/2)/k_\alpha R_\alpha. \tag{2.24}$$

Then in obvious matrix notation

$$\underset{\sim}{h}^+(R_\alpha) = \underset{\sim}{n}(R_\alpha) + i\underset{\sim}{j}(R_\alpha), \tag{2.25}$$

and we see that (2.6) may be written as

$$\begin{aligned} \underset{\sim}{U}^J(R_\alpha) = \underset{\sim}{j}(R_\alpha) + \frac{2\mu}{\hbar^2}\underset{\sim}{k}\int_0^{R_\alpha} dR'_\alpha {R'_\alpha}^2 [\underset{\sim}{j}(R_\alpha)\underset{\sim}{n}(R'_\alpha) \\ + i\underset{\sim}{j}(R_\alpha)\underset{\sim}{j}(R'_\alpha) - \underset{\sim}{n}(R_\alpha)\underset{\sim}{j}(R'_\alpha) - i\underset{\sim}{j}(R_\alpha)\underset{\sim}{j}(R'_\alpha)]\underset{\sim}{V}^J(R'_\alpha)\underset{\sim}{U}^J(R'_\alpha). \end{aligned} \tag{2.26}$$

Clearly the imaginary terms cancel, leaving

$$\begin{aligned} \underset{\sim}{U}^J(R_\alpha) = \underset{\sim}{j}(R_\alpha) + \frac{2\mu}{\hbar^2}\underset{\sim}{k}\int_0^{R_\alpha} dR'_\alpha {R'_\alpha}^2 [\underset{\sim}{j}(R_\alpha)\underset{\sim}{n}(R'_\alpha) \\ - \underset{\sim}{n}(R_\alpha)\underset{\sim}{j}(R'_\alpha)]\underset{\sim}{V}^J(R'_\alpha)\underset{\sim}{U}^J(R'_\alpha). \end{aligned} \tag{2.27}$$

showing that, in fact, $\underset{\sim}{U}^J(R_\alpha)$ is a *real* function. Thus, the recursion algorithm can be formulated in a way which involves complex arithmetic only when the $\underset{\sim}{T}^J$ matrix is constructed, at the end of the calculation.

2.2. Basis function method. We now discuss an alternative method for solving collision problems by noniteratively solving the Lippmann-Schwinger equation (1.46) for $\chi^J(R_\alpha)$, Again, it is generally convenient to convert (1.46) into an equation for a related, real function using the relationship (2.23). We may then write

$$\begin{aligned} \underset{\sim}{\chi}^J(R_\alpha) = \underset{\sim}{j}(R_\alpha)[\underset{\sim}{1} - \frac{2\mu i}{\hbar^2}\underset{\sim}{k}\int_0^\infty dR'_\alpha {R'_\alpha}^2 \underset{\sim}{j}(R'_\alpha)\underset{\sim}{V}^J(R'_\alpha)\underset{\sim}{\chi}^J(R'_\alpha)] \\ - \frac{2\mu}{\hbar^2}\underset{\sim}{k}\int_0^\infty d{R'_\alpha}^2 {R'_\alpha}^2 \underset{\sim}{j}(R^<_\alpha)\underset{\sim}{n}(R^>_\alpha)\underset{\sim}{V}^J(R'_\alpha)\underset{\sim}{\chi}^J(R'_\alpha), \end{aligned} \tag{2.28}$$

or using (2.10)

$$\begin{aligned} \underset{\sim}{\chi}^J(R_\alpha) = \underset{\sim}{j}(R_\alpha)[\underset{\sim}{1} - \frac{1}{2}\underset{\sim}{k}^{1/2}\underset{\sim}{T}^J\underset{\sim}{k}^{-1/2}] \\ - \frac{2\mu}{\hbar^2}\underset{\sim}{k}\int_0^\infty dR'_\alpha {R'_\alpha}^2 \underset{\sim}{j}(R^<_\alpha)\underset{\sim}{n}(R^>_\alpha)\underset{\sim}{V}^J(R'_\alpha)\underset{\sim}{\chi}^J(R'_\alpha). \end{aligned} \tag{2.29}$$

We define the real function $\underset{\sim}{W}^J(R_\alpha)$ by

$$\underset{\sim}{\chi}^J(R_\alpha) = \underset{\sim}{W}^J(R_\alpha)[\underset{\sim}{1} - \frac{1}{2}\underset{\sim}{k}^{1/2}\underset{\sim}{T}^J\underset{\sim}{k}^{-1/2}], \tag{2.30}$$

where

$$W^J(R_\alpha) = j(R_\alpha) - \frac{2\mu}{\hbar^2} k \int_0^\infty dR'_\alpha {R'_\alpha}^2 j(R^<_\alpha) n(R^>_\alpha) V^J(R'_\alpha) W^J(R'_\alpha). \tag{2.31}$$

We then introduce an auxiliary function $\zeta^J(R_\alpha)$, termed the "amplitude density." (The amplitude density for nonreactive scattering was introduced by Johnson and Secrest [53].) The amplitude density is defined by

$$\zeta^J(R_\alpha) = V^J(R_\alpha) W^J(R_\alpha). \tag{2.32}$$

We remark that the amplitude density is quadratically integrable ($\mathcal{L}^2$) as a function of R_α and can be expanded in terms of $\mathcal{L}^2$ basis functions of R_α. We derive an integral equation for the $\zeta(R_\alpha)$ by applying $V^J(R_\alpha)$ to both sides of (2.31), and use (2.32) to obtain

$$\zeta^J(R_\alpha) = V^J(R_\alpha) j(R_\alpha) - \frac{2\mu}{\hbar^2} V^J(R_\alpha) k \int_0^\infty dR'_\alpha {R'_\alpha}^2 j(R^<_\alpha) n(R^>_\alpha) \zeta^J(R'_\alpha). \tag{2.33}$$

We expand

$$\zeta^J(n_\alpha|k^0_\alpha, n^0_\alpha|R_\alpha) = \sum_{t_\alpha} f(t_\alpha|R_\alpha) a^J(t_\alpha, n_\alpha|k^0_\alpha, n^0_\alpha), \tag{2.34}$$

where the $f(t_\alpha|R_\alpha)$ can be any convenient, complete set of functions (*e.g.*, harmonic oscillator functions [54, 55], sinusoidal functions [56–58], gaussians [59–63], or Bessel functions [64]),substitute into (2.34), multiply by the various $f^*(t_\alpha|R_\alpha)R^2_\alpha$, and integrate over R_α:

$$\begin{aligned} &\sum_{t'_\alpha} \mathcal{O}(t_\alpha|t'_\alpha) a^J(t'_\alpha, n_\alpha|k^0_\alpha, n^0_\alpha) \\ &= \sum_{n'_\alpha t'_\alpha} \int_0^\infty dR_\alpha R^2_\alpha f^*(t_\alpha|R_\alpha) V^J(n_\alpha|n'_\alpha|R_\alpha) \\ &\times [\delta_{v'_\alpha v^0_\alpha} \delta_{j'_\alpha j^0_\alpha} \delta_{\ell'_\alpha \ell^0_\alpha} j_{\ell^0_\alpha}(k^0_\alpha R_\alpha) - \frac{2\mu}{\hbar^2} k_{v'_\alpha j'_\alpha} \int_0^\infty dR'_\alpha {R'_\alpha}^2 \\ &\times j_{\ell'_\alpha}(k_{v'_\alpha j'_\alpha} R^<_\alpha) n_{\ell'_\alpha}(k_{v'_\alpha j'_\alpha} R^>_\alpha) f(t'_\alpha|R'_\alpha) a^J(t'_\alpha, n'_\alpha|k^0_\alpha, n^0_\alpha). \end{aligned} \tag{2.35}$$

Here $\mathcal{O}(t_\alpha|t'_\alpha)$ is an overlap matrix,

$$\mathcal{O}(t_\alpha|t'_\alpha) = \int_0^\infty dR_\alpha R^2_\alpha f^*(t_\alpha|R_\alpha) f(t'_\alpha|R_\alpha), \tag{2.36}$$

since the basis functions need not be orthogonal. Equation (2.35) is a set of linear, inhomogeneous algebraic equations whose solution is highly suited to modern high-speed vector supercomputers. We note that once (2.35) has been solved for the

$a^J(t_\alpha n_\alpha|k^0_\alpha n^0_\alpha)$, the transition amplitudes may be calculated as follows. By (2.10) and (2.30),

$$\underset{\sim}{T}^J = \frac{4\mu i}{\hbar^2}\underset{\sim}{k}^{1/2}\int_0^\infty dR'_\alpha {R'_\alpha}^2 \underset{\sim}{j}(R'_\alpha)\underset{\sim}{V}(R'_\alpha) \times \underset{\sim}{W}^J(R'_\alpha)[\underset{\sim}{1} - \frac{1}{2}\underset{\sim}{k}^{1/2}\underset{\sim}{T}^J\underset{\sim}{k}^{-1/2}]\underset{\sim}{k}^{1/2}. \tag{2.37}$$

But using (2.32), this yields

$$\underset{\sim}{T}^J = \frac{4\mu i}{\hbar^2}\underset{\sim}{k}^{1/2}\int_0^\infty dR'_\alpha {R'_\alpha}^2 \underset{\sim}{j}(R'_\alpha)\underset{\sim}{\zeta}^J(R'_\alpha)[\underset{\sim}{1} - \frac{1}{2}\underset{\sim}{k}^{1/2}\underset{\sim}{T}^J\underset{\sim}{k}^{-1/2}]\underset{\sim}{k}^{1/2}. \tag{2.38}$$

It is clear from (2.34) that to evaluate the integral in (2.38), one must calculate integrals of the sort

$$\int_0^\infty dR'_\alpha {R'_\alpha}^2 j_{\ell_\alpha}(k_{v_\alpha j_\alpha}R'_\alpha)f(t_\alpha|R'_\alpha), \tag{2.39}$$

and weight them by the expansion coefficients $a^J(t_\alpha n_\alpha|k^0_\alpha n^0_\alpha)$. Then (2.38) is solved for the $\underset{\sim}{T}^J$. If we define a so called "reactance matrix" $\underset{\sim}{R}^J$ as

$$\underset{\sim}{R}^J = \frac{4\mu}{\hbar^2}\underset{\sim}{k}^{1/2}\int_0^\infty dR_\alpha R_\alpha^2 \underset{\sim}{j}(R_\alpha)\underset{\sim}{\zeta}^J(R_\alpha)\underset{\sim}{k}^{1/2}, \tag{2.40}$$

then (2.38) can be written as

$$\underset{\sim}{T}^J = i[\underset{\sim}{R}^J - \frac{1}{2}\underset{\sim}{R}^J\underset{\sim}{T}^J], \tag{2.41}$$

or

$$\underset{\sim}{T}^J = i[\underset{\sim}{1} + \frac{i}{2}\underset{\sim}{R}^J]^{-1}\underset{\sim}{R}^J \tag{2.42}$$

$$= iR^J[1 + \frac{i}{2}R^J]^{-1}. \tag{2.43}$$

The above relations (2.41)–(2.43) are examples of the well known Heitler damping equation [42].

It is important to discuss briefly how the computational labor involved in applying the $\mathcal{L}^2$-amplitude-density moment method introduced above will vary with the number of quantum states and translational basis functions included. In the crudest form of application, one might include the same number M of translational basis functions in expanding every $\zeta^J(n|k^0, n^0|R)$. Then if there are N possible sets of quantum states n_α, the total dimension of the coefficient matrix in the algebraic equation for the $a^J(t_\alpha, n_\alpha|k^0_\alpha, n^0_\alpha)$ will be $(MN)\times(MN)$. The computational effort required for the inversion of the coefficient matrix will then vary as $(MN)^3$. This is in contrast to the N^3 variation of the Volterra equation method. However, the Volterra method involves an N^3 step at each quadrature point. If there are N_Q

quadrature points, the total work will scale as $N_Q N^3$. If $N_Q \gtrsim M^3$, the algebraic method will be more efficient. If $N_Q \lesssim M^3$, the Volterra method will be more efficient.

If desired, one may also use the $\mathcal{L}^2$ method of moments for the amplitude density to directly solve for the T^J-amplitudes. To do this one does not make the substitution of $\underset{\sim}{n}(R_\alpha) + i\underset{\sim}{j}(R_\alpha)$ for the $\underset{\sim}{h}^+(R_\alpha)$ in Eq.(1.46). Then in place of Eq.(2.32) defining the amplitude density, one defines the complex amplitude density $\tilde{\zeta}^J(R_\alpha)$ by

$$\tilde{\underset{\sim}{\zeta}}^J(R_\alpha) = \underset{\sim}{V}^J(R_\alpha)\underset{\sim}{\chi}^J(R_\alpha). \tag{2.44}$$

This new amplitude density is also $\mathcal{L}^2$, and its integral equation,

$$\begin{aligned}\tilde{\underset{\sim}{\zeta}}^J(R_\alpha) &= \underset{\sim}{V}^J(R_\alpha)\underset{\sim}{j}(R_\alpha) \\ &- \frac{2\mu}{\hbar^2}\underset{\sim}{V}^J(R_\alpha)\underset{\sim}{k}\int_0^\infty dR'_\alpha R'^{\,2}_\alpha \underset{\sim}{j}(R_\alpha^<)\underset{\sim}{h}^+(R_\alpha^>)\tilde{\underset{\sim}{\zeta}}(R_\alpha),\end{aligned} \tag{2.45}$$

may be solved by precisely the same technique as was used for Eq.(2.33). Furthermore, because one no longer has to utilize the damping equation to obtain $\underset{\sim}{T}^J$ from the $\tilde{\zeta}^J(R_\alpha)$, but rather

$$\underset{\sim}{T}^J = \frac{4\mu i}{\hbar^2}\underset{\sim}{k}^{1/2}\int_0^\infty dR_\alpha R_\alpha^2 \underset{\sim}{j}(R_\alpha)\tilde{\underset{\sim}{\zeta}}^J(R_\alpha)\underset{\sim}{k}^{1/2}, \tag{2.46}$$

it is possible to solve for a single initial condition (as can be done, *e.g.*, in the wavepacket method). However, the direct solution would still involve calculating the inverse of a coefficient matrix, which is now complex, and the work would still scale as $(MN)^3$ just as in the reactance matrix case. However, if a convergent iterative solution can be found, then the work for a single initial condition would scale as $(MN)^2 N_I$ where N_I the number of iterations. If the required number of iterations is small enough, the iterative procedure would then be more efficient. If one employed such an iterative method with real algebra to calculate the whole reactance matrix, the computational effort for N_O open channels would scale as $(MNN_O)^2 N_I$ which may be less favorable than $(MN)^3$.

Finally, we remark that very similar $\mathcal{L}^2$ expansions for the amplitude density can be combined with variational principles for the $\underset{\sim}{R}^J$ or $\underset{\sim}{T}^J$ matrices, and this has in fact been found to greatly reduce the number of basis functions required for convergence [65–68]. Thus, the factor M is reduced in many cases by an order of magnitude or more, resulting in a three-orders-of-magnitude decrease in the computational labor. As a result, variationally correct versions of the $\mathcal{L}^2$ expansions of amplitude densities promise to be an extremely powerful tool for treating the most demanding inelastic and reactive molecular collision systems.

2.3. Multi-arrangement-channel integral equations. Unlike the Volterra equation procedure, the above $\mathcal{L}^2$-amplitude-density method is equally as applicable to reactive (rearrangement) scattering as nonreactive scattering, and to see this, we now consider generalizing the Lippmann-Schwinger integral equation to a situation where atoms can be interchanged so as to lead to new molecular species in the collision. (The amplitude density was originally generalized to reactive systems using uncoupled equations for the reactive transition operator [69] and using a permutative arangement-channel coupling scheme [70]. The generalization to the Fock coupling scheme used here was first presented by some of the present authors and Shima [63].) The basic problem in reactive scattering has already been alluded to in Section 1.1, and it consists in the fact that more than one way of forming subclusters (molecules and atoms) can be realized, as indicated in the nonuniqueness of the unperturbed Hamiltonian H_α and corresponding perturbation V_α of (1.4). The total state function $|\chi^J\rangle$ must produce outgoing scattered waves in all limits $R_\alpha \to \infty$, $\alpha = 1, 2, 3$. We might try to solve for $|\chi^J\rangle$ as being made up of 3 pieces, each of which support scattering in one (and only one) arrangement. Then we write

$$\text{(2.47)} \qquad \Big|\chi^J(\alpha_0, n^0_{\alpha_0})\Big\rangle = \sum_{\alpha=1}^{3} \Big|\chi^J(\alpha|\alpha_0, n^0_{\alpha_0})\Big\rangle$$

and

$$\text{(2.48)} \qquad (E - H)\sum_{\alpha=1}^{3} \Big|\chi^J(\alpha|\alpha_0, n^0_{\alpha_0})\Big\rangle = 0$$

Equation (2.47) is clearly nonunique. The three arrangement components are made unique by defining a coupling scheme and introducing finite basis expansions for the 3 pieces of $|\chi^J\rangle$. To do this, we rearrange eq. (2.48) into three equations, in each of which a different arrangement is singled out on the left-hand side:

$$\text{(2.49)} \qquad (E - H_\alpha)\Big|\chi^J(\alpha|\alpha_0, n^0_{\alpha_0})\Big\rangle = V_\alpha\Big|\chi^J(\alpha|\alpha_0, n^0_{\alpha})\Big\rangle + \sum_{\alpha' \neq \alpha}(H - E)\Big|\chi^J(\alpha'|\alpha^0_{\alpha_0})\Big\rangle.$$

This is called the Fock coupling scheme [31, 63] because it can be shown [65] to be a generalization of the continuum multiconfiguration Hartree-Fock method mentioned in Sect. 1.1. (Other schemes for coupling the arrangement components in an expansion of the form (2.47) have also been studied [31, 55, 57, 63, 69, 70, 72, 73].) Just as in deriving (1.19), we write [57]

(2.50)
$$\Big|\chi^J(\alpha|\alpha_0, n^0_{\alpha_0})\Big\rangle = \delta_{\alpha\alpha_0}\Big|\Phi^J(\alpha_0, n^0_{\alpha_0})\Big\rangle + (E - H_\alpha + i\epsilon)^{-1}V_\alpha\Big|\chi^J(\alpha|\alpha_0, n^0_{\alpha_0})\Big\rangle + \sum_{\alpha' \neq \alpha}(E - H_\alpha + i\epsilon)^{-1}(H - E)\Big|\chi^J(\alpha'|\alpha_0, n^0_{\alpha_0})\Big\rangle,$$

It is also convenient to rewrite this using matrices in arrangement channel (α, α') space,

$$\text{(2.51)} \qquad \Big|\underset{\sim}{\chi}^J(\alpha_0, n^0_{\alpha})\Big\rangle = \Big|\underset{\sim}{\Phi}^J(\alpha_0, n^0_{\alpha_0})\Big\rangle + \underset{\sim}{G}{}^+_0\underset{\sim}{\mathcal{V}}\Big|\underset{\sim}{\chi}^J(\alpha_0, n^0_{\alpha_0})\Big\rangle.$$

Here, the vectors (in arrangement channel space) are

$$\left[\left|\underset{\sim}{\chi}^{J}(\alpha_0, n^0_{\alpha_0})\right\rangle\right]_\alpha = \left|\chi^J(\alpha|\alpha_0, n^0_{\alpha_0})\right\rangle, \tag{2.52}$$

$$\left[\left|\underset{\sim}{\Phi}^{J}(\alpha_0, n^0_{\alpha_0})\right\rangle\right]_\alpha = \delta_{\alpha\alpha_0}\left|\Phi^J(\alpha_0, n^0_{\alpha_0})\right\rangle, \tag{2.53}$$

and the matrices are

$$[\underset{\sim}{G}_0^+]_{\alpha\alpha'} = \delta_{\alpha\alpha'}[E - H_\alpha + i\epsilon]^{-1} \tag{2.54}$$

$$[\underset{\sim}{\mathcal{V}}]_{\alpha\alpha'} = \delta_{\alpha\alpha'}V_\alpha + (1 - \delta_{\alpha\alpha'})(H - E). \tag{2.55}$$

If (2.51) is put in the coordinate representation, it is complicated compared to the single-arrangement version considered previously by the fact that the scattering variable $\underset{\sim}{R}_\alpha$ in arrangement α depends on the scattering variable $\underset{\sim}{\boldsymbol{R}}_{\alpha'}$ and the molecular internal vector $\underset{\sim}{r}_{\alpha'}$ in arrangement α', and similarly for $\underset{\sim}{r}_\alpha$. As a result, integrals over $\underset{\sim}{r}_\alpha$ will involve knowledge of functions of $\underset{\sim}{R}_{\alpha'}$, $\alpha' \neq \alpha$, over all $\underset{\sim}{R}_{\alpha'}$ space, in general. This eliminates the possibility to convert (2.51) into a Volterra integral equation [47–50, 74] in any simple fashion. However, in analogy to the discussion of the amplitude density in Sect. 2.2, we define a generalized "reactive amplitude density" $|\underset{\sim}{\zeta}^J(\alpha_0, n^0_{\alpha_0} j^0_{\alpha_0} l^0_{\alpha_0})\rangle$ by

$$\left|\underset{\sim}{\zeta}^J(\alpha_0, n^0_{\alpha_0})\right\rangle = \underset{\sim}{\mathcal{V}}\left|\underset{\sim}{\chi}^J(\alpha_0, n^0_{\alpha_0})\right\rangle. \tag{2.56}$$

The equation satisfied by the $|\underset{\sim}{\zeta}^J(\alpha_0, n^0_{\alpha_0})\rangle$ is obtained by applying $\underset{\sim}{\mathcal{V}}$ to (2.51) from the left, using (2.56), to obtain

$$\left|\underset{\sim}{\zeta}^J(\alpha_0, n^0_{\alpha_0})\right\rangle = \underset{\sim}{\mathcal{V}}\left|\underset{\sim}{\Phi}^J(\alpha_0, n^0_{\alpha_0})\right\rangle + \underset{\sim}{\mathcal{V}}\underset{\sim}{G}_0^+\left|\underset{\sim}{\zeta}^J(\alpha_0, n^0_{\alpha_0})\right\rangle. \tag{2.57}$$

This is the abstract vector form of the reactive amplitude density integral equation.

The derivation given above of the Fock coupling scheme amplitude density equations, (2.50), can be made mathematically precise using the techniques of injection operators. This approach to few-body collision equations has been used by Chandler and Gibson [75] in their "Two-Hilbert-Space" formalism, and subsequently by Evans [76] in discussing coupled arrangement channel wavefunction equations. The derivation using injection operators will be given elsewhere [77].

It is of particular interest to examine, for a moment, the action of $\underset{\sim}{\mathcal{V}}$ on $\underset{\sim}{G}_0^+$. We note from (2.55) that off the diagonal ($\alpha \neq \alpha'$), the operators comprising $\underset{\sim}{\mathcal{V}}$ are all $(H - E)$. This operator involves the kinetic energy, which in the coordinate representation is a differential operator, and is therefore inconvenient. In addition, the energy E is a constant. We can simplify it by noting that [63]

$$(\underset{\sim}{\mathcal{V}}\underset{\sim}{G}_0^+)_{\alpha\alpha'} = (H - E)(E - H_{\alpha'} + i\epsilon)^{-1} \tag{2.58}$$

for $\alpha \neq \alpha'$. This is readily expressed as

$$(\underset{\sim}{\mathcal{V}}\underset{\sim}{G}_0^+)_{\alpha\alpha'} = -1 + V_{\alpha'}(E - H_{\alpha'} + i\epsilon)^{-1} \tag{2.59}$$

or

$$\underset{\sim}{\mathcal{V}}\underset{\sim}{G}_0^+ = -\underset{\sim}{N} + \underset{\sim}{V}\underset{\sim}{G}_0^+, \tag{2.60}$$

where

$$(\underset{\sim}{N})_{\alpha\alpha'} = (1 - \delta_{\alpha\alpha'}), \tag{2.61}$$

$$(\underset{\sim}{V})_{\alpha\alpha'} = V_{\alpha'}. \tag{2.62}$$

Finally, we note that, because $H_{\alpha_0}\Phi(\alpha_0) = E\Phi(\alpha_0)$,

$$\left(\underset{\sim}{\mathcal{V}}\left|\underset{\sim}{\Phi}^J(\alpha_0, n^0_{\alpha_0})\right\rangle\right)_\alpha = V_{\alpha_0}\left|\underset{\sim}{\Phi}^J(\alpha_0, n^0_{\alpha_0})\right\rangle, \tag{2.63}$$

and

$$\underset{\sim}{\mathcal{V}}\left|\underset{\sim}{\Phi}^J(\alpha_0, n^0_{\alpha_0})\right\rangle = \underset{\sim}{V}\left|\underset{\sim}{\Phi}^J(\alpha_0, n^0_{\alpha_0})\right\rangle. \tag{2.64}$$

Using these results, we can write our reactive amplitude density equation as [63]

$$\left|\underset{\sim}{\zeta}^J(\alpha_0, n^0_{\alpha_0})\right\rangle = \underset{\sim}{V}\left|\underset{\sim}{\Phi}^J(\alpha_0, n^0_{\alpha_0})\right\rangle - \underset{\sim}{N}\left|\underset{\sim}{\zeta}^J(\alpha_0, n^0_{\alpha_0})\right\rangle + \underset{\sim}{V}\underset{\sim}{G}_0^+\left|\underset{\sim}{\zeta}^J(\alpha_0, n^0_{\alpha_0})\right\rangle \tag{2.65}$$

which involves no differential operators. This can be put into the coordinate representation by projecting the αth component onto the coordinate state $\langle \underset{\sim}{R}_\alpha, \underset{\sim}{r}_\alpha | (\equiv \langle R_\alpha \theta_\alpha \phi_\alpha r_\alpha \gamma_\alpha \xi_\alpha |)$, to obtain

(2.66)

$$\begin{aligned}
&\zeta^J(\alpha|\alpha_0, n^0_{\alpha_0}|\underset{\sim}{R}_\alpha, \underset{\sim}{r}_\alpha) = V_{\alpha_0}\left[\underset{\sim}{R}_{\alpha_0}(\underset{\sim}{R}_\alpha, \underset{\sim}{r}_\alpha), \underset{\sim}{r}_{\alpha_0}(\underset{\sim}{R}_\alpha, \underset{\sim}{r}_\alpha)\right] \\
&\times \Phi^J(\alpha_0, n^0_{\alpha_0}|\underset{\sim}{R}_{\alpha_0}(\underset{\sim}{R}_\alpha, \underset{\sim}{r}_\alpha), \underset{\sim}{r}_{\alpha_0}(\underset{\sim}{R}_\alpha, \underset{\sim}{r}_\alpha)) \\
&- \sum_{\alpha' \neq \alpha} \zeta^J(\alpha'|\alpha_0, n^0_{\alpha_0}|\underset{\sim}{R}_{\alpha'}(\underset{\sim}{R}_\alpha, \underset{\sim}{r}_\alpha), \underset{\sim}{r}_{\alpha'}(\underset{\sim}{R}_\alpha, \underset{\sim}{r}_\alpha)) \\
&+ \sum_{\alpha'} V_{\alpha'}\left[\underset{\sim}{R}_{\alpha'}(\underset{\sim}{R}_\alpha, \underset{\sim}{r}_\alpha), \underset{\sim}{r}_{\alpha'}(\underset{\sim}{R}_\alpha, \underset{\sim}{r}_\alpha)\right] \\
&\times \int d\underset{\sim}{R}'_{\alpha'} d\underset{\sim}{r}'_{\alpha'} G^+_{\alpha'}\left[\underset{\sim}{R}_{\alpha'}(\underset{\sim}{R}_\alpha, \underset{\sim}{r}_\alpha), \underset{\sim}{r}_{\alpha'}(\underset{\sim}{R}_\alpha, \underset{\sim}{r}_\alpha)|\underset{\sim}{R}'_{\alpha'}, \underset{\sim}{r}'_{\alpha'}\right]\zeta^J(\alpha'|\alpha_0, n^0_{\alpha_0}|\underset{\sim}{R}'_{\alpha'}, \underset{\sim}{r}'_{\alpha'}),
\end{aligned}$$

where by $\underset{\sim}{R}_{\alpha'}(\underset{\sim}{R}_\alpha, \underset{\sim}{r}_\alpha)$ and $\underset{\sim}{r}_{\alpha'}(\underset{\sim}{R}_\alpha, \underset{\sim}{r}_\alpha)$ we mean the point $\underset{\sim}{R}_\alpha, \underset{\sim}{r}_\alpha$ expressed in terms of the α' arrangement coordinates. Thus, $\underset{\sim}{R}_{\alpha'}(\underset{\sim}{R}_\alpha, \underset{\sim}{r}_\alpha)$, $\underset{\sim}{r}_{\alpha'}(\underset{\sim}{R}_\alpha, \underset{\sim}{r}_\alpha)$ is simply an "alias" for the point $\underset{\sim}{R}_\alpha, \underset{\sim}{r}_\alpha$. However, $\underset{\sim}{R}'_{\alpha'}$ and $\underset{\sim}{r}'_{\alpha'}$ are integration variables and are not restricted by $\underset{\sim}{R}_\alpha, \underset{\sim}{r}_\alpha$. It is the fact that the radial variable $R_{\alpha'}$ depends on

$\underset{\sim}{R}_\alpha$ and $\underset{\sim}{r}_\alpha$ that prevents one from converting (2.66) into a Volterra equation, so as to employ the recursion method discussed earlier.

In this connection it is instructive to again consider the MCHF method for electron scattering [6, 9–29], which was discussed briefly in Section 1.1. In that case all V_α are sums of coulomb operators. Furthermore consider the three-body problem of $e_1^- + p^+e_2^-$, where e_1^- and e_2^- denote electrons and p^+e^- is an H atom. Then, because the mass of a nucleus is so much larger than that of an electron, we may take the $m_B \to \infty$ limit in eqs. (1.12) and (1.13) with negligible error. This plus the form of V_α makes the integral kernel in eq. (2.66) particularly simple [21] so that a noniterative integral equation propagation method may still be applied [47, 50, 78], but this is not possible for more general arrangements such as considered here. (Other methods for treating chemical reactions by propagation methods may be based on natural collision coordinates or hyperspherical coordinates [31, 79–82], but it is beyond the scope of the present chapter to discuss those approaches or yet other approaches based on matching local solutions obtained in disjoint regions of configuration space.) Thus previous work on the MCHF approach to chemical reactions was cast in terms of coupled integrodifferential equations [32, 33, 83–87]. Equation (2.66) provides an equivalent formulation in terms of coupled integral equations.

If one projects (2.66) onto a particular α-arrangement internal state by multiplying the equation by some $\langle Z^J_{n_\alpha} | \underset{\sim}{r}_\alpha, \hat{R}_\alpha \rangle$ and integrating over $\underset{\sim}{r}_\alpha$, effectively this also integrates the $R_{\alpha'}$ variable in the Green's function, and obviates the steps (2.1)–(2.3) required to obtain Volterra equations. The procedure then is to expand the $|\zeta^J(\alpha'|\alpha_0, n^0_{\alpha_0} | \underset{\sim}{R}_{\alpha'}, \underset{\sim}{r}_{\alpha'}\rangle$ in terms of basis functions appropriate for each particular arrangement α', for all α'. Projecting (2.66) with specific basis functions then leads to simultaneous linear inhomogeneous algebraic equations which may be very efficiently solved utilizing vector-processing supercomputers. The α'-arrangement basis set is taken to be comprised of the $\Lambda^{JM}_{j_{\alpha'}\ell_{\alpha'}}(\theta_{\alpha'}, \phi_{\alpha'}, \gamma_{\alpha'}, \xi_{\alpha'})\Delta(n_{\alpha'}, j_{\alpha'}|r_{\alpha'})f(t_{\alpha'}|R_{\alpha'})$, just as would be used in nonreactive scattering in the same arrangement. Because there are no external torques, the equations are uncoupled in the J, M quantum numbers (and are independent of M) [4, 9, 10]. Thus, we write

$$\zeta^J(\alpha'|\alpha_0, n^0_{\alpha_0}|\underset{\sim}{R}_\alpha, \underset{\sim}{r}_\alpha) = \sum_{\substack{n_{\alpha'} \\ t_{\alpha'}}} \Lambda^{JM}_{j_\alpha \ell_\alpha}(\theta_{\alpha'}, \phi_{\alpha'}, \gamma_{\alpha'}, \xi_{\alpha'}) \times \Delta(v_{\alpha'}, j_{\alpha'}|r_{\alpha'}) f(t_{\alpha'}|R_{\alpha'}) a^J(t_{\alpha'}, n_{\alpha'}|k^0_{\alpha_0}, n^0_{\alpha_0}) \tag{2.67}$$

where

$$\begin{aligned} \sum_{t_\alpha} \mathcal{O}(t_\alpha|t'_\alpha) a^J(t'_\alpha, n_\alpha|k^0_{\alpha_0}, n^0_{\alpha_0}) &= \int d\underset{\sim}{R}_\alpha d\underset{\sim}{r}_\alpha \Lambda^{JM*}_{j_\alpha \ell_\alpha}(\hat{R}_\alpha, \hat{r}_\alpha)\Delta^*(v_\alpha, j_\alpha|r_\alpha) \\ &\times f^*(t_\alpha|R_\alpha) V_{\alpha_0}\left[\underset{\sim}{R}_{\alpha_0}(\underset{\sim}{R}_\alpha, \underset{\sim}{r}_\alpha), \underset{\sim}{r}_{\alpha_0}(\underset{\sim}{R}_\alpha, \underset{\sim}{r}_\alpha)\right] \Phi^J[\alpha_0, n^0_{\alpha_0}|\underset{\sim}{R}_{\alpha_0}(\underset{\sim}{R}_\alpha, \underset{\sim}{r}_\alpha), \underset{\sim}{r}_{\alpha_0}(\underset{\sim}{R}_\alpha, \underset{\sim}{r}_\alpha)] \\ &- \sum_{\alpha' \neq \alpha} \sum_{n_{\alpha'} t_{\alpha'}} \left[\int d\underset{\sim}{R}_\alpha d\underset{\sim}{r}_\alpha \, \Lambda^{JM*}_{j_\alpha \ell_\alpha}(\hat{R}_\alpha, \hat{r}_\alpha)\Delta^*(v_\alpha, j_\alpha|r_\alpha) f^*(t_\alpha|R_\alpha) \Lambda^{JM}_{j_{\alpha'}\ell_{\alpha'}}(\hat{R}_{\alpha'}, \hat{r}_{\alpha'}) \right. \end{aligned} \tag{2.68}$$

$$\times \Delta[v_{\alpha'}, j_{\alpha'}|r_{\alpha'}(\underset{\sim}{R}_\alpha, \underset{\sim}{r}_\alpha)] f(t_{\alpha'}|R_{\alpha'}(\underset{\sim}{R}_\alpha, \underset{\sim}{r}_\alpha)) \Big] a^J(t_{\alpha'}, n_{\alpha'}|k^0_{\alpha_0}, n^0_{\alpha_0})$$

$$- \frac{2\mu}{\hbar^2} \sum_{\alpha'} \sum_{n_{\alpha'} t_{\alpha'}} \Big[\iint d\underset{\sim}{R}_\alpha d\underset{\sim}{r}_\alpha \int_0^\infty d\underset{\sim}{R}'_{\alpha'} \underset{\sim}{R}'^{\,2}_{\alpha'} \Lambda^{JM*}_{j_\alpha \ell_\alpha}(\hat{R}_\alpha, \hat{r}_\alpha) \Delta^*(v_\alpha, j_\alpha|r_\alpha) f^*(t_\alpha|R_\alpha)$$

$$\times V_{\alpha'} \Big[\underset{\sim}{R}_{\alpha'}(\underset{\sim}{R}_\alpha, \underset{\sim}{r}_\alpha), \underset{\sim}{r}_{\alpha'}(\underset{\sim}{R}_\alpha, \underset{\sim}{r}_\alpha) \Big] k_{v_{\alpha'} j_{\alpha'}} h^+_{\ell_{\alpha'}}(k_{v_{\alpha'} j_{\alpha'}} R^>_{\alpha'}) j_{\ell_{\alpha'}}(k_{v_{\alpha'} j_{\alpha'}} R^<_{\alpha'})$$

$$\times f(t_{\alpha'}|R'_{\alpha'}) \Lambda^{JM}_{j_{\alpha'} \ell_{\alpha'}} \Big[\hat{R}_{\alpha'}(\underset{\sim}{R}_\alpha, \underset{\sim}{r}_\alpha) \hat{r}_{\alpha'}(\underset{\sim}{R}_\alpha, \underset{\sim}{r}_\alpha) \Big] \Delta^*(v_{\alpha'}, j_{\alpha'}|r_{\alpha'}) \Big]$$

$$\times a^J(t_{\alpha'}, n_{\alpha'}|k^0_{\alpha_0}, n^0_{\alpha_0}).$$

The integrals over all angles (except for the γ_α) can be done analytically by use of rotational symmetry properties [63, 88, 89], leaving integrals over the R_α, r_α, γ_α, and $R'_{\alpha'}$. To achieve this, we rewrite (1.25) as

(2.69)
$$\Lambda^{JM}_{j_{\alpha'} \ell_{\alpha'}}(\hat{R}_{\alpha'}, \hat{r}_{\alpha'}) = \sqrt{\frac{2\ell_{\alpha'} + 1}{4\pi}} \sum_{m_{\alpha'}} \langle \ell_{\alpha'} o j_{\alpha'} m_{\alpha'} | J m_{\alpha'} \rangle Y_{j_{\alpha'} m_{\alpha'}}(\gamma_{\alpha'}, 0) \mathcal{D}^{J*}_{M m'_{\alpha'}}(\phi_{\alpha'}, \theta_{\alpha'}, \xi_{\alpha'}),$$

and

(2.70)
$$\mathcal{D}^{J*}_{M m_{\alpha'}}(\phi_{\alpha'}, \theta_{\alpha'}, \xi_{\alpha'}) = \sum_{m'_{\alpha'}} d^J_{m'_{\alpha'}, m_{\alpha'}}(\Delta_{\alpha\alpha'}) \mathcal{D}^{J*}_{M m'_{\alpha'}}(\phi_\alpha, \theta_\alpha, \xi_\alpha),$$

where $\Delta_{\alpha\alpha'}$ is the angle between $\hat{R}_{\alpha'}$ and $\hat{R}_\alpha$, and $\mathcal{D}^J_{Mm}$ and d^J_{Mm} are the rotation matrices of Wigner [5, 63]. Then because the potentials depend only on the size and shape of the 3-atom triangle and not its orientation, the integrals over $d\hat{R}_\alpha$ and $d\xi_\alpha$ can be done using

(2.71)
$$\int_0^{2\pi} d\xi_\alpha \int_0^{2\pi} d\phi_\alpha \int_0^\pi d\theta_\alpha \sin\theta_\alpha \mathcal{D}^J_{M m_\alpha}(\phi_\alpha \theta_\alpha \xi_\alpha), \times \mathcal{D}^{J*}_{M m'_{\alpha'}}(\phi_\alpha \theta_\alpha \xi_\alpha) = \frac{8\pi^2}{(2J+1)} \delta_{m_\alpha m'_{\alpha'}}.$$

Then we find

(2.72)
$$\sum_{t'_\alpha} \mathcal{O}(t_\alpha|t'_\alpha) a^J(t'_\alpha, n_\alpha|k^0_{\alpha_0}, n^0_{\alpha_0}) =$$

$$2\pi \int_0^\infty dR_\alpha R^2_\alpha \int_0^\infty dr_\alpha r^2_\alpha \int^\pi d\gamma_\alpha \sin\gamma_\alpha \; \frac{\sqrt{(2\ell_\alpha + 1)(2\ell^0_{\alpha_0} + 1)}}{(2J+1)}$$

$$\times \sum_{m_\alpha m_{\alpha_0}} \langle \ell_\alpha 0 j_\alpha m_\alpha | J m_\alpha \rangle Y^*_{j_\alpha m_\alpha}(\gamma_\alpha, 0) \Delta^*(v_\alpha, j_\alpha|r_\alpha)$$

$$\times f^*(t_\alpha|R_\alpha) V_{\alpha_0}[R_{\alpha_0}(R_\alpha, r_\alpha, \gamma_\alpha), r_{\alpha_0}(R_\alpha, r_\alpha, \gamma_\alpha), \gamma_{\alpha_0}(R_\alpha, r_\alpha, \gamma_\alpha)]$$

$$\times \langle \ell^0_{\alpha_0} 0 j^0_{\alpha_0} m_{\alpha_0} | J m_{\alpha_0} \rangle Y_{j^0_{\alpha_0} m_{\alpha_0}}(\gamma_{\alpha_0}, 0) \Delta(v^0_{\alpha_0}, j^0_{\alpha_0}|r_{\alpha_0}) d^J_{m_\alpha m_{\alpha_0}}(\Delta_{\alpha\alpha_0}) j_{\ell^0_{\alpha_0}}(k^0_{\alpha_0} R_{\alpha_0})$$

$$
\begin{aligned}
&-2\pi \sum_{\substack{\alpha'\neq\alpha \\ m_\alpha m_{\alpha'}}} \sum_{n_{\alpha'} t_{\alpha'}} \frac{\sqrt{(2\ell_\alpha+1)(2\ell_{\alpha'}+1)}}{(2J+1)} \langle \ell_\alpha 0 j_\alpha m_\alpha | J m_\alpha\rangle \langle \ell_{\alpha'} 0 j_{\alpha'} m_{\alpha'} | J m_{\alpha'}\rangle \\
&\times \Bigg[\int_0^\infty dR_\alpha R_\alpha^2 \int_0^\infty dr_\alpha r_\alpha^2 \int_0^\pi d\gamma_\alpha \sin\gamma_\alpha \, Y^*_{j_\alpha m_\alpha}(\gamma_\alpha, 0)\Delta^*(v_\alpha, j_\alpha | r_\alpha) \\
&\times f^*(t_\alpha|R_\alpha) Y_{j_{\alpha'} m_{\alpha'}}(\gamma_{\alpha'}, 0) d^J_{m_\alpha m_{\alpha'}}(\Delta_{\alpha\alpha'})\Delta(v_{\alpha'}, j_{\alpha'}|r_{\alpha'}) f(t_{\alpha'}|R_{\alpha'})\Bigg] \\
&\times a^J(t_{\alpha'}, n_{\alpha'}|k^0_{\alpha_0}, n^0_{\alpha_0}) - \frac{4\pi\mu}{\hbar^2} \sum_{\alpha' m_\alpha m_{\alpha'}} \sum_{n_{\alpha'} t_{\alpha'}} k_{v_{\alpha'} j_{\alpha'}} \frac{\sqrt{(2\ell_\alpha+1)(2\ell_{\alpha'}+1)}}{(2J+1)} \\
&\quad \langle \ell_\alpha 0 j_\alpha m_\alpha | J m_\alpha \rangle \\
&\times \langle \ell_{\alpha'} 0 j_{\alpha'} m_{\alpha'} | J m_{\alpha'}\rangle \int_0^\infty dR_\alpha R_\alpha^2 \int_0^\infty dr_\alpha r_\alpha^2 \int_0^\pi d\gamma_\alpha \sin\gamma_\alpha \\
&\times \int_0^\infty dR'_{\alpha'} R'^{\,2}_{\alpha'} Y^*_{j_\alpha m_\alpha}(\gamma_\alpha, 0)\Delta^*(v_\alpha, j_\alpha|r_\alpha) f^*(t_\alpha|R_\alpha) \\
&\times V_{\alpha'},(R_{\alpha'}, r_{\alpha'}, \gamma_{\alpha'}) h^+_{\ell_{\alpha'}}(k_{v_{\alpha'} j_{\alpha'}} R^>_{\alpha'}) j_{\ell_{\alpha'}}(k_{v_{\alpha'} j_{\alpha'}} R^<_{\alpha'}) \\
&\times f(t_{\alpha'}|R'_{\alpha'}) Y_{j_{\alpha'} m_{\alpha'}}(\gamma_{\alpha'}, 0) d^J_{m_\alpha m_{\alpha'}}(\Delta_{\alpha\alpha'})\Delta(v_{\alpha'} j_{\alpha'}|r_{\alpha'}) a^J(t_{\alpha'}, n_{\alpha'}|k^0_{\alpha_0}, n^0_{\alpha_0}).
\end{aligned}
$$

The remaining integrals over R_α, r_α, γ_α, and $R'_{\alpha'}$ are done numerically using either Gaussian quadratures or Newton-Cotes quadratures. The final result again is a set of algebraic equations for the expansion coefficients $a^J(t_\alpha n_\alpha|k^0_{\alpha_0} n^0_{\alpha^0})$ whose structure is basically the same as in the nonreactive case. This leads then to a very robust method which may be used for general collision problems. Applications have been made for the fully 3-dimensional collisions systems: I + H_2 nonreactive [63], H + H_2 and D + H_2 reactive [63, 90], O + H_2 reactive [71, 91]. In addition we are currently studying Br + H_2 reactive [92], Cl + H_2 reactive, and F + H_2 reactive collisions. (The latter two systems are currently under study in a collaboration involving the Minnesota, Houston, and Ames research groups.)

3. Time-dependent wavepacket methods. We conclude our discussion of methods for solving molecular collision problems by outlining a practical method for obtaining cross sections by solving the time-dependent Schrödinger equation. The detailed time-dependent Schrödinger equation for atom-diatom inelastic scattering, in matrix notation, is

(3.1)

$$
i\hbar\frac{\partial}{\partial t}\underset{\sim}{\Psi}^J(n^0_\alpha|R_\alpha, t) = \left[\underset{\sim}{1}\left(-\frac{\hbar^2}{2\mu R_\alpha}\frac{\partial^2}{2R^2_\alpha}R_\alpha\right) + \frac{\hbar^2}{2\mu R^2_\alpha}\underset{\sim}{\ell}^2 + \underset{\sim}{E} + \underset{\sim}{V}^J(R_\alpha)\right]\underset{\sim}{\Psi}^J(n^0_\alpha|R_\alpha, t)
$$

(3.2)

$$
= \underset{\sim}{H}\underset{\sim}{\Psi}^J(n^0_\alpha|R_\alpha, t),
$$

where

$$(\underset{\sim}{1})_{n_\alpha n_\alpha^0} = \delta_{v_\alpha v_\alpha^0}\delta_{j_\alpha j_\alpha^0}\delta_{\ell_\alpha \ell_\alpha^0}, \tag{3.3}$$

$$(\underset{\sim}{\ell^2})_{n_\alpha n_\alpha^0} = \delta_{v_\alpha v_\alpha^0}\delta_{j_\alpha j_\alpha^0}\delta_{\ell_\alpha \ell_\alpha^0}\ell_\alpha^0(\ell_\alpha^0+1), \tag{3.4}$$

$$(\underset{\sim}{E})_{n_\alpha n_\alpha^0} = \delta_{v_\alpha v_\alpha^0}\delta_{j_\alpha j_\alpha^0}\delta_{\ell_\alpha \ell_\alpha^0}E_{v_\alpha j_\alpha}, \tag{3.5}$$

$$[\underset{\sim}{V}^J(R_\alpha)]_{n_\alpha n_{\alpha'}} = V^J(n_\alpha|n'_\alpha|R_\alpha), \tag{3.6}$$

and

$$[\underset{\sim}{\Psi}^J(n_\alpha^0|R_\alpha,t)]_{n_\alpha} = \Psi^J(n_\alpha|n_\alpha^0|R_\alpha,t). \tag{3.7}$$

A convenient initial condition for $\underset{\sim}{\Psi}^J$ has already been discussed in Section 1. (In fact, other initial conditions are possible which can reduce the amount of labor required to calculate physically significant scattering amplitudes [93].) In order to propagate the coupled equations (3.1) in time, we note that the formal solution is

$$\underset{\sim}{\Psi}^J(n_\alpha^0|R_\alpha t) = \exp(-i\underset{\sim}{H}t/\hbar)\underset{\sim}{\Psi}^J(n_\alpha^0|R_\alpha 0). \tag{3.8}$$

This requires repeated evaluations of the radial derivatives of the $\underset{\sim}{\Psi}^J(n_\alpha^0|R_\alpha,t)$. Furthermore, we require a sufficiently accurate evaluation of the action of $\exp(-i\underset{\sim}{H}t/\hbar)$ so that long propagation times can be treated, since many cases of interest involve long-lived resonance complexes. Once the long time packet has been determined, the final state analysis may be carried out as discussed in Section 1. The action of $[-\frac{\hbar^2}{2\mu R_\alpha}\frac{\partial^2}{\partial R_\alpha^2}R_\alpha + \frac{\hbar^2}{2\mu R_\alpha^2}\ell_\alpha(\ell_\alpha+1)]$ on $\Psi^J(n_\alpha|n_\alpha^0|R_\alpha,t)$ can be evaluated by means of the Fast Fourier Transform (FFT) method. (The use of the FFT for molecular scattering calculations was originally suggested by Kosloff and Kosloff [94].) Here we review a technique developed by three of the authors [37]. It is based on the fact that the spherical Bessel function satisfies

$$\frac{\hbar^2}{2\mu}\left(-\frac{1}{R_\alpha}\frac{\partial^2}{\partial R_\alpha^2}R_\alpha + \frac{\ell_\alpha(\ell_\alpha+1)}{R_\alpha^2}\right)j_{\ell_\alpha}(kR_\alpha) = \frac{\hbar^2k^2}{2\mu}j_{\ell_\alpha}(kR_\alpha), \tag{3.9}$$

and the completeness relation

$$\int_0^\infty dR_\alpha R_\alpha^2 j_{\ell_\alpha}(kR_\alpha)j_{\ell_\alpha}(k'R_\alpha) = \frac{\pi}{2}\frac{\delta(k-k')}{k^2}. \tag{3.10}$$

Thus, we write

$$\Psi^J(n_\alpha|n_\alpha^0|R_\alpha,t) = \int_0^\infty dk k^2 j_{\ell_\alpha}(kR_\alpha)\overline{\Psi}^J(n_\alpha|n_\alpha^0|k,t), \tag{3.11}$$

so that by (3.9),

$$\left(-\frac{1}{R_\alpha}\frac{\partial^2}{\partial R_\alpha^2}R_\alpha + \frac{\ell_\alpha(\ell_\alpha(\ell_\alpha+1)}{R_\alpha^2}\right)\Psi^J(n_\alpha|n_\alpha^0|R,t) = \int_0^\infty dk k^4 j_{\ell_\alpha}(kR_\alpha)\overline{\Psi}^J(n_\alpha|n_\alpha^0|k,t). \tag{3.12}$$

To obtain the $\overline{\Psi}^J(n_\alpha|n_\alpha^0|kt)$ to use in (3.12), we employ (3.10) to write

$$\overline{\Psi}^J(n_\alpha|n_\alpha^0|k,t) = \frac{2}{\pi}\int_0^\infty dR_\alpha R_\alpha^2 j_{\ell_\alpha}(kR_\alpha)\Psi^J(n_\alpha|n_\alpha^0|R_\alpha,t). \tag{3.13}$$

In order to use these expressions, we need to convert them into a form involving the FFT. To illustrate, we focus on an integral of the form

$$I = \int_0^\infty dRR^2 j_\ell(kr)\Psi(R), \tag{3.14}$$

where $\Psi(R)$, as a prototypical wavepacket, is assumed to be basically gaussian in shape, behaving as $\exp(-\alpha R^2)$ for large R. Using the relation

$$j_\ell(kR) = \left(\frac{\pi k}{2R}\right)^{1/2} J_{\ell+1/2}(kR), \tag{3.15}$$

we write I as

$$I = \left(\frac{\pi k}{2}\right)^{1/2}\int_0^\infty dRR^{3/2}J_{\ell+1/2}(kR)\Psi(R), \tag{3.16}$$

where $J_{\ell+1/2}(kR)$ is a cylindrical Bessel function of order $\ell+\frac{1}{2}$. We then insert unity in the form

$$1 = \exp(-\delta R)\exp(\delta R), \tag{3.17}$$

so that

$$I = \left(\frac{\pi k}{2}\right)^{1/2}\int_0^\infty dRJ_{\ell+1/2}(kR)\exp(-\delta R)R^{3/2}\Psi(R)\exp(\delta R). \tag{3.18}$$

At this stage, δ is arbitrary but we will find it must be greater than zero, and not too large, for the method to work. We now expand $[J_{\ell+\frac{1}{2}}(kR)\exp(-\delta R)]$ and $[\exp(\delta R)R^{\frac{3}{2}}\Psi(R)]$ in terms of sine functions:

$$J_{\ell+1/2}(kR)\exp(-\delta R) = \int_0^\infty dKC_1(K)\sin(KR), \tag{3.19}$$

$$\exp(\delta R)R^{3/2}\Psi(R) = \int_0^\infty dKC_2(K)\sin(KR), \tag{3.20}$$

where

$$C_1(K) = \frac{2}{\pi}\int_0^\infty dR\exp(-\delta R)J_{\ell+1/2}(kR)\sin(KR) \tag{3.21}$$

and

$$C_2(K) = \frac{2}{\pi}\int_0^\infty dR \exp(\delta R) R^{3/2}\Psi(R)\sin(KR). \tag{3.22}$$

Then one easily finds that

$$I = \frac{\pi}{2}\left(\frac{\pi k}{2}\right)^{1/2}\int_0^\infty dK C_1(K) C_2(K). \tag{3.23}$$

Furthermore, $C_1(K)$ is readily evaluated analytically to be

$$C_1(K) = \frac{2}{\pi}\text{Im}\left\{\frac{[\sqrt{1+(\delta - iK)^2} - (\delta - iK)]^{\ell+1/2}}{k^{\ell+1/2}\sqrt{(\delta - iK)^2 + k^2}}\right\}. \tag{3.24}$$

It is here that $\delta > 0$ is required. Numerical tests show that large δ values lead to accurate results, but if δ is too large, then $\exp(\delta R)R^{\frac{3}{2}}\Psi(R)$ may extend too far for the FFT grid used, thereby necessitating use of a larger grid. This then increases the computation time, so that too large a δ makes the method become inefficient.

Another possibility is to include the centrifugal potential $\frac{\hbar^2}{2\mu R_\alpha^2}\underset{\sim}{\ell}^2$ in the potential and evaluate only the action of $-\frac{\hbar^2}{2\mu}\left(\frac{1}{R_\alpha}\frac{\partial^2}{\partial R_\alpha^2}R_\alpha\right)$ on Ψ^J. This can be done directly using the spherical Bessel functions $j_0(kR_\alpha)$ rather than $j_{\ell_\alpha}(kR_\alpha)$ [95]. This leads directly to sine type integrals which can be evaluated by FFT. However, in *any* case, the final state analysis requires evaluation of integrals of the sort given in (3.13), so that the above analysis is important even if not used in propagating the wavepacket. Another alternate procedure, discussed by Bisseling and Kosloff [95], is to use Fast Hankel Transforms; however, these have been found to be unstable in applications to gas phase atom-diatom collisions [37].

The above FFT procedures are accurate and efficient for repeatedly applying the radial kinetic energy operator to the wavepacket. The final step in developing a method for solving gas phase 3-dimensional atom-diatom collisions is to specify how (3.10) will be carried out. An efficient method was suggested by Tal-Ezer and Kosloff [96]. In this procedure, $\exp(-i\underset{\sim}{H}t/\hbar)$ is developed in an expansion using Chebychev polynomials, according to [37, 39, 96]

$$\exp(-i\underset{\sim}{H}dt/\hbar) = \sum_{t=0}^{\infty} a_t T_t(\underset{\sim}{W}), \tag{3.25}$$

where T_t is the Chebychev polynomial of order t, and the operator $\underset{\sim}{W}$ is defined by

$$(\underset{\sim}{W}) = (1/F)(-\underset{\sim}{H}\frac{dt}{\hbar} + (F+G)\underset{\sim}{1}), \tag{3.26}$$

$$F = dt(E_{\max} - E_{\min})/2\hbar, \tag{3.27}$$

$$G = E_{\min}dt/\hbar, \tag{3.28}$$

and

$$a_t = i^t(2 - \delta_{t,0}) \exp[-i(F + G)] J_t(F). \tag{3.29}$$

Here, J_t is a cylindrical Bessel function of order t. When the expansion index t is larger than F, $J_t(F) \sim 0$ and higher terms in the sum are negligible. Thus F determines the number of terms needed in the series. The energies $E_{\max}$ and $E_{\min}$ are the largest and smallest eigenvalues of $\underset{\sim}{H}$, so that the eigenvalues of $\underset{\sim}{W}$ lie between ± 1. In practice, $E_{\max}$ and $E_{\min}$ may be estimated as [39, 96]

$$E_{\max} = \frac{\pi^2 \hbar^2}{2\mu(\Delta R)^2} + (E_{v_\alpha j_\alpha})_{\max} + V_{\max}, \tag{3.30}$$

$$E_{\min} = (E_{v_\alpha j_\alpha})_{\min} + V_{\min}, \tag{3.31}$$

where ΔR is the grid size in the radial variable, $V_{\max}$ is the maximum value of the potential on the grid, $V_{\min}$ is its smallest (non-negative) value, $(E_{v_\alpha j_\alpha})_{\max}$ is the largest vib-rotor energy included in the basis set, and $(E_{v_\alpha j_\alpha})_{\min}$ is the smallest. Note that by the uncertainty principle, $\pi^2\hbar^2/2\mu(\Delta R)^2$ is the largest radial momentum describable in the wavepacket. We point out that (3.25) enables one to propagate the wavepacket in a single time step by an amount dt. The *larger* dt is, the more terms are required in the sum in (3.25). If one is uninterested in analysis at intermediate times, one may propagate the packet in one step from the initial region to the asymptotic region, simply by making dt large enough, and then including enough terms in (3.25).

It is important to note that $\exp(-i\underset{\sim}{H}t/\hbar)$ is applied to a *vector* $\underset{\sim}{\Psi}^J(n_\alpha^0|R_\alpha, t)$, with a number of components equal to the number N of basis functions $\Lambda^{JM}_{j_\alpha \ell_\alpha} \Delta(v_\alpha, j_\alpha)$ included in the calculation. Thus, $\underset{\sim}{H}$ (or $\underset{\sim}{W}$) is an $N \times N$ matrix and $\underset{\sim}{\Psi}^J$ is an $N \times 1$ matrix. Application of $\exp(-i\underset{\sim}{H}t/\hbar)$ then involves multiplications that scale as N^2, rather than as the cube, N^3, as do the Volterra and $\mathcal{L}^2$-amplitude-density methods. This means one can handle many more channels in the time-dependent approach than in the time-independent methods. The price one pays for this is that one only gets information about scattering from a single initial state, although by judicious choice of the wavepacket, results for many energies can be obtained in a single propagation. The time-independent methods we have described, on the other hand, give results for *all* possible initial states. If, however, one could solve the complex $\mathcal{L}^2$-amplitude density equations for a single initial state by iteration, then the work would scale as the number of iterations times N^2. If the method converged in a few iterations, this might be more efficient than the wavepacket method. To date, the largest numerically exact quantum scattering calculation for full 3-dimensional scattering has been done by means of the wavepacket procedure; it has successfully been used to solve a problem with 18,711 quantum states, simulating the collision of an N_2 diatom with a corrugated (but nonvibrating) model of a LiF crystal surface [97]. It is of considerable interest to extend these calculations to other types of collisions. Initial work on strictly gas-phase collisions is underway. Finally, we note that wavepacket propagation in more than the radial variable may be more efficient than using basis function expansions in those variables (see, *e.g.*, [44, 45, 94, 95, 98–100]), and this is also under further study. In summary, the prognosis for new time-dependent methods is very bright. A continuing challenge, though, is to find the most efficient formulation for various problems with larger and larger N.

REFERENCES

[1] P.A.M. DIRAC, *Quantum Mechanics*, fourth edition, Oxford U.P., London, 1958.

[2] F. T. SMITH, *Generalized angular momentum in many-body collisions,*, Phys. Rev., 120 (1960), pp. 1058–1069.

[3] B. A. LIPPMANN AND J. SCHWINGER, *Variational principles for scattering processes, I.*, Phys. Rev., 79 (1950), pp. 469–480.

[4] A. M. ARTHURS AND A. DALGARNO, *The theory of scattering by a rigid rotator*, Proc. Roy. Soc., London A256 (1960), pp. 540–551.

[5] M. E. ROSE, *Angular Momentum*, Wiley, New York, 1957.

[6] H. S. W. MASSEY, *Theory of the scattering of slow electrons*, Rev. Mod. Phys., 28 (1956), pp. 199–213.

[7] D. SECREST, *Rotational excitation I: The quantal treatment* and, *Vibrational excitation I: The quantal treatment*, in *Atom-Molecule Collision Theory*, edited by R. B. Bernstein, Plenum, New York, 1979, pp. 265–299 and 377–390.

[8] J. R. OPPENHEIMER, *On the quantum theory of electronic impacts*, Phys. Rev., 32 (1928), pp. 361–376;
D. R. BATES, A. FUNDAMINSKY, AND H. S. W. MASSEY, *Excitation and Ionization of Atoms by Electron Impact—The Born and Oppenheimer Approximations,* Phil. Trans. Roy. Soc. Lond. A243 (1951), pp. 93–117.

[9] M. J. SEATON,, *The Hartree-Fock equations for continuous states with applications to electron excitation of the ground configuration terms of OI*, Phil. Trans., A245 (1953), pp. 469–499.

[10] I. C. PERCIVAL AND M. J. SEATON, *The partial wave theory of electron-hydrogen atom collisions*, Proc. Camb. Phil. Soc., 53 (1957), pp. 654–662.

[11] L. CASTILLEJO, I. C. PERCIVAL AND M. J. SEATON, *On the theory of elastic collisions between electrons and hydrogen atoms*, Proc. Roy. Soc., London, A254 (1960), pp. 259–272.

[12] P. G. BURKE AND K. SMITH, *The low-energy scattering of electrons and positrons by hydrogen atoms*, Rev. Mod. Phys., 34 (1962), pp. 458–502.

[13] P. G. BURKE AND H. M. SCHEY, *Elastic scattering of low-energy electrons by atomic hydrogen*, Phys. Rev., 126 (1962), pp. 147–162.

[14] P. G. BURKE, H. M. SCHEY AND K. SMITH, *Collisions of slow electrons and positrons with atomic hydrogen*, Phys. Rev., 129 (1963), pp. 1258–1274.

[15] N. F. MOTT AND H. S W. MASSEY, *The Theory of Atomic Collisions*, 3rd ed., Oxford University Press, London, 1965, Chapt. XV.

[16] P. G. BURKE, *Resonances in electron scattering and photon absorption*, Adv. Phys., 14 (1965), pp. 521–567.

[17] P. G. BURKE AND A. J. TAYLOR, *Correlation in the elastic and inelastic S-wave scattering of electrons by H and* He^+, Proc. Phys. Soc., 88 (1966), p. 549–562.

[18] K. SMITH, R. J. W. HENRY AND P. G. BURKE, *Scattering of electrons by atomic systems with configurations* $2p^q$ *and* $3p^q$, Phys. Rev., 147 (1966), pp. 21–28.

[19] K. SMITH, M. J. CONNELY AND L. A. MORGAN, *Trial wave functions in the close-coupling approximation*, Phys. Rev, 177 (1969), pp. 196–203.

[20] H. SARAPH, M. J. SEATON, AND J. SCHEMMING, *Excitation of forbidden lines in gaseous nebulae: I. Formulation and calculations for* $2p^q$ *ions*, Phil. Trans. Roy. Soc. London, A264 (1969), pp. 77–105.

[21] P. G. BURKE AND M. J. SEATON, *Numerical solutions of the integro-differential equations of electron-atom collisions theory*, Methods Comp. Phys., 10 (1971), pp. 1–80.

[22] F. E. HARRIS AND H. H. MICHELS, *Expansion method for electron-atom scattering*, Methods Comp. Phys., 10 (1971), pp. 143–210.

[23] K. SMITH, *The Calculation of Atomic Collision Processes*, Wiley-Interscience, New York, 1971.

[24] D. G. TRUHLAR, J. ABDALLAH, JR. AND R. L. SMITH, *Algebraic variational methods in scattering theory*, Adv. Chem. Phys., 25 (1974), pp. 211–293.

[25] P. G. BURKE AND W. D. ROBB, *The R-matrix theory of atomic processes*, Advan. At. Molec. Phys., 11 (1975), p. 143–214.

[26] D. G. TRUHLAR, *Electron scattering*, in *Semiempirical Methods of Electronic Structure Theory, Part B*, G. A. Segal, ed., Plenum, New York, 1977, pp. 247–288.

[27] G. GALLUP AND J. MACEK, *The separable approximation in multichannel electron-molecule collisions*, in *Electron-Molecule and Photon-Molecule Collisions*, T. Rescigno, V. McKoy, and B. I. Schneider, eds., Plenum, New York, 1977, pp. 109–121.

[28] R. K. NESBET, *Variational Methods in Electron-Atom Scattering Theory*, Plenum, New York, 1980.

[29] B. H. BRANSDEN AND C. J. JOACHAIN, *Physics of Atoms and Molecules*, Longman Group, Ltd., London, 1983, p. 510–511.

[30] K. SMITH, *Partial wave theory of positron-hydrogen atom collisions*, Proc. Phys. Soc., London, 78 (1961), pp. 549–553.

[31] D. W. SCHWENKE, D. G. TRUHLAR, AND D. J. KOURI, J. Chem. Phys., 86 (1987), pp. 2772–2786.

[32] D. A. MICHA, *A quantum mechanical model for simple molecular reactions*, Arkiv Fysik, 30 (1965), pp. 411–423.

[33] W. H. MILLER, *Coupled equations and the minimum principle for collisions of an atom and a diatomic molecule including arrengements*, J. Chem. Phys., 50 (1969), pp. 407–418.

[34] D. EYRE AND T. A. OSBORN, *Cluster expansions of the three-body problem: nonseparable interactions*, Phys. Rev. C, 26 (1982), pp. 1369–1378. and references therein.

[35] Y. HAHN, *Orthogonality properties of the coupled equations for rearrangement collision*, Phys. Lett., 30B (1969), pp. 595–596.

[36] Y HAHN, *Generalized variational bounds for multichannel scatterings*, Phys. Rev. C, 1 (1970), pp. 12–16.

[37] Y. SUN, R. C. MOWREY AND D. J. KOURI, *Spherical wave close coupling wavepacket formalism for gas phase nonreactive atom-diatom collisions*, J. Chem. Phys., 87 (1987), pp. 339–349.

[38] R. C. MOWREY AND D. J. KOURI, *On a hybrid close-coupling wave-packet approach to molecular scattering*, Chem. Phys. Lett., 119 (1985), pp. 285–289.

[39] R. C. MOWREY AND D. J. KOURI, *Close coupling wavepacket approach to numerically exact molecule-surface scattering calculations*, J. Chem. Phys., 84 (1986), pp. 6466–6473.

[40] B. JACKSON AND H. METIU, *A multiple gaussian wave packet theory of H_2 diffraction and rotational excitation by collision with solid surfaces*, J. Chem. Phys, 85 (1986), pp. 4129–4139.

[41] R. D. LEVINE, *Quantum mechanics of melecular rate processes*, Oxford U. P., London, 1969. Section 2.7.0

[42] R. G. NEWTON, *Scattering theory of waves and particles*, 2nd ed., Springer-Verlag, New York, 1982.

[43] E. J. HELLER, *Time-dependent approach to semiclassical dynamics*, J. Chem. Phys., 62 (1975), pp. 1544–1555.

[44] K. C. KULANDER, *Collision induced dissociation in collinear $H + H_2$: Quantum mechanical probabilities using the time-dependent wavepacket approach*, J. Chem Phys., 69 (1978), pp. 5064–5072.

[45] J. C. GRAY, G. A. FRASER, D. G. TRUHLAR AND K. C. KULANDER, *Quasiclassical trajectory calculations and quantal wave packet calculations for vibrational energy transfer at energies above the dissociation threshold*, J. Chem. Phys., 73 (1980), pp. 5726–5733.

[46] R. T. SKODJE AND D. G. TRUHLAR, *Localized gaussian wave packet methods for inelastic collisions involving anharmonic oscillations*, J. Chem. Phys., 80 (1984), pp. 3123–3136.

[47] W. N. SAMS AND D. J. KOURI, *Noniterative solutions of integral equations for scattering. I. Single channels and II. Coupled channels*, J. Chem. Phys., 51 (1969), pp. 4809–4819.

[48] D. SECREST, *Amplitude densities in molecular scattering*, in *Methods in Computational Physics*, eds. B. Alder, S. Fernbach, and M. Rotenberg, Vol 10, Academic Press, New York, 1983, pp. 243–286.

[49] R. A. WHITE AND E. F. HAYES, *Quantum mechanical studies of the vibrational excitation of H_2 by Li^+*, J. Chem. Phys., 57 (1972), pp. 2985–2993.

[50] E. R. SMITH AND R. J. W. HENRY, *Noniterative integral equation approach to scattering problems*, Phys. Rev. A, 7 (1973), pp. 1585–1590.

[51] M. E. RILEY AND A. KUPPERMANN, *Vibrational energy transfer in collisions between diatomic molecules*, Chem. Phys. Lett., 1 (1968), pp. 537–538.

[52] R. G. GORDON, *New method for constructing wave functions for bound states and scattering*, J. Chem. Phys., 51 (1969), p. 14–25.

[53] B. R. JOHNSON AND D. SECREST, *The solution of the nonrelativistic quantum scattering problem without exchange*, J. Math. Phys., 7 (1966), pp. 2187–2195.

[54] Z. C. KUROUGLU AND D. A. MICHA, *Diatomic transition operators: Results of L^2 basis expansions*, J. Chem. Phys., 72 (1980), pp. 3327–3336.

[55] Y. SHIMA AND M. BAER, *Arrangement channel approach to reactive systems: Theory and numerical algorithms (as applied to the HFH system)*, J. Phys. B, 16 (1983), pp. 2169–2184.

[56] D. J. ZVIJAC AND J. C. LIGHT, *R-matix theory for collinear chemical reactions*, Chem. Phys., 12 (1976), pp. 237–251.

[57] Y. SHIMA, D. J. KOURI, AND M. BAER, *BKLT equations for reactive scattering. I. Theory and application to three finite mass atom systems*, J. Chem. Phys., 78 (1983), pp. 6666–6679.

[58] C. J. BOCCHETTA AND J. GERRATT, *The application of the Wigner R-matrix theory to molecular collisions*, J. Chem. Phys., 82 (1985), pp. 1351–1362.

[59] O. H. CRAWFORD, *Calculation of chemical reaction rates by the R-matrix method*, J. Chem. Phys, 55 (1971), pp. 2563–2570.

[60] R. DER, O. GEBHARDT, AND R. HABERLANDT, *R-matrix theory of reactive atom-diatom collisions*, Chem. Phys. Lett., 27 (1974), pp. 107–110.

[61] I. P. HAMILTON AND J. C. LIGHT, *On distributed gaussian bases for simple model multidimensional vibrational problems*, J. Chem. Phys., 84 (1986), pp. 306–317.

[62] Z. BAČIĆ AND J. C. LIGHT, *Highly excited vibrational levels of "floppy" tratomic molecules: A discrete variable representation-distributed Gaussian basis approach*, J. Chem. Phys., 85 (1986), pp. 4594–4604.

[63] J. Z. H. ZHANG, D. J. KOURI, K. HAUG, D. W. SCHWENKE, Y. SHIMA AND D. G. TRUHLAR, *L^2-amplitude density method for multichannel inelastic and rearrangement collisions*, J. Chem. Phys. (in press).

[64] G. H. RAWITSCHER AND G. DELIC, *Sturmian eigenvalue equations with a Bessel function basis*, J. Math. Phys., 27 (1986), pp. 816–823.

[65] G. STASZEWSKA AND D. G. TRUHLAR, *Rapid convergence of discrete-basis representations of the amplitude density for quantal scattering calculations*, Chem. Phys. Lett., 130 (1986), pp. 341–345.

[66] G. STASZEWSKA AND D. G. TRUHLAR, *Convergence of L^2 methods for scattering problems*, J. Chem. Phys., 86 (1987), pp. 2793–2804.

[67] D. W. SCHWENKE, K. HAUG, D. G. TRUHLAR, Y. SUN, J. Z. H. ZHANG AND D. J. KOURI, J. Phys. Chem. (to be published).

[68] D. W. SCHWENKE, K. HAUG, M. ZHAO, D. G. TRUHLAR, Y. SUN, J. Z. H. ZHANG AND D. J. KOURI, J. Phys. Chem. (to be published).

[69] D. J. KOURI, *Theory of reactive scattering. I. Homogeneous integral solution formalism for the rearrangement T-operator integral equation*, J. Chem. Phys., 51 (1969), pp. 5204–5215.

[70] M. BAER AND D. J. KOURI, *Rearrangement channel operator approach to models for three body reactions I.*, Phys. Rev. A, 4 (1971), pp. 1924–1934.

[71] J. Z. H. ZHANG, Y. ZHANG, D. J. KOURI, B. C. GARRETT, K. HAUG, D. W. SCHWENKE AND D. G. TRUHLAR, *L^2 calculations of accurate quantal dynamical reactive scattering transition probabilities and their use to test semiclassical applications*, Faraday Discussions Chem. Soc., 84 (in press).

[72] D. J. KOURI, H. KRUGER AND F. S. LEVIN, *Arrangement channel quantum mechanics: A general time-dependent formalism for multiparticle scattering*, 15 (1977), pp. 1156–1171.

[73] D. J. KOURI AND M. BAER, *Arrangement channel quantum mechanical approach to reactive scattering, in The Theory of chemical reaction dynamics*, ed. D. C. Clary, D. Reidel, Dordrecht, Holland, 1986, pp. 359–381.

[74] D. J. KOURI, *Theory of reactive scattering. VI. Volterra equation formalism for coupled channel amplitude densities and modified wavefunctions*, Mol. Phys., 22 (1971), pp. 421–431.

[75] C. CHANDLER AND A. G. GIBSON, *Transition from time-dependent to time-independent multichannel quantum scattering theory*, J. Math. Phys., 14 (1973), pp. 1328–1335.

[76] J. W. EVANS, *Two-Hilbert space formulations of the quantum statistical mechanics of reactive fluids: Dimer formation and decay*, J. Chem. Phys., 85 (1986), pp. 5991–6003.

[77] D. J. KOURI, D. G. TRUHLAR, AND D. K. HOFFMAN, *manuscript in preparation.*.

[78] D. L. Knirk, E. F. Hayes, and D. J. Kouri, *Close coupled equations of electron-hydrogen atom scattering using a noniterative integral equation technique*, J. Chem. Phys., 57 (1972), pp. 4770–4781.

[79] R. E. Wyatt, *Direct-mode chemical reactions. I. Methodology for accurate quantal calculations*, in *Atom-Molecule collision Theory*, R. B. Bernstein, ed., Plenum, New York, 1979, pp. 567–594.

[80] A. Kuppermann, *A useful mapping of triatomic potential energy surfaces*, Chem. Phys. Lett., 82 (1975), pp. 374–375.

[81] B. R. Johnson, *The quantum dynamics of three particles in hyperspherical coordinates*, J. Chem. Phys., 79 (1983), pp. 1916–1925.

[82] J. Linderberg, *Basis for coupled channel approach to reactive scattering*, J. Quantum Chem. Symp., 19 (1986), pp. 467–476.

[83] G. Wolken, Jr. and M. Karplus, *Theoretical studies of* $H + H_2$ *reactive scattering*, J. Chem. Phys., 60 (1974), pp. 351–367.

[84] B. C. Garrett and W. H. Miller, *Quantum mechanical reactive scattering via exchange kernels: Application to the collinear* $H + H_2$ *reaction*, J. Chem. Phys., 68 (1978), pp. 4051–4055.

[85] J. E. Adams and W. H. Miller, *Expansion of exchange kernels for reactive scattering*, J. Phys. Chem., 83 (1979), pp. 1505–1508.

[86] P. S. Dardi, S. Shi, and W. H. Miller, *Quantum mechanical reactive scattering via exchange kernels: Infinite order exchange on a grid*, J. Chem. Phys., 83 (1985), pp. 575–583.

[87] M. R. Hermann and W. H. Miller, *Quantum mechanical reactive scattering via exchange kernels: Comparison of grid versus basis set expansion of the exchange interaction*, Chem. Phys., 109 (1986), pp. 163–172.

[88] W. H. Miller, *Distorted-wave theory for collisions of an atom and a diatomic molecule*, J. Chem. Phys., 49 (1968), pp. 2373–2381.

[89] B. H. Choi and K. T. Tang, *Theory of distorted-wave Born approximation for reactive scattering of an atom and a diatomic molecule*, J. Chem. Phys., 61 (1974), pp. 5147–5157.

[90] K. Haug, D. W. Schwenke, Y. Shima, D. G. Truhlar, J. Zhang and D. J. Kouri, L^2*-solution of the quantum mechanical reactive scattering problem. The threshold energy for* $D + H_2(v = 1) \rightarrow HD + H$, J. Phys. Chem., 90 (1986), pp. 6757–6759.

[91] K. Haug, D. W. Schwenke, D. G. Truhlar, Y. Zhang, J. Z. H. Zhang and D. J. Kouri, *Accurate quantum mechanical reaction probabilities for the reaction* $O + H_2 \rightarrow OH + H$, J. Chem. Phys., 87 (1987), pp. 1892–1894.

[92] Y. C. Zhang, J. Z. H. Zhang, D. J. Kouri, K. Haug, D. W. Schwenke and D. G. Truhlar, *Three-dimensional quantum mechanical reactive scattering. Numerically exact* L^2*-amplitude density study of the* H_2Br *reactive system*, in preparation.

[93] Y. Sun and D. J. Kouri, *Wavepacket study of gas phase atom-diatom scattering*, J. Chem. Phys. (submitted).

[94] R. Kosloff and D. Kosloff, *A Fourier method solution for the time-dependent Schrödinger equation: A study of the reaction* $H^+ + H_2$, $D^+ + HD$ *and* $D^+ + H_2$, J. Chem. Phys., 72 (1983), pp. 1823–1833.

[95] R. Bisseling and R. Kosloff, *The fast Hankel transform as a tool in the solution of the time-dependent Schrödinger equation*, J. Comput. Phys., 59 (1985), pp. 136–151.

[96] H. Tal-Ezer and R. Kosloff, *An accurate and efficient scheme for propagating the time-dependent Schrödinger equation*, J. Chem. Phys., 81 (1984), pp. 3967–3971.

[97] R. C. Mowrey, H. F. Bowen and D. J. Kouri, *Molecule-corrugated surface scattering calculations using the close coupling wavepacket method*, J. Chem. Phys., 86 (1987), pp. 2441–2442; and to be published.

[98] E. A. McCullough, Jr. and R. E. Wyatt, *Dynamics of the collinear* $H + H_2$ *reaction. I. Probability density and flux*, J. Chem. Phys., 54 (1971), pp. 3578–3583.

[99] R. B. Gerber, R. Kosloff and M. Berman, *Time-dependent wavepacket calculations of molecular scattering from surfaces*, Computer Phys. Repts., 5 (1986), pp. 59–114.

[100] C. Leforestier, *Competition between dissociation and exchange processes in a collinear A+BC collision. I. Exact quantum results*, Chem. Phys., 87 (1984), p. 241–261.

DYNAMICAL GROUPS

LIE ALGEBRAIC APPROACH TO MOLECULAR STRUCTURE AND DYNAMICS

R.D. LEVINE†

Abstract. The applications of Lie algebraic techniques to molecular physics are discussed with special reference to the open problems.

1. Introduction. The study of Lie Groups and Algebras is highly developed in Minnesota. The basic theory and its application has been described in important research papers as well as in books [1-4]. I could, of course, discuss additional and more specialized applications. It is unlikely however that I can improve upon the original presentations. Furthermore, as far as I can tell from *ca.* 130° longitude away, that is not the purpose of the workshop. Rather, what I shall try to do is describe what it is that some of us, coming from the applications side, are trying to do and what are the technical problems that we face.

I should also state explicitly that my discussion and judgement is biased in that I am first and foremost interested in dynamics. I need not only to know the structure of a molecule for its own sake but also as a key ingredient in the understanding of how the molecule responds to a perturbation, be it due to an electromagnetic field, or a collision with another molecule, etc. These are not idle remarks as there are algebraic procedures, which are of interest in their own sake, but which do not easily meet my requirements. Specifically, let H_0 be the Hamiltonian operator for the isolated atom or molecule. It is not enough for my purpose, to be able to solve the time independent Schrödinger equation

$$H_0\psi = E\psi \tag{1.1}$$

for the eigenvalue E and eigenfunction ψ. That can sometimes be done by rewriting the equation as

$$F(E - H_0)\psi = 0 \tag{1.2}$$

for some operator F. But such an approach will not, in a way that I can see, be immediately useful when the problem is to solve the time dependent equation

$$i\partial\psi/\partial t = (H_0 + V)\psi \tag{1.3}$$

where V is the perturbation. The celebrated 'factorization method' of Infeld and Hull [5] is often of limited use for my purposes because of these reasons. There are, of course, still some options, as we shall discuss below, such as using time

---

†The Fritz Haber Research Center for Molecular Dynamics, The Hebrew University, Jerusalem 91904, Israel. The work reported was supported by the Air Force Office of Scientific Research, Grant AFOSR 86-0011 and the Stiftung Volkswagenwerk. The Fritz Haber Research Center is supported by the Minerva Gesellschaft für die Forschung, mbH, München, BRD.

dilation [6,7] or writing $W = H_0 - i\partial/\partial t$ so that the stationary problem for W is equivalent to the time dependent Schrödinger equation for H_0 [8]. All I intended in the paragraph above is to state my bias.

My emphasis on the dynamics also means that I am interested (like most people these days), in large amplitude motions. It is these motions (which were inaccessible using conventional light sources) that we are now able to probe using lasers and it is inevitably such motion which takes place when chemical bonds break and reform during a chemical reaction. The harmonic approximation, valid near the bottom of the potential well, is not adequate even when corrected by low order perturbation theory, for such problems.

The recent surge in activity in both nuclear and molecular physics [9–19] is due to the (in part, phenomenological) result that Hamiltonians for systems of realistic structure can be written as a linear combination of a few terms provided these are *bilinear* in the generators of some Lie Group. This observation is the basis for the two central technical issues which stand in the way of further progress. The first is the 'derivation' of the algebraic Hamiltonian. Physicists and Chemists very strongly prefer to begin with a Hamiltonian which has a clear geometrical interpretation. In part, this is due to the historical development where the quantum mechanical Hamiltonian operator is fashioned from the corresponding classical kinetic and potential energies. There is however another and equally important reason. Our physical and, in particular, our chemical, intuition is very much geometrical. No matter how useful a purely algebraic Hamiltonian proves to be, the practicing chemists would insist on providing for it a geometrical interpretation. As we shall discuss, the route between the algebraic Hamiltonian operator and its geometrical realization is not well paved. Starting with an algebra one can find a realization but it need not be the realization one wants for the system of interest. The other direction, and one which is more common, is that one starts with a Hamiltonian as a differential operator in coordinate space. As an example, Miller ([2], table V page 74) quotes all one dimensional potentials for which the Schrödinger equation admits of a non trivial symmetry. I will discuss one example (the Morse potential) and mention others (*e.g.* the Poschl-Teller potential) which have been extensively studied (*e.g.* [20]) by algebraic procedures and which are not in the table. Even the most standard text book example, that of a square-well potential, for which an algebraic Hamiltonian, bilinear in the generator of $SU(1,1)$ is known [21], is not in the table. The reason all these example are not in the table is presumably that they do not belong in the table in that they do not possess a linear differential operator L as defined on page 74 of [2]. However, it is very useful to study such problems by algebraic means. What I am asking for is another table. As far as I can tell, the essential generalization is a simple one. The coordinate y which appears in the realization of the algebra need not be the same as the coordinate x in terms of which the physical problem is stated. Rather, $y = y(x)$. I do not however know how to do this in a systematic fashion.

The second essential difficulty with Hamiltonians which are bilinear in the generator is in evaluating the time propagation. The time displacement operator in quantum mechanics is $U(t) \equiv \exp(-iHt)$. For most purposes what is necessary to compute is of the generic form

$$U(t)XU^{\dagger}(t) \equiv I(t) \tag{1.4}$$

where X is a generator [22,23]. Since $U(t)$ is unitary, knowing $I(t)$ is sufficient to propagate in time any analytic function, $f(X)$, of X,

$$U(t)f(X)U^{\dagger}(t) = f(I(t)). \tag{1.5}$$

When the Hamiltonian is given as a linear combination of the generators $U(t)$ is an element of the corresponding group and hence $I(t)$ as defined in (4.1) is immediately known to be a linear combination of generators. In most realistic problems where the Hamiltonian is bilinear in the generators, there is not much that we can say, in general, about $I(t)$. One can argue that this is hardly surprising. Hamiltonians which are bilinear in the generators cover a truly extensive range of diverse physical behavior including systems whose classical motion is chaotic [24]. Knowing the general structure of $I(t)$ is knowing a lot about the possible dynamics. Perhaps one should be content with less, such as under what conditions on the Hamiltonian can one say something about $I(t)$. For example, say the Hamiltonian is a linear combination of the bilinear Casimir invariants of subgroups, what is the restriction of $I(t)$.

The operator $I(t)$ is closely related to the symmetry operator L as defined by Miller [2]. This can be seen by replacing the definition (1.4) by

$$\left[i\frac{\partial}{\partial t} - H, I\right] = 0 \tag{1.6}$$

and the boundary condition $I(0) = X$. The problem is that if H is bilinear in the generator, the definition (1.6) does not, in an obvious way, offer clues as to the structure of $I(t)$. In particular, taking $I(t)$ to be a linear combination of operators with time dependent coefficients does not, for me, offer a systematic approach.

Molecular physicists do not expect to obtain closed analytic solution for problem of realistic complexity. We have indeed provided a variational solution for $I(t)$ which effective 'linearizes' the Hamiltonian, [25]. This is a, so called, self consistency procedure. Even more special cases include, for example, an adiabatic procedure [26]. Chemists thrive on practical approximations and more work is required towards computational schemes. The one method that every theorist learns at the freshperson year is to represent the Hamiltonian as a matrix in some suitable basis and to solve (1.3) or (1.4) numerically as a matrix equation. Such a method is not realistic for many problems of practical interest since the matrix size for acceptable accuracy is far too large, unless the basis is chosen with considerable foresight.

For the purpose of starting the discussion I have chosen to illustrate all the point above in the context of one particular example.

2. Example: The Morse Potential. The Morse [27] potential is

$$V(r) = D[\exp(-2ax) - 2\exp(-ax)]. \tag{2.1}$$

Here x is the displacement, r, of the system from its equilibrium, r_e, point $x = r - r_e$, a is the range parameter and D is the (finite) dissociation energy. The potential $V(r)$, which is practically harmonic about $x = 0$ where $V(r_e) = -D$, tends to a finite (i.e., zero) value as $x \to \infty$. The one dimensional Schrödinger equation, is, as usual

$$H_0\psi \equiv -\frac{1}{2\mu}\frac{\partial^2\psi}{\partial r^2} + V(r)\psi = i\frac{\partial\psi}{\partial t} \tag{2.2}$$

or, for a stationary solution, $\psi(t) = \exp(-Et)\psi(0)$, $H_0\psi = E\psi$. Here μ is the mass and, as elsewhere, we put $\hbar = 1$. The Morse potential is a simple example of what I refer to as a 'realistic' problem. The motion can sample regions where the harmonic approximation ($V(r) = -D + Da^2(r - r_e)^2$) is quite poor. Indeed, the spectrum of the Morse Hamiltonian is discrete only for $E < 0$. For $E > 0$ the spectrum is continuous and the motion is unbound in the $r \to \infty$ direction.

There have been numerous papers dealing with an algebraic approach to the Morse Hamiltonian. Let me begin with a procedure *a la* (1.2). For this purpose start with the change of variable

$$y(x) = N\exp(-ax) \tag{2.3}$$

where N, $N^2 = 8\mu D/a^2$, will turn out to have the significance of the maximal number of bound states. We also note that the vibration frequency ω, in the harmonic regime is $\omega = 2N(a^2/2\mu) = 4AN$ where $A = a^2/8\mu$ is an often used parameter. Consider now the generators of $SU(1,1)$, $[J_1, J_2] = -iJ_3$, $[J_2, J_3] = iJ_1, [J_3, J_1] = iJ_2$ with the realization ($p \equiv i\partial/\partial x$)

$$\begin{aligned} J_1 &= \frac{1}{a^2 y(x)}(p^2 - 2\mu E) - \frac{1}{4}y(x) \\ J_2 &= i[i\frac{p}{a} + \frac{1}{ax} - \frac{1}{2}] \\ J_3 &= \frac{1}{a^2 y(x)}(p^2 - 2\mu E) + \frac{1}{4}y(x). \end{aligned} \tag{2.4}$$

The Morse Hamiltonian can now be written as

$$\begin{aligned} H_0 &\equiv (p^2/2m) + V(r) \\ &= 4Ay(x)(J_3 - N/2) + E. \end{aligned} \tag{2.5}$$

The time-independent Schrödinger equation $(H_0 - E)\psi = 0$ is then of the form (cf. (1.2))

$$4Ay(x)(J_3 - N/2)\psi = 0. \tag{2.6}$$

The eigenvalue equation for E has thereby been converted to an eigenvalue problem for the maximal number N of bound states. That the substitution (2.3) will reduce the stationary Schrödinger equation for the Morse potential to a confluent hypergeometric equation with y as the variable has been known for quite a while [27]. The factorization method [5] can then be applied. The problem with this approach, which is being periodically rediscovered, is that the eigenvalue equation is for N at a fixed E. For a given physical system N has a given value. The group generators connect states of fixed E (where the zero of E is common to all states) and variable N. These are states of the same binding energy of different Morse potentials. What we really want is states of different energy of the same Morse potential (i.e., different E but same N). That in a given representation the energy is constant can be seen from the (bilinear) Casimir invariant of $SU(1,1)$ [28]

$$\begin{aligned} C &\equiv J_3^2 - J_1^2 - J_2^2 \\ &= -(E/4A) - 1/4 \end{aligned} \tag{2.7}$$

whose eigenvalues in a given representation are $-k(k+1)$, *e.g.*

$$\begin{aligned} C|n,k> &= -k(k+1)|n,k> \\ J_0|n,k> &= (n+lk)|n,k> \end{aligned} \tag{2.8}$$

where, from (2.7), $k - 1/2 = (-E/4A)^{\frac{1}{2}}$. ($E$ is negative for the bound states). The number n labels the bound states where, since (cf. (2.8)) $n + k = N/2$

$$E_n = -4A[(N/2) - (n + 1/2)]^2 \tag{2.9}$$

and $-k \leq n \leq k$.

The operators

$$J_\pm = (J_1 \pm iJ_2)/\sqrt{2} \tag{2.10}$$

act as step-up and down operators for N at a given E

$$J_\pm|n,k> \propto |n \pm 1, k> . \tag{2.11}$$

The shift operators which connect different eigenstates of the same Morse potential correspond to shifting k (and hence, E) for given $n+k$ (i.e., for given number N, cf. (2.9), of bound states). Such operators can be shown to act as generators of SU(2). It is possible to combine both types of generators by considering, *e.g.* Sp(4). One then has both types of shift operators in one common structure. In terms of two boson operators a, b

$$\begin{aligned} J_+ &= \frac{1}{2}(a^\dagger a^\dagger + b^\dagger b^\dagger) \\ J_- &= \frac{1}{2}(aa + bb) \\ J_3 &= \frac{1}{2}(a^\dagger a + b^\dagger b + 1). \end{aligned} \tag{2.12}$$

The operators which shift among the physical states (i.e., of the same well) are

(2.13)
$$\begin{aligned} L_+ &= a^\dagger b \\ L_- &= b^\dagger a \\ L_3 &= \frac{1}{2}(a^\dagger a - b^\dagger b). \end{aligned}$$

Amongst the 10 linearly independent generators of Sp(4) there is yet a third SU(1,1) subgroup

(2.14)
$$\begin{aligned} K_+ &= a^\dagger b^\dagger \\ K_- &= ab \\ K_3 &= \frac{1}{2}(a^\dagger a + b^\dagger b + 1). \end{aligned}$$

One can show [12,13] that these act amongst the states of the continuous spectrum of a given Morse potential. So we have three, three parameter Lie groups. One for the discrete spectrum, one for the continuous spectrum and one for shifting the value of N. The list is clearly not complete since the Morse eigenvalue problem depends on two physical parameters A and N. It is indeed possible to write down a symmetry group which includes the generator for dilating A, [29]. I do not know whether the shift of A is included amongst the generators of Sp(4). The direct product algebra for SU(2)⊗SU(2) discussed in [29] suggests that this may be feasible. Hence we have a:

- Technical point: Use Sp(4), or some other group, to fully classify the one dimensional Morse spectrum, including the dilation symmetries* in N and A.

What is very important and still missing both for the Morse and in other, more general cases, is how to have the bound and the continuous spectrum in the same representation. This is a purely technical problem but one whose solution will promote real and rapid progress.

The physical origin of the problem introduced in the paragraph above is that, in the laboratory and also in the geometrical (differential) approach to quantum mechanics one readily constructs operators that connect bound (i.e. $E < 0$) and continuous ($E > 0$) eigenstates of the Morse Hamiltonian. One does not have to search far. The operator r will do it. Yet one has no simple algebraic analogue. One way or another, one switches representations to go from the bound states to the continuum [12,13,28]. I then no longer know how to construct matrix elements of operators since the states belong to different representations. In summary, we have

- Open problem: Construct an irreducible unitary representation which spans both the discrete and continuous spectrum of the physical Hamiltonian. If that is not possible, provide alternative means for computing matrix element of the generators between states of different representations.

*Chemists know them as 'the law of corresponding states' which states that, to realistic accuracy, two scale parameters specify the potential.

For the bound states the bilinear Morse Hamiltonian is typically written as

$$H_0 = -4AL_3^2 \tag{2.15}$$

where L_3 is a generator of the su(2) algebra (2.13). Using angular momentum quantum numbers, we have the spectrum $-4Am^2$ so that $l = (N-1)/2$ and the vibrational quantum number n (cf. (2.9)) is $n = 1 - m$. The degeneracy is due to the invariance of the Morse Hamiltonian to reflection about $x = 0$. ($r = r_e$). The advantage of using SU(2) for the bound states is that the finite number of such states is immediately guaranteed by the finite value of l. This remains the case also for the Morse oscillator in three dimensions [10] where the corresponding group is SU(4). While the inherent cut-off of the spectrum of the three dimensional Morse oscillator is quite realistic, it is not fully quantitative. A study of the realization shows a possible deficiency. In SU(4), the quantum mechanical kinetic energy operator (which is the Laplacian) is not well realized. As is well known, after separation of variables one acquires the centrifugal term $l(l+1)/r^2$, where r is variable. In SU(4), this is approximated by $l(l+1)/r_0^2$ where r_0 is constant. Even in the geometric, differential equations approach, one does not know an exact solution for the three dimensional Morse potential [30]. In the algebraic approach one can obtain improved approximations by including higher orders (beyond bilinear) in the generators [31]. So far, however, I do not know of an exact non-trivial algebraic Hamiltonian whose realization is $-\Delta_3 + V(r)$. Hence we have

- Technical point: Provide an algebraic Hamiltonian for a three dimensional Schrödinger equation with a realistic potential $V(r)$ where r is a scalar radial coordinate.

The bosonic representation (2.13) of the SU(2) generators provides for a two-dimensional realization

$$\begin{aligned} a^\dagger &= \frac{1}{\sqrt{2}}(q - \frac{\partial}{\partial q}) \\ b^\dagger &= \frac{1}{\sqrt{2}}(p - \frac{\partial}{\partial p}). \end{aligned} \tag{2.16}$$

Upon change of variables

$$q = y\cos\phi, \quad p = y\sin\phi \tag{2.17}$$

and separation in the y and ϕ coordinates, the eigenvalue equations (N is the number operator)

$$\begin{aligned} H_0\psi(y,\phi) &= E\psi(y,\phi) \\ N\psi(y,\phi) &= N\psi(y,\phi) \end{aligned} \tag{2.18}$$

lead finally to the Morse potential problem with $y = N\exp(-x)$, cf. (2.3). With so many changes of variable along the way, it is not surprising that the algebraists have a tendency to construct useful Hamiltonians in terms of generators and not to worry about the geometrical realization. In many ways this has led to useful progress but it has created a gap with the conventional point of view. Hence, while the problem is technical, its magnitude is such that it is a real issue

- Open problem: We need a more systematic route between algebraic Hamiltonians and their geometric realization.

The bosonic representation is also very convenient for taking the limit to the problem where the motion is purely harmonic [11,32]. As a first step we note that since in this limit the dissociation energy, D, is infinite, one chooses the zero of energy at the bottom of the potential. Hence we add $A(N^2-1)$ to the Hamiltonian (2.15) so that with respect to the new zero, $H_0 = -4AL_3^2 + A(N^2-1)$. Now, with $l = (N-1)/2$, the eigenvalues $l(l+1)$ of the Casimir of SU(2) are $(N^2-1)/4$ and an equivalent Hamiltonian operator is

$$H_0 = -4AL_3^2 + 4AC = 4A(J_+J_- + \frac{1}{2}[J_-, J_+]). \tag{2.19}$$

The final form in (2.19) already looks remarkably similar to the Hamiltonian for the harmonic oscillator. The similarity can be made closer by factoring out the harmonic frequency $\omega = 4AN$

$$H = \omega(Q_+Q_- + \frac{1}{2}[Q_-, Q_+]) \tag{2.20}$$

with [11]

$$Q_+ = N^{-1/2}J_+, \quad Q_- = N^{-1/2}J_-. \tag{2.21}$$

It is now only a matter of group contraction to show that [32] in the $N \to \infty$, $A \to 0$, ($AN^2 = D \to \infty$, $AN = \omega \to$ finite) limit $Q_+ \to a^\dagger$, $Q_- \to a$, where a and $a^\dagger$ are the annihilation and creation operators of the harmonic oscillator.

The harmonic-like form (2.19) is not unique to the Morse problem. Other harmonic potentials (*e.g.*, the square well [21]) can be cast in a similar form. For all these cases the Hamiltonian bilinear in the generators is a Casimir invariant of a subgroup. Is this the reason for this ubiquitous form? In the factorization method ([5],[1], ch. 7) a form very reminiscent of (2.19) is derived where, however, the raising and lowering operators are of the type we discussed earlier (shifting the potential at a given E). Hence we have a

- Technical point: Is there an alternative, complementary factorization method where the raising and lowering operators shift among the states of a given potential. If there is, can one develop it in a systematic fashion.

3. Time Evolution. What this typically means is that in addition to the secular motion under the influence of H_0 the system is subject to an additional perturbation V, which may itself be time dependent. The corresponding Schrödinger equation is now (1.3). In many problems of interest, the perturbation V is realistically given as a linear combination of generators. Unfortunately, as we discussed, realistic molecular Hamiltonians are typically bilinear in the generators. The standard way of solving (1.3), that of eliminating the secular variation by solving for $\psi_I(t) = \exp(iH_0 t)\psi(t)$

$$i\frac{\partial}{\partial t}\psi_I(t) = V_I(t)\psi_I(t) \tag{3.1}$$

$$V_I(t) \equiv \exp(iH_0 t)V\exp(-iH_0 t) \tag{3.2}$$

is not useful since $V_I(t)$ is not necessarily a linear combination of generators even if V is. As a simple example, take $V = L_+ + L_-$ (as per (2.13)) and H_0 as the Morse Hamiltonian (2.15). Then use of the Baker-Campbell-Hausdorff formula leads to [32]

$$V_I(t) = \exp[4iAt(1 - 2L_3)]L_+ + L_-\exp[-4iAt(1 - 2L_3)]. \tag{3.3}$$

Beyond this point, the only way I know how to handle the solution of (3.1) by algebraic means is via the Magnus [33] expansion. Past the first order this requires evaluating integrals over commutators of $V_I(t)$ at different times. For most molecular problems $N < 100$. Hence faced with a real need to solve (3.1) a brute force numerical procedure would be the sensible choice. For more complicated problems, say a triatomic molecule where the matrix representation is of order N^2 (at least) and even a numerical solution is taxing, one simply has no viable options. Of course, the whole problem stems from H_0 being bilinear while V is linear. Mathematically, one can suggest eliminating V in (1.3) by the substitution $\psi_{IV} = \exp(iVt)\psi(t)$ and treating H_0 as a perturbation. For very intense laser fields that would be not unreasonable. In general however H_0 is stronger than V and one would need to solve the problem exactly (or at least to high order) in H_0 in order to get sensible results.

So far we had in mind an external field as a perturbation. Many of us, myself included, are interested in molecular collisions where the perturbation is the intermolecular force which, like the intramolecular one, often requires terms beyond first order in the generators. Some time ago we have introduced a phenomenological analysis aimed at uncovering the symmetries of such processes (*e.g.*, [22]). One can show that such symmetries are indeed given by (1.6). At least approximately, these symmetries are quite ubiquitous and correspond to simple operators. The problem is explaining and ultimately, predicting these symmetries starting from the Hamiltonian. The best we can presently do with bilinear Hamiltonians is a variational procedure [25], as discussed in section 5. This is based on approximating $I(t)$ by

$$I(t) = \sum_r i^r(t)X_r \tag{3.4}$$

where the X_r's are the generators, in terms of which H is expressed and the i_r's are time dependent coefficients. Such a form can be exact when H is linear in the generators [23] as will be shown in section 4. Otherwise, it is but a variational approximation. We do not necessarily have to have a solution of the simple form

(3.4). In particular, the coefficients i_r can depend on the space as well as on the time variables. What we require however is that $I(t)$ is a symmetry in the special sense appropriate for the dynamics, namely

$$\text{Tr}\{\rho dI(t)/dt\} = 0 \tag{3.5}$$

where ρ is the von Neumann-Liouville density matrix

$$id\rho/dt \equiv i\partial\rho/\partial t - [H, \rho] = 0. \tag{3.6}$$

Hence the next technical problem

- Technical point. Can one determine exact symmetries, which are linear in the generators, for Hamiltonians which are bilinear in the generators.

My conjecture that this should be possible and a proof for pure states (i.e., when ρ is idempotent) is reasonably clear. If however I insist also on, so called, mixed states for which one must use (3.6) as the equation of motion (rather than (1.3) with $\rho = \psi\psi^*$) then it appears necessary to generalize (1.6), for example to

$$[I, i\frac{d}{dt}]\rho = 0 \tag{3.7}$$

where $id/dt \equiv i\partial/\partial t - [H,]$ is the total time derivative and where ρ satisfies (3.6). Then the condition on I is

$$[I, i\frac{d}{dt}] = Ri\frac{d}{dt}. \tag{3.8}$$

Here R is chosen so that both sides of (3.8) are of the same order in the adjoint representation.

4. Exact Results for the Dynamics. A simple but nontrivial case for which exact results are available is which the set of generators of a Lie algebra is closed under communication with the Hamiltonian

$$[H, X_r] = \sum_s \alpha_r^s X_s \tag{4.1}$$

Here the α's are a set of numerical coefficients which are independent of the state of the system. In section 5 we derive a self-consistent approximate Hamiltonian, denoted there by h, for which (4.1) remains valid (except that the α's will then depend on the state). The reason why such a generalization is necessary is that, as it stands, (4.1) allows for only a restricted set of Hamiltonians.

To see how (4.1) is useful, consider solving for a time-dependent constant of motion $I(t)$ by a separation of variables [22,23]

$$I(t) = \sum_r i^r(t) X_r. \tag{4.2}$$

Here the i's are numerical coefficients which depend on time only and the X_r's are (time independent) generators. From $dI/dt = 0$ we have a set of coupled equations

(4.3a) $$\sum_s (di^s(t)/dt) X_s = i \sum_r i^r [H, X_r]$$

and using (4.1) we obtain

(4.3b) $$\sum_s (di^s(t)/dt) X_s = i \sum_s X_s \sum_r i^r(t) \alpha_r^s.$$

Equating the coefficients of each generator on both sides of (4.3), we see that (4.1) is indeed a sufficient condition and that there will be up to m linearly independent constants of motion given by the solutions of

(4.4) $$di^s/dt = i \sum_r i^r \alpha_r^s.$$

5. A Self-Consistent Procedure. In general, to satisfy (4.1) the Hamiltonian need be linear in the generators. But as we have seen, this puts a severe restriction on the eigenfrequencies of the system. A practical procedure for determining approximate dynamical symmetries for Hamiltonians bilinear in the generators will be discussed in this section. The procedure will be motivated first on intuitive grounds of self-consistency and derived later from a variational principle. The procedure will also determine an effective Hamiltonian, linear in the generators, for which the dynamical symmetry is exact. To achieve a compact derivation, it is convenient to work in a notation where the state of the system is described in terms of a density matrix ρ. This is however not essential. In any case, ρ need not be a mixture but can represent a pure state, and such a pure state will remain a pure state under our approximation.

In general, when ρ evolves under an approximate Hamiltonian, it will not be true that

(5.1) $$\begin{aligned} \partial \mathrm{Tr}(\rho X_r)/\partial t &\equiv \mathrm{Tr}(x_r \partial\rho/\partial t) \\ &= i\mathrm{Tr}(\rho[H, X_r]) \end{aligned}$$

where H is the true Hamiltonian of the system. In other words, time propagation of the state and time propagation of the observables need not yield the same results. Our requirement is that they do, i.e., that (5.1) be correct at least for the generators of the algebra. A sufficient condition is that a Hamiltonian h, linear in the generators

(5.2) $$h = \sum_h h^r X_r$$

can be defined such that

(5.3) $$\mathrm{Tr}(\rho[H, X_r] = \mathrm{Tr}(\rho[h, X_r]).$$

It should be noted that (5.3) is only required to hold for the generators and that, in general, h will depend on the state ρ of the system.

To determine the coefficients h^r in (5.2), one introduces the matrix σ

$$i \sum_s \mathrm{Tr}\{\rho[X_r, X_s]\}\sigma^{st} = \delta_{rt} \tag{5.4}$$

so that from (5.4) and (5.3)

$$h^r = i \sum_s \sigma^{sr} \mathrm{Tr}\{\rho[H, X_s]\}. \tag{5.5}$$

Introducing the generator G, $G = \sum g^r(t)X_r$, we see, using (5.5) and (5.3), that the condition that G is a constant of the motion of h for the state ρ

$$\mathrm{Tr}\{\rho(\partial G/\partial t - i[G, h])\} = 0 \tag{5.6}$$

is equivalent to the self-consistency condition (5.1). Indeed a variational principle can be readily based on this condition. Introduce a Lagrangian L.

$$L = \mathrm{Tr}\{\rho(t)dG(t)/dt\} \tag{5.7}$$

where both ρ and G are evaluated at time t. The corresponding action integral is

$$I = \int_{t_o}^{t_1} L(t)dt - \mathrm{Tr}\{\rho(t_1)G(_1)\}. \tag{5.8}$$

Since both ρ and G satisfy equations of motion which are first order in time, one can only specify their boundary values at one time point. The simplest choice is ρ at t_0 and G at t_1. Subject to these mixed boundary conditions the stationary action, $\delta I = 0$, condition yields

$$\mathrm{Tr}\{\delta\rho(t)dG(t)/dt\} = 0 \tag{5.9}$$

and using the invariance of the trace to cyclic permutation and integrating by parts, we obtain

$$\mathrm{Tr}\{\delta G(t)d\rho(t)/dt\} = 0. \tag{5.10}$$

If the variation $\delta\rho(t) \propto \rho(t)$ is allowed, it follows from (5.9) that $\mathrm{Tr}(\rho dG/dt) = 0$. Of course, if G is an exact constant of the motion, then (5.9) vanishes for all possible variations and the action is at its minimum (and equal to $\mathrm{Tr}(\rho(t)G(t))$ which is time independent). We shall vary G by varying the g^r coefficients. Using $\delta G = \sum \delta g^r X_r$, one readily verifies that for independent variations of the coefficients δg^r we recover the self-consistency conditions (5.3) that were imposed on intuitive grounds.

As in section 4, the generators are closed under commutation with the self-consistent Hamiltonian h

$$[h, X_r] = \sum_s \alpha_r^s X_s \tag{5.11}$$

except that here the coefficients α do depend on the state

$$\alpha_r^s = \sum_t h^t C_{tr}^s \tag{5.12}$$

so that the equation of motion

$$dg^s/dt = i \sum_r g^r \alpha_r^s \tag{5.13}$$

need be solved in a self consistent fashion as discussed in [25].

6. Group Dynamics. To implement the self-consistent procedure of section 5, we write the generator as $UX_rU^\dagger$ (cf. (1.4)), where U is the evolution operator for the self consistent Hamiltonian h (5.2). Since this Hamiltonian is linear in the generators, its evolution operator is an element of the Lie group and can be written as

$$U(t) = \exp(i \sum_r \zeta^r(t) X_r). \tag{6.1}$$

To prove that U is the evolution operator, differentiate it with respect to the group parameters

$$\partial U/\partial \zeta^r = i\hat{X}_r U. \tag{6.2}$$

Here $\hat{X}_r = X_r$ only for an Abelian group and is, in general, a linear combination of generators, reflecting the need to take care of the special noncommutation of the X_r's. To determine the necessary linear combination, one can, for example, note the relation

$$UX_rU^\dagger = \sum_s g_r^s(\zeta) X_s \tag{6.3}$$

valid for any algebra, and differentiate both sides with respect to ζ. We shall be particularly concerned here with the situation where the Hamiltonian H is given by $H = H_0 + V(t)$, where the perturbation $V(t)$ vanished prior to some initial time, say t_0. If one chooses ψ_0 as an eigenstate of $H_0, \psi = U\psi_0$ is a solution of the Schrödinger equation (1.3).

Since the time dependence of U is via the group parameters

$$\begin{aligned} \partial U/\partial t &= i \sum_r (\partial \zeta^r/\partial t)\hat{X}_r U \\ &= i \sum_s (\sum_r D_r^s \partial \zeta^r/\partial t) X_s U \end{aligned} \tag{6.4}$$

where $\hat{X}_r - = \sum D_r^s X_s$ and the matrix $\mathbf{D}(t)$, determined via (6.3), is time dependent. The equation of motion of the group parameters is therefore

$$\sum_r D_r^s(t) \partial \zeta^r/\partial t = h^s \tag{6.5}$$

where h^s is defined by (5.5).

The coefficients h^s are themselves functions of the group parameters. To see this and to introduce a suggestive way of writing (6.5), define

$$H(\zeta) = < \psi_0 | U^\dagger H U | \psi_0 > \tag{6.6}$$

where $\psi = U\psi_0$ is the current state of the system. Then

$$\begin{aligned} \partial H/\partial \zeta^r &= i < \psi_0 | U^\dagger [H, X_r] U | \psi_0 > \\ &= i \sum_s D_r^s < \psi_0 | U^\dagger [H, X_r] U | \psi_0 > . \end{aligned} \tag{6.7}$$

Defining

$$\partial H/\partial\hat{\zeta}^r = \sum_s (D^{-1})^r_s \partial H/\partial\zeta^r \tag{6.8}$$
$$= i < \psi_0|U^\dagger[H, X_r]U|\psi_0 >$$

it follows from (6.7) and (5.5) that

$$h^s = \sum_r \sigma^{rs} \partial H/\partial\hat{\zeta}^r \tag{6.9}$$

or from (6.5)

$$\sum_s \sigma_{rs} \partial\zeta^s/\partial t = \partial H/\partial\zeta^r. \tag{6.10}$$

Having brought the equations of motion of the group parameters to a Hamiltonian form, the only thing left is to compute the matrix elements $\partial H/\partial\zeta^r$. Now if H is linear in the generators, that is immediate, using the g matrix introduced in (6.3). But suppose H is not. Say it is bilinear or even higher order in the generators. The order of the commutator $[H, X]$ will be the same as the order of H. One can now invoke the group automorphism (3.9) to conclude that $U^\dagger[H, X]U$ will also be of the same order. As a specific example, consider the bilinear term $X_r X_s$; then

$$U^\dagger X_r X_s U = U^\dagger X_r U U^\dagger X_s U \tag{6.11}$$
$$= \sum_{\alpha,\beta} G^\alpha_r G^\beta_s X_\alpha X_\beta.$$

For higher powers, simpler results apply. The only matrix elements over ψ_0 that need be computed are thus of products of generators. If the Hamiltonian is only bilinear, then only bilinear terms, i.e. $\langle\psi_0|X_\alpha X_\beta|\psi_0\rangle$, need be computed. All else is done by the group automorphism (6.3). Of course, the final stage is the (numerical) solution of the "classical" equations of motion (6.10).

7. Coherent time evolution. When the Hamiltonian is linear in the generators $\{X_r\}$ of some group, the evolution operator $U(t)$ can be written explicitly as an element of this group as in (6.1), $U = U(\zeta(t))$. Here $\zeta(t)$ is the set of group parameters whose time dependence is given by the solution of (6.10). Say now $U(\zeta)$ with a given set ζ of group parameters is some other element of the group. It follows from the group property that

$$U(\zeta') = U(\zeta(t))U(\zeta) \tag{7.1}$$

where ζ' is a set of group parameters determined by $\zeta(t)$ and the ζ via the group composition law,

$$\zeta' \equiv \phi(\zeta(t), \zeta). \tag{7.2}$$

The group property (7.1) can now be used to define a special type of states. These states are parametrized by the group parameters and their time evolution is such that the values of the parameters can change with time but the functional form of the state remains unchanged. A familiar example is a Gaussian state which in the coordinate representation is [36]

$$< x|\alpha(t) >= \exp[-(x-x(t))^2/2 + ip(t)x + i\gamma]. \tag{7.3}$$

Here $x(t)$ and $p(t)$ are the parameters of the state (7.3) and for a motion in a harmonic potential, (7.3) is an exact solution for all values of t.

To generalise (7.3), let $|0>$ be the ground state of H_0 and define the 'coherent state' [34,35] $|\zeta>$ by

$$|\zeta >= U(\zeta)|0> . \tag{7.4}$$

Here $|\zeta>$ is a state parametrized by the set of parameters ζ, which are time independent. The point is that a coherent state remains coherent under time evolution generated by H. The proof is immediate. Using (7.1)

$$\begin{aligned} U(\zeta(t))|\zeta> &= U(\zeta(t))U(\zeta)|0> \\ &= U(\zeta')|0> \\ &\equiv |\zeta'> . \end{aligned} \tag{7.5}$$

For every time t, there is a coherent state with parameters ζ' determined by the group composition law (7.2).

Coherent evolution as hereby discussed is restricted however to Hamiltonians linear in the generators of some group. It can be extended [25] as a self consistent scheme for more general Hamiltonians, where now it is h, (cf. (5.2)) which acts as the generator of time displacements.

One can clearly define coherent time evolution also for mixed states. A particular class of such coherent states, and one to which we have given much attention [37,22,25] because of the implications for analysis of scattering data is as follows. Say that prior to the action of perturbation, the (mixed) state of the system has the (maximal entropy, [22]) form

$$\rho_{\text{in}} = \exp(-\sum_r \lambda^r_{\text{in}} X_r). \tag{7.6}$$

Here λ^{in}_r are a set of parameters. It follows on using (6.3) that upon time evolution the functional form is retained and only the values of the parameters evolve with time

$$\begin{aligned} \rho(t) &\equiv U(t)\rho_{\text{in}}U^\dagger(t) \\ &= \exp(-\sum_r \lambda^r_{\text{in}} U X_r U^\dagger) \\ &= \exp(-\sum_s \lambda^s(t) X_s) \end{aligned} \tag{7.7}$$

where (cf. (6.3))

$$\lambda^s(t) \sum_r \lambda^r_{\text{in}} g^s_r(t). \tag{7.7}$$

Acknowledgement.

During my studies I had the benefit of discussions with many coworkers including Y. Alhassid, I. Benjamin, I.L. Cooper, R. Gilmore, F. Iachello, S. Kais, S.K. Kim, J.L. Kinsey and C.E. Wulfman.

REFERENCES

[1] W. MILLER, JR., *Lie theory and special function*, Academic Press, New York, 1968.
[2] W. MILLER, JR., *Symmetry and separation of variables*, Addison-Wesley, Reading, 1977.
[3] P.J. OLVER, *Applications of Lie group to differential equations*, Springer Verlag, New York, 1986.
[4] D.H. SATTINGER AND O.L. WEAVER, *Lie group and algebras with applications to physics, geometry and mechanics*, Springer Verlag, New York, 1986.
[5] L. INFELD AND T.E. HULL, *The Factorization Method*, Rev. Mod. Phys., 23 (1951), p. 21.
[6] R.L. ANDERSON, S. KUMEI AND C.E. WULFMAN, *Generalization of the Concept of Invariance of Differential Equations*, Phys. Rev. Letts., 28 (1972), p. 988; R. Anderson, S. Kumei and C.E. Wulfman, Invariants of the Equations of Wave Mechanics, Rev. Fiz. Mex. 21, 1(1972) p. 35.
[7] C.E. WULFMAN, *Dynamical Groups in Atomic and Molecular Physics*, in *Recent advances in group theory and their applications to spectroscopy*, J.C. Donini, ed., Plenum, New York, 1979.
[8] P. PFEIFER AND R.D. LEVINE, *A Stationary Formulation of Time Dependent Problems in Quantum Mechanics*, J. Chem. Phys, 79 (1983), p. 5512.
[9] R.D. LEVINE AND C.E. WULFMAN, *Energy Transfer to a Morse Oscillator*, Chem. Phys. Lett, 60 (1979), p. 372.
[10] F. IACHELLO AND R.D. LEVINE, *Algebraic Approach to Molecular Rotation-Vibration Spectra I. Diatomic Molecules*, J. Chem. Phys., 77 (1982), p. 3046.
[11] R.D. LEVINE, *Harmonizing the Morse Oscillator*, Chem. Phys. Letts., 95 (1983), p. 87.
[12] C.E. WULFMAN AND R.D. LEVINE, *A Unified Algebraic Approach to Bound and Continuum States of Anharmonic Potentials*, Chem. Phys. Lett, 97 (1983), p. 361.
[13] Y. ALHASSID, F. GURSEY AND F. IACHELLO, *Group Theory of the Morse Oscillator*, Chem. Phys. Lett., 99 (1983), p. 27.
[14] O.S. VAN ROOSMALEN, F. IACHELLO, R.D. LEVINE AND A.E.L. DIEPERINK, *Algebraic Approach to Molecular Rotation-Vibration Spectra. II. Tri- and Polyatomic Molecules*, J. Chem. Phys., 79 (1983), p. 2515.
[15] O.S. VAN ROOSMALEN, I. BENJAMIN, R.D. LEVINE, *A Unified Algebraic Model Description for Interacting Vibrational Modes in ABA Molecules*, J. Chem. Phys, 81 (1984), p. 5986.
[16] A. ARIMA AND F. IACHELLO, *The Interacting Boson Model*, Adv. Nucl. Phys., 13 (1984), p. 139.
[17] Y. ALHASSID, F. IACHELLO, R.D. LEVINE, *Resonance Widths and Positions by an Algebraic Approach*, Phys. Rev. Lett., 54 (1985), p. 1746; I. Benjamin and R.D. Levine, Complex Scaling and Algebraic Calculation of Resonances, Phys. Rev., A33 (1986) p. 2833.
[18] A. FRANK, F. IACHELLO AND R. LEMUS, *Algebraic Methods for Molecular Electronic Spectra*, Chem. Phys. Lett, 131 (1986), p. 380.
[19] I. BENJAMIN AND R.D. LEVINE, *Potential Energy Surfaces for Stable Triatomic Molecules Using an Algebraic Hamiltonian*, Chem. Phys. Lett., 117 (1985), p. 314.
[20] Y. ALHASSID, F. GURSEY AND F. IACHELLO, *Group Theory Approach to Scattering*, Ann. Phys., 148 (1983), p. 346.
[21] S. KAIS AND R.D. LEVINE, *The Square Well Potential by an Algebraic Approach*, Phys. Rev., A34 (1986), p. 4615.
[22] R.D. LEVINE, *The Information Theoretic Approach to Intramolecular Dynamics*, in *Photoselective Chemistry*, J. Jortner, R.D. Levine and S.A. Rice, eds., Wiley, New York, 1980.

[23] R.D. LEVINE, *Time Dependent Constants of the Motion*, in *New Horizons of Quantum Chemistry*, p. 135, P.-O. Löwdin and B. Pullman, eds., Reidel, Dordrecht, 1982; R.D. Levine, J. Phys. Chem. 89 (1985) p. 2122.

[24] V.I. ARNOLD AND A. AVEZ, *Ergodic Problems of Classical Mechanics*, Benjamin, New York, 1968.

[25] N.Z. TISHBY AND R.D. LEVINE, *Time Evolution via A Self Consistent Maximal Entropy Propagation: The Reversible Case*, Phys. Rev., A30 (1984), p. 1477.

[26] C.E. WULFMAN AND R.D. LEVINE, *An Adiabatic Approximation for Time Dependent Constants of Motion*, Chem. Phys. Lett, 84 (1981), p. 13.

[27] P.M. MORSE, *Diatomic Molecules According to the Wave Mechanics. II. Vibrational Levels*, Phys. Rev, 34 (1929), p. 57; D. Ter Haar, The Vibrational Levels of an Anharmonic Oscillator, Phys. Rev., 70 (1946) p. 222.

[28] B.G. WYBOURNE, *Classical groups for physicists*, Wiley, New York, 1974.

[29] C.E. WULFMAN AND R.D. LEVINE, *Isotropic Substitution as a Symmetry Operation in Molecular Vibrational Spectroscopy*, Chem. Phys. Lett, 104 (1984), p. 9.

[30] I.R. ELSUN AND R.G. GORDON, *Accurate Analytical Approximations for the Rotating Morse Oscillator*, J. Chem. Phys, 76 (1982), p. 5452; C.L. Pekeris, The Rotation-Vibration Coupling in Diatomic Molecules, Phys. Rev., 45 (1934) p. 98.

[31] S.K. KIM, I.L. COOPER AND R.D. LEVINE, *The Algebraic Hamiltonian for Diatomic Molecules in the Vibron Model*, Chem. Phys, 106 (1986), p. 1.

[32] R.D. LEVINE, *Algebraic Approach to Molecular Structure and Dynamics*, in *Intramolecular Dynamics*, p. 17, J. Jortner and B. Pullman. eds., Reidel, Dordrecht, 1982.

[33] W. MAGNUS, *On the Exponential Solution of Differential Equations for a Linear Operator*, Comm. Pure Appl. Math., 7 (1954), p. 649.

[34] R. GILMORE, *On the Properties of Coherent States*, Rev. Mex. Fis, 23 (1974), p. 143.

[35] A. PERELOMOV, *Generalized Coherent States and Their Applications*, Springer-Verlag, Berlin, 1986.

[36] M. SARGENT, III., M.O. SCULLY AND W.E. LAMB, JR., *Laser Physics*, Addison-Wesley, Mass, 1974.

[37] Y. ALHASSID AND R.D. LEVINE, *Connection Between the Maximal Entropy and the Scattering Theoretic Analyses of Collision Processes*, Phys. Rev, A18 (1978), p. 89.

LIE ALGEBRAIC APPROACH TO THE MANY-ELECTRON CORRELATION PROBLEM

JOSEF PALDUS†*

Abstract. A brief outline of the unitary group approach (UGA) to the many-electron correlation problem, its basic principles, development and present and future exploitation are presented. After a concise formulation of basic quantum chemical models and methods which are currently employed to study the molecular electronic structure, a short overview of basic concepts of UGA is given. A special attention is payed to recent exploitations of Green-Gould representation theory and to the Clifford algebra UGA, to which separate sections are devoted.

Key words. many-electron correlation problem, shell-model, configuration interaction, coupled cluster approach, unitary group approach, Clifford algebras

1. Introduction. The orthodox exploitation of group theory in quantum mechanics is based on the concept of an invariance (symmetry, degeneracy) group of the Hamiltonian of the studied system, and provides quantum numbers (conservation laws, integrals of motion) for the state labeling and associated selection rules. This is the case, for example, of the rotation group for central systems or of the permutation group S_N for systems consisting of N indistinguishable particles. During the last two decades it became clear, however, that much larger groups, which are not necessarily the invariance groups for the studied system, can provide a very useful formalism for the description of quantum mechanical systems [1-3]. These non-invariance groups—invariably various non-compact or compact Lie groups—are nowadays referred to as dynamical groups or, their Lie algebras, as spectrum generating algebras [3]. A dynamical group is usually required to be sufficiently large to contain all the bound states of a system in a single irreducible representation (irrep). In this sense, for example [4], the dynamical group of the hydrogen atom is SO(4,2). However, various smaller subgroups can be equally useful in describing a given system [4] [*e.g.*, the group SO(2,1) for the H atom]. We thus prefer a more general view, regarding any group as a dynamical group (in a weaker sense) of the system, if the Hamiltonian can be expressed in terms of the generators (infinitesimal operators) of this group [5]. Letting such groups act on spaces spanned by the states of the studied system provides a useful organization and rules for the corresponding matrix representative of the Hamiltonian or, in simple enough cases, directly for its eigenstates.

We shall see that an appropriate dynamical group for an n-level (or orbital) model of an N-electron system, which can be described by the spin-independent

---

†Institute for Advanced Study Berlin, Wallotstraβe 19, D - 1000 Berlin 33, FRG; permanent address: Department of Applied Mathematics, University of Waterloo, Waterloo, Ontario, N2L 3G1, Canada. Also at: Department of Chemistry, University of Waterloo and Guelph-Waterloo Centre for Graduate Work in Chemistry.

*Senior Killam Scholar 1987-88.

Hamiltonian, is the unitary group $U(2^n)$ [6, 7]. However, the various subgroups in the chain [8, 9]

$$U(2^n) \supset \text{Spin}(m) \supset SO(m) \supset U(n) , \tag{1}$$

where $m = 2n+1$ or $m = 2n$, can be equally useful in the many-electron correlation problem. Particularly the smallest group in this chain, the unitary group $U(n)$, has led to the so-called unitary group approach [10-16] (UGA), which is being exploited in large scale configuration interaction (CI) calculations [17-23]. This approach and its various applications have been described in numerous reviews [11, 16, 24-27] and even in monographs [28-30], and will also be dealt with in this volume by Shavitt [31]. We will thus restrict ourselves to general aspects of UGA, while concentrating on more recent developments, which have not yet received such coverage. In particular we wish to stress the basic role played by Clifford algebras in the general many-fermion problem [7] and the possibilities offered by the so-called Clifford algebra UGA (CAUGA) [8, 9, 32]. We also wish to point out the possibilities for the standard UGA offered by the Green-Gould approach [33], recently exploited by Gould, Chandler and others [34-36] as well as to briefly mention other developments [37-39] and open problems.

2. Many-Electron Correlation Problem. In spite of its fundamental significance and the excellent total energies it can provide, the independent particle model (IPM) is unable to correctly describe many chemical phenomena and properties. To go beyond the IPM, the motion of individual electrons must be properly correlated, which leads to very demanding computational schemes. Out of the many approaches which have been developed to handle the many-electron (or, generally, many-fermion) correlation problem, the most often exploited one is the simplest variational CI (or shell model) approach, followed by coupled cluster (CC) and perturbative (MBPT - many-body perturbation theory) approaches. The CI matrix, representing the Hamiltonian $\hat{H}$ in a chosen finite-dimensional N-particle subspace, can also serve as a starting point in CC and PT approaches, although these are traditionally based on one- and two-electron integral representations of $\hat{H}$. All these approaches suffer from a poor convergence of the expansions employed (even when "optimal" natural orbitals [40] are used) and require large scale computations in order to achieve the desired accuracy. The dimension of the relevant N-electron spaces can be considerably reduced by exploiting various symmetry or invariance properties of the electronic Hamiltonian. Exploitation of the dynamical group or some of its subgroups, Eq. (1), for an n-orbital N-electron model, can achieve simultaneous adaptation to these various symmetries as well as provide an efficient and versatile algorithm for the computation of the actual or effective matrix representative of $\hat{H}$. We now give a concise formulation to this problem.

2.1. Models. *1.* All quantum-chemical orbital models exploit a *finite* dimensional subspace $\mathcal{V}$ of the general one-electron Hilbert space $\mathcal{H}$ and construct the appropriate N-electron space as an antisymmetric component of the N-th rank tensor power of $\mathcal{V}$, $\mathcal{V}^{\wedge N}$. Although $\mathcal{V}$ is usually defined as the span of a non- orthogonal set of atomic orbitals (AO's), in going beyond the IPM one often starts from an equivalent orthonormal basis of (usually SCF or HF) molecular orbitals (MO's). We thus restrict ourselves to orthogonal spinorbital sets, *i.e.*, to the MO-type or the so-called $\overline{\text{VB}}$-type approaches [41] (note: the valence bond-bar approach exploits orthogonalized AO's). In view of the spin independence of the Hamiltonian considered (see below), our spinorbitals $|I\rangle$ can be written as tensor products of the orbital and spin parts, $|I\rangle = |i\rangle|\sigma\rangle$.

2. In the second-quantization formalism [42] we associate with a given orthonormal spinorbital basis $\{|I\rangle\}$ a Grassman-type algebra of creation ($X_I^\dagger$) and annihilation (X_I) operators, which satisfy the anticommutation relations

$$\{X_I, X_J\} = \{X_I^\dagger, X_J^\dagger\} = 0, \;\; \{X_I, X_J^\dagger\} = \delta_{IJ} \,, \tag{2}$$

where

$$\{A, B\} \equiv AB + BA \,, \tag{3}$$

as well as the vacuum property

$$X_I|0\rangle = 0 \,. \tag{4}$$

These operators are defined on the Fock space $\mathcal{F}$,

$$\mathcal{F} = \bigoplus_{k=0}^{n} \mathcal{V}^{\wedge k} \,, \tag{5}$$

where $n = \dim \mathcal{V}$, and the zero-particle component $\mathcal{V}^0$ is isomorphic with $\mathbb{C}$ (the field of complex numbers) and spanned by the vacuum state $|0\rangle$. The N-particle states then automatically possess the correct permutational symmetry required of the system of identical fermions. We also define the number operators N_I ,and their complements $\bar{N}_I$,

$$N_I = X_I^\dagger X_I \,, \quad \bar{N}_I = X_I X_I^\dagger \,, \qquad (I = 1, ..., 2n) \tag{6}$$

which form a system of mutually commuting Hermitean idempotents

$$\begin{aligned} &N_I^2 = N_I, \;\; \bar{N}_I^2 = \bar{N}_I, \;\; N_I^\dagger = N_I, \;\; \bar{N}_I^\dagger = \bar{N}_I \,, \\ &[N_I, N_J] = [\bar{N}_I, \bar{N}_J] = [N_I, \bar{N}_J] = 0 \,, \end{aligned} \tag{7}$$

where

$$[A, B] \equiv AB - BA \,. \tag{8}$$

Note that also

$$N_I + \bar{N}_I = e \quad \text{and} \quad N_I \bar{N}_I = 0 \,, \tag{9}$$

where e is the identity of the algebra.

3. The molecular system studied is modeled by the spin-independent electronic Hamiltonian $\hat{H}$ in the Born-Oppenheimer approximation,

$$\hat{H} = \sum_{i,j} \sum_{\sigma} \langle i|\hat{z}|j\rangle \; X^{\dagger}_{i\sigma} X_{j\sigma} + \frac{1}{2} \sum_{i,j,k,\ell} \sum_{\sigma,\tau} \langle ij|\hat{v}|k\ell\rangle X^{\dagger}_{i\sigma} X^{\dagger}_{j\tau} X_{\ell\tau} X_{k\sigma} \,, \tag{10}$$

where $\hat{z}$ represents the kinetic energy and the potential due to the clamped-nuclei framework, while $\hat{v}$ is the interelectronic Coulomb potential. Since all the forces involved are spin independent, $\hat{H}$ commutes with total spin operators,

$$[\hat{H}, \hat{S}^2] = [\hat{H}, \hat{S}_z] = 0\,. \tag{11}$$

4. One distinguishes *ab initio* models, which are defined by the chosen AO spanning set for $\mathcal{V}$, $n \equiv \dim \mathcal{V} \geq N$, and *semi-empirical* models where $n \equiv \dim \mathcal{V} = N$. The latter are often directly defined through a judicious parametrization (based on experimental data) of one- and two-electron integrals $\langle i\,|\hat{z}|j\rangle$ and $\langle ij|\hat{v}|k\ell\rangle$, respectively, after their number has been drastically reduced using various simplifying assumptions. In either case the objective is to find the eigenvalues and the eigenstates of $\hat{H}$ (generally as a function of the geometry of the nuclear framework). Since the dimension of the N-electron space increases rapidly with increasing n, only an approximate determination of the eigenvalues and eigenvectors can be contemplated.

2.2. Methodology. *1.* The simplest, most universal and most widely employed method is the CI approach [43], in which the "exact"(*i.e.*, within the model) wavefunction $|\Psi\rangle$ is represented as a linear combination of ordered N-electron configuration states $|\Phi_K\rangle$,

$$|\Psi\rangle = \sum_k C_k |\Phi_k\rangle \;, \tag{12}$$

the linear coefficients C_k being determined by the diagonalization of $\hat{H}$ represented in the basis $\{|\Phi_k\rangle\}$ using the variation principle. We can also write

$$|\Psi\rangle = \hat{C}|\Phi_0\rangle \;, \; \hat{C} = \sum_{k=0}^{N} \hat{C}_k, \qquad \hat{C}_0 \equiv \hat{1}, \tag{13}$$

assuming intermediate normalization, and

$$\hat{C}_k = \sum_{[I'']\,[I']} \langle [I'']|\hat{c}_k|[I']\rangle X^{\dagger}_{[I'']} X_{[I']}, \quad (k > 0) \tag{13'}$$

where $[J]$ designates the ordered spinorbital set

$$[J] \equiv [J_k] \equiv [J_1, J_2, ..., J_k]; \;\; J_1 < J_2 < ... < J_k, \tag{14}$$

and

$$X^{\dagger}_{[J]} = X^{\dagger}_{J_1} \cdots X^{\dagger}_{J_k} \,,$$
$$X_{[J]} = X_{J_k} \cdots X_{J_1} \,. \tag{15}$$

The spinorbital set was further subdivided into occupied ($[I_0]$ or primed indices) and unoccupied (doubly primed indices) disjoint subsets, the occupied spinorbitals defining the reference configuration $|\Phi_0\rangle$,

$$|\Phi_0\rangle = X^{\dagger}_{[I_0]}|0\rangle \,, \tag{16'}$$

while, generally

$$|[J]\rangle = X^{\dagger}_{[J]}|0\rangle \,. \tag{16}$$

The cardinality k of the sets $[I']$ and $[I'']$ defines the excitation order of the configurations constituting $\hat{C}_k$. In view of the spin-independency of $\hat{H}$, it is advantageous to spin-adapt the configuration states (see below).

The key problem in the implementation of the CI approach is to efficiently construct the matrix representative of $\hat{H}$, given the one- and two-electron integrals and the chosen n-electron configuration basis. In the so-called direct CI approaches [44] only the action of $\hat{H}$ on some trial vector $\|C_k\|$ is computed and the lowest eigenvalue is iteratively determined, using a diagonalization algorithm based on the Krylov sequence. Presently, spaces of dimension exceeding 10^6 are employed, although most applications use dimensions $10^2 - 10^5$. In spite of these large dimensions (or heuristic extrapolation procedures [45]) only very limited subspaces of $\mathcal{V}^{\wedge N}$, usually spanned by singly and doubly excited configurations (with respect to one or more reference states), can be achieved. This truncation by excitation order results in size-consistency (extensivity) defects (*i.e.*, non-linear energy dependence on the particle number when the interaction is switched off, *e.g.*, a single reference CI wave-function restricted to biexcited configurations can be shown to be orthogonal to the exact ground state when $N \to \infty$ [46].

2. The main advantage of the MBPT is its size-extensivity, since each connected energy diagram gives a contribution with a correct particle number dependence. However, it is very difficult to go beyond the fourth order of PT as well as to handle general open shells [47]. Nevertheless, this approach is capable of providing very good results, since the higher order contributions, when required, can often be simulated by denominator shifting [48].

3. The coupled cluster approach [49] exploits the connected cluster theorem [50] asserting the exponential character of the wavefunction in terms of connected cluster components $\hat{T}_k$,

$$|\Psi\rangle = \exp(\hat{T})|\Phi_0\rangle, \;\; \hat{T} = \sum_{k=1}^{N} \hat{T}_k \,,$$
$$\hat{T}_k = \sum_{[I''],[I']} \langle [I'']|\hat{t}_k|\,[I']\rangle X^{\dagger}_{[I'']} X_{[I']}. \tag{17}$$

A comparison with the CI expansion, Eq. (13), shows that

$$\hat{C}_k = \sum_{\mathcal{P}_K} \prod_{j=1}^{k} (n_j!)^{-1}\ \hat{T}_j^{n_j}\ , \tag{18}$$

where the sum extends over all partitions $\mathcal{P}_k$ of k with

$$k = \sum_{j=1}^{k} j n_j\ , \qquad 0 \leq n_j \leq k\ , \qquad \hat{T}_j^0 = \hat{1}\ . \tag{19}$$

The term consisting of a single factor, $\hat{T}_k$, represents the *connected* k-times excited cluster component, while other terms, involving a product of at least two lower-order clusters, represent disconnected components. Since in the closed-shell case $\hat{T}_1 \approx 0$ and $\hat{T}_4 \ll \frac{1}{2}\hat{T}_2^2$, we can approximate $\hat{T}$ by pair clusters $\hat{T}_2$, assuming that $\hat{T}_3$ is also small, obtaining coupled-pair many electron theory [49] (CPMET or CC-D). Considering also $\hat{T}_1$ and/or $\hat{T}_3$, we obtain extended CC approaches [51] (the latter being computationally very demanding due to a large number of tri-excited clusters).

The CC equations determining various cluster components are generally non-linear algebraic equations, representing essentially the recurrence formulas for the generation of higher order PT diagrams of a certain type. Thus, solving these equations we automatically generate these diagrams and sum their contribution to an infinite order. If no quasi-degeneracy is present, then the linear approximation (LCC) yields very good results [52] (note that LCC-D corresponds to CI-D with its unlinked clusters eliminated, thus becoming size-extensive [53]). A close connection between LCC-D and CI-D enables one to exploit CI programs in CC calculations (*cf.*, *e.g.*, [52]). This connection might be particularly useful for multi-reference (MR) schemes, as originally suggested in [52] and recently exploited by Bartlett's group [54].

2.3 Spin-adaptation. The spin-independence of the Hamiltonian (10) can be exploited to significantly reduce the dimension of the CI problem. Numerous techniques of spin-adaptation have been developed since the early days of quantum mechanics, exploiting either the symmetric group S_N or the spin group SU(2) [55]. Particularly suitable in various applications proved to be so called geminally antisymmetric spin functions [56, 57], which include Yamanouchi-Kotani [58], Serber [59] and spin-bonded [60] functions. The latter are closely related with Rumer-Weyl functions of the VB approach [41]. The symmetric group enables a formal elimination of the spin-variables when appropriate irreps of S_N are considered. This so-called spin-free approach was particularly championed by Matsen [61]. In the late sixties it was realized [62] that the same goal can be achieved by exploiting the unitary group, which can play the role of a dynamical group (in a weaker sense, see Section 1) for many-fermion systems. In the many-electron case this is achieved by exploiting the chain [63]

$$\mathrm{U}(2n) \supset \mathrm{U}(n) \times \mathrm{U}(2)\,, \tag{20}$$

where n is the number of orbitals defining the model (cf. Sec. 2.1.4).

Needless to say, there is a close relationship between the S_N and U(n) approaches, since the symmetric group plays a fundamental role in the representation theory of U(n). Particularly powerful and elegant is the so-called boson calculus [64], which is based on the generalized Jordan-Schwinger maps [65], providing a general representation theory of U(n). The general expression for the generator matrix elements, Wigner coefficients, etc. are quite formidable and their practical exploitation may not be immediately obvious. For the N-electron case, however, this formalism can be drastically simplified [10] and exploited in actual applications [11-31]. In this respect also the relationship with the SU(2) angular momentum theory, particularly the so-called graphical methods of spin algebras [66, 67], proved very useful. In turn, the UGA developments provided a stimulus for novel developments based on the symmetric group approach [25].

Recently, Gould and Chandler [40] exploited another formalism based on U(n) polynomial identities developed by Green and Gould (GG) [33]. This approach is capable of providing the segment level formalism for the generator matrix element evaluation and is very versatile, enabling the extension to spin-dependent operators [35], system partitioning [36] or various orbital partitionings [68]. Since this approach received only a very cursory treatment in existing reviews, we shall devote a special section to it.

Finally, following the work of Nikam and Sarma [69], we have recently realized the fundamental role played by Clifford algebras and the related U(2^n) dynamical group for the many-electron correlation problem [7-9]. In fact, the proposed formalism is not restricted to spin-$\frac{1}{2}$ particles, but easily generalizes for arbitrary fermions [37-39], and is amenable to system partitioning [9] and general point group symmetry adaptation [32]. Very recently we have also examined some rather general aspects of CAUGA and its relationship to general spin-adaptation techniques used in various many-body approaches [7]. We shall briefly describe these developments in the last section.

2.4. Implementation. For small CI it is indeed quite unessential which procedure is employed. Since, however, the dimension rapidly increases with increasing n and N, so that a truncation of the CI space is essential, and since in large scale computations one has to handle very large sparse matrices, the elements of which are functions of a large number of one-and two-electron integrals which cannot be in a direct access memory, the structuring and the flexibility of the CI algorithm employed is very crucial. In the basic diagonalization step one has to handle generally the five-fold summations (four orbital labels and one configuration label) of the product of four terms: a two-electron integral, two generator matrix elements and a trial vector component. The UGA provides much desired flexibility for this crucial step and enables us to construct algorithms driven either by integrals (direct CI [44]), matrix elements, loops of branching diagrams, etc. Since this aspect of UGA will be covered in detail by Shavitt [31], we shall not elaborate on it any further. In closing, let us only mention that the UGA formalism provides the needed insight even when spin-non-adapted basis of Slater determinants is employed, as has been done by Knowles and Handy [23] in their full CI totally vectorized approach. We shall also see that their approach can be viewed as a special case of CAUGA, which in turn can provide additional insight [7].

3. Unitary Group Approach (UGA). In this section we present a brief overview of the unitary group approach as it pertains to the many-electron correlation problem. After pointing out the relevant facts of the unitary group representation theory, we concentrate on the formalism which is appropriate to the many-electron case. We discuss briefly the structure and construction of the spin-adapted many-electron Gel'fand-Tsetlin basis [70] and the problem of an evaluation of the matrix representative of the Hamiltonian in this basis, particularly as it pertains to large-scale shell-model or CI calculations, where this formalism has been most often exploited [16-23].

1. The unitary group U(n) is a (matrix) Lie group of $n \times n$ unitary matrices over $\mathbb{C}$. Any $U \in \mathrm{U}(n)$ can be written as $U = \exp(A)$, where A is skew-Hermitean. All $n \times n$ skew-Hermitean matrices over $\mathbb{C}$ form a *real* Lie algebra $u(n)$. Any $A \in u(n)$ can also be written as $A = iS$ with S Hermitean. Clearly, $u(n)$ is a (Lie) subalgebra of the general linear Lie algebra $g\ell(n, \mathbb{C})$ of all $n \times n$ matrices over $\mathbb{C}$. In fact, the complexification of $u(n)$ gives $g\ell(n, \mathbb{C})$. A standard basis for $g\ell(n, \mathbb{C})$ consists of the so-called matric units $\mathbf{e}^{ij}$, whose (k, ℓ)-entry is

$$(\mathbf{e}^{ij})_{k\ell} = \delta_{ik}\delta_{j\ell}, \tag{21}$$

so that

$$[\mathbf{e}^{ij}, \mathbf{e}^{k\ell}] = \delta_{jk}\mathbf{e}^{i\ell} - \delta_{i\ell}\mathbf{e}^{kj}, \tag{22}$$

defining the structure constants of $g\ell(n, \mathbb{C})$ and its various subalgebras. The basis for $u(n)$ can be obtained by defining appropriate linear combinations of $g\ell(n, \mathbb{C})$ matric units, say $\mathbf{e}_{ij}$, which in addition to satisfying (22) also possess the following Hermitean property

$$(\tilde{\mathbf{e}}_{ij})^\dagger = \tilde{\mathbf{e}}_{ji}. \tag{23}$$

2. For a connected Lie group G there is a one-to-one correspondence between the reps (representations) and irreps (irreducible reps) of G and of associated Lie algebra g. The possible irreps of g are then determined as irreducible sets of endomorphisms E_{ij} of some finite-dimensional inner product space (the carrier space of the irrep), which possess the same structure constants as the matric units of g. For unitary representations we require in addition that

$$E_{ij}^\dagger = E_{ji}\,. \tag{24}$$

The E_{ij}'s are referred to as the generators (infinitesimal operators) of the rep. (Note that in physics texts also the basis elements of the algebra are referred to as generators, since both sets possess the same structure constants.) We shall use the same symbol E_{ij} to designate the matric units and their representations $\Gamma(E_{ij})$ unless a confusion can arise.

3. U(n) is a "universal" compact Lie group in the sense that every compact Lie group is isomorphic to a subgroup of some U(n) [compare with the "universality" of S_N for finite groups as expressed by Cayley's theorem]. The representation theory of compact Lie groups is very similar to that of finite groups. In fact, if G is compact, then

(a) G has a faithful linear rep,

(b) all irreps of G are finite dimensional and can be realized as tensor reps over a linear space which provides a faithful rep of G,

(c) all finite-dimensional reps of G are equivalent to unitary reps and are decomposable (fully reducible), and

(d) the number of irreps of G (up to equivalence) is denumerable.

4. Any irrep Γ of U(n) is uniquely labeled by its highest weight vector $\mathbf{m}_n$,

$$\mathbf{m}_n = (m_{1n}, m_{2n}, \cdots, m_{nn}), \tag{25}$$

with non-increasing integer components. When $m_{nn} \geq 0$, the integers m_{in} can be regarded as a partition of some integer N,

$$N = \sum_{i=1}^{n} m_{in} \quad , \tag{26}$$

and we can associate a Young pattern with Γ having m_{in} boxes in its $i-th$ row. Γ and $\tilde{\Gamma}$ are mutually conjugate if their Young patterns are conjugate (*i.e.*, obtainable one from the other by interchanging the rows with columns).

5. Subduction of Γ of U(n) to U($n-1$) [or, more precisely, to $U(n-1)\dot{+}$ (1)] is governed by the Weyl branching rule

$$\Gamma(\mathbf{m}_n) \downarrow \mathrm{U}(n-1) \;=\; \bigoplus_{\mathbf{m}_{n-1}} \Gamma'(\mathbf{m}_{n-1}), \tag{27}$$

the sum extending over the irreps of U($n-1$), all with unit multiplicity, such that

$$m_{in} \geq m_{i,n-1} \geq m_{i+1,n} \;, \qquad (i = 1, \cdots, n-1) \;. \tag{28}$$

Applying recursively Weyl's branching rule, we obtain the canonical (orthonormal) Gel'fand-Tsetlin basis [70] for the carrier space of any irrep $\Gamma(\mathbf{m}_n)$ of U(n), which is associated with the subgroup chain

$$\mathrm{U}(n) \supset \mathrm{U}(n-1) \supset \;\cdots \supset \mathrm{U}(2) \supset \mathrm{U}(1), \tag{29}$$

since U(1) is Abellian. The vectors of this basis are uniquely labeled by the triangular Gel'fand tableaux $[m]$ whose rows are the irrep labels $(\mathbf{m}_k)$, $k = 1, \cdots, n$

$$[m] = \begin{bmatrix} \mathbf{m}_n \\ \mathbf{m}_{n-1} \\ \cdots \\ \cdots \\ \cdots \\ \mathbf{m}_2 \\ \mathbf{m}_1 \end{bmatrix} = \begin{bmatrix} m_{1n} \quad m_{2n} & \cdots & m_{nn} \\ m_{1,n-1} & \cdots & m_{n-1,n-1} \\ & \cdots & \\ & \cdots & \\ & \cdots & \\ m_{12} & & m_{22} \\ & m_{11} & \end{bmatrix}. \tag{30}$$

In view of (28) the allowed (lexical) Gel'fand tableau entries m_{ij} satisfy at each level the "betweenness conditions"

$$m_{ij} \geq m_{i,j-1} \geq m_{i+1,j} \; . \tag{28'}$$

Linearizing the array $[m]$ we define the *lexical* order of the basis vectors $|[m]\rangle$. The dimension of $\Gamma(\mathbf{m}_n)$ is given by Weyl's or Robinson's formulas [71].

6. The generators E_{ij} of U(n) are classified into the raising $(i < j)$, weight $(i = j)$ and lowering $(i > j)$ generators according to whether they raise, preserve or lower the weight, respectively. Their matrix representatives (in the canonical basis) are strictly upper triangular, diagonal and strictly lower triangular, respectively. The real representation matrices of E_{ij} and E_{ji} are mutually transposed in view of Eq. (24), *i.e.*,

$$\langle [m'] | E_{ij} | [m] \rangle = \langle [m] \, | E_{ji} | [m'] \rangle \, . \tag{31}$$

The diagonal matrix representatives of weight generators E_{ii} are

$$\langle [m'] | E_{ii} | [m] \rangle = \delta([m'], [m])(k_i - k_{i-1}) \, , \tag{32}$$

where

$$k_i = \sum_{j=1}^{i} m_{ji}, \qquad (i = 1, \cdots, n); \; k_0 \equiv 0 \, , \tag{32'}$$

and

$$\delta([m'], [m]) = \prod_{i \leq j} \delta(m'_{ij}, m_{ij}) \, . \tag{32''}$$

For typographical reasons we write $\delta(i, j) \equiv \delta_{ij}$ for the Kronecker deltas. The matrix elements of raising (lowering) generators were given for a general U(n) irrep $\Gamma(\mathbf{m})$ by Gel'fand and Tsetlin [71] and Baird and Biedenharn [72]. The commutation relations (22) for E_{ij}'s imply that

$$E_{i,j\pm 1} = [E_{ij}, \; E_{j,j\pm 1}] \, , \tag{33}$$

indicating that the elementary (primitive) generators $E_{i,i\pm 1}$ can be used to determine all the generators. However, this procedure can be directly employed when the full irrep carrier space is considered and cannot be generally used for truncated spaces, which are required in the limited CI approaches. We shall next outline how the general representation theory simplifies in the many-electron case.

7. To facilitate a better understanding of the U(n) representation theory, Table I provides a schematic comparison of important concepts for the familiar [64] rotation group SO(3), for the SU(2) group which is closely related to it, as well as the U(2) and general U(n) groups. In fact, only the relevant Lie algebras are important for

Table I.*

Schematic comparison of the SO(3), SU(2) and U(n) groups and corresponding Lie algebras.

Lie group	SO(3) $\approx$	SU(2) $\subset$	U(2)	U(n)
Lie algebra	$so(3)$	$su(2)$ $\subset$	$u(2)$	$u(n)$
Generators	J_k or iJ_k $(k = 1,2,3)$ or $J_3,\ J_\pm = J_1 \pm iJ_2$	$E_{11} - E_{22}$, E_{12}, E_{21}	$E_{11},\ E_{22}$, E_{12}, E_{21}	E_{ij} $(i,j = 1,\dots,n)$
Structure constants	$[J_1, J_2] = iJ_3$ (cycl.)			$[E_{ij}, E_{k\ell}] = \delta_{jk}E_{i\ell} - \delta_{i\ell}E_{kj}$ or $[E_{rs}, E_{st}] = E_{rt}\ (r \neq t)$, $[E_{rs}, E_{sr}] = E_{rr} - E_{ss}$
Hermitean property	$J_3^\dagger = J_3$, $J_\mp^\dagger = J_\pm$			$E_{ij}^\dagger = E_{ji}$
Casimir* operators (irrep invariants)	$J^2 = J_1^2 + J_2^2 + J_3^2$			$I_k = tr(\mathbf{E}^k)$, $\mathbf{E} = [E_{ij}]$
Irrep labels	$\mathcal{D}^{(j)}$ $[\dim \mathcal{D}^{(j)} = 2j+1]$			$[\mathbf{m}_n] = [m_{1n} m_{2n} \dots m_{nn}]$ $m_{1n} \geq m_{2n} \geq \cdots \geq m_{nn}$
Group chain and irrep carrier space basis labels	$SO(3) \supset SO(2)$ $\lvert j, m\rangle$	$\begin{bmatrix} 2j & & 0 \\ & j+m & \end{bmatrix}$	$\begin{bmatrix} m_{12} & & m_{22} \\ & m_{11} & \end{bmatrix}$	$U(n) \supset U(n-1) \supset \cdots \supset U(1)$ $\begin{bmatrix} m_{1n}\, m_{2n} \dots m_{nn} \\ m_{1,n-1} \dots m_{n-1,n-1} \\ \dots \\ m_{12}\, m_{22} \\ m_{11} \end{bmatrix}$
"Betweenness" conditions	$j \geq m \geq -j$			$m_{ij} \geq m_{i,j-1} \geq m_{i+1,j}$
Highest weight state (HWS)	$\lvert j, j\rangle$			$\lvert [m^{(0)}]\rangle \equiv \left\lvert \begin{bmatrix} m_{1n}\, m_{2n} \dots m_{nn} \\ m_{1n} \dots m_{n-1,n} \\ \dots \\ m_{1n}\, m_{2n} \\ m_{1n} \end{bmatrix} \right\rangle$
HWS property	$J_+ \lvert j,j\rangle = 0$, $J_3 \lvert j,j\rangle = j\lvert j,j\rangle$			$E_{ij}\lvert [m^{(0)}]\rangle = 0 \quad (i<j)$, $E_{ii}\lvert [m^{(0)}]\rangle = m_{in}\lvert [m^{(0)}]\rangle$
Weight generator property	$J_3\lvert j,m\rangle = m\lvert j,m\rangle$			Eq. (32)
Matrix elements	$\langle jm\lvert J_3 \rvert jm\rangle = m$, $\langle jm \pm 1 \lvert J_\pm \rvert jm\rangle = [(j \mp m)(j \pm m + 1)]^{1/2}$			$\langle [m'] \lvert E_{ij} \rvert [m]\rangle$ (see following sections)

*See Sec. 4.

our applications, and we note that both groups SO(3) and SU(2) possess isomorphic Lie algebras, resulting when only traceless matrices are retained in $u(2)$ (*cf.*, *e.g.*, [11] for an elementary introduction). For the algebra generators and irrep basis labels, Table 1 has an entry for each of the four group headings; for the other rows the first entry is for SO(3), and the second entry corresponds to SU(2), U(2), and U(n).

8. Any spin-independent Hamiltonian (10) can be expressed as a two-form in terms of U(n) orbital generators E_{ij},

$$\hat{H} = \sum_{i,j} \langle i|\hat{z}|j\rangle E_{ij} + \frac{1}{2} \sum_{i,j,k,\ell} \langle ij|\hat{v}|k\ell\rangle (E_{ik}E_{j\ell} - \delta_{jk}E_{i\ell}). \tag{34}$$

Thus, instead of U($2n$) we can consider its proper subgroup U(n) × U(2), Eq. (20). Subducing the totally antisymmetric (one-column) irreps of U($2n$) to this subgroup, we obtain a direct sum of U(n) × U(2) irreps, which must be of the form $\Gamma \times \tilde{\Gamma}$. Thus, at most two-column irreps of U(n) can occur. Moreover, any U(2) irrep is uniquely specified by the total electron number N and the total spin quantum number S [labeling SU(2) irreps], the individual vectors being labeled by S_z quantum number. Thus, the relevant Γ ($\mathbf{m}_n$) irrep of U(n) must satisfy the conditions

$$0 \le m_{in} \le 2, \qquad \text{(Pauli principle)}\ , \tag{35}$$

and

$$\sum_{i=1}^{n} m_{in} = N, \qquad \text{(electron number conservation)}\ , \tag{36}$$

and is fully determined by n, N and S since

$$\sum_{i=1}^{n} \delta(1, m_{in}) = 2S. \tag{37}$$

Since $\hat{H}$ is spin-independent, we can use U(n) irreps to provide us with the desired spin-adapted bases for the system with N electrons and multiplicity $(2S+1)$. These irreps are labeled by at most 2-column Young frames $\{2^a 1^b\}$ or, equivalently, by the triple of integers a, b and c, *i.e.*, $\Gamma(a, b, c)$, where

$$a = \frac{1}{2}N - S, \quad b = 2S \quad \text{and } c = n - a - b = n - \frac{1}{2}N - S. \tag{38}$$

9. Exploiting this simplified irrep labeling at each level of the Gel'fand-Tsetlin chain (29), we can replace Gel'fand tableau $[m]$, Eq. (30), by an $n \times 3$ tableau* (called ABC [10, 11, 14-16] or Paldus [12, 13, 16-22, 24, 28-31] tableau)

$$[\mathrm{m}] \equiv [a_i\ b_i\ c_i]\,, \tag{39}$$

*We call these matrices tableaux to remind us the unconventional labeling of rows (from the bottom upwards).

where

$$a_i + b_i + c_i = i\,. \tag{40}$$

We also define the difference tableaux ΔABC with entries

$$\Delta x_i = x_i - x_{i-1} \qquad (x = a, b, c; \quad i = 1, \ldots, n; \quad x_0 \equiv 0), \tag{41}$$

in which case

$$\Delta a_i + \Delta b_i + \Delta c_i = 1\,. \tag{42}$$

Obviously,

$$x_i = \sum_{j=1}^{i} \Delta x_i \qquad (x = a, b, c; \quad i = 1, \ldots, n). \tag{43}$$

Clearly, any two-column subtableau (*e.g.*, $AB, \Delta AC$, etc.) also uniquely labels our spin-adapted states. Since Δa_i and Δc_i can only take on binary values 0 and 1 (while $\Delta b_i = 0, \pm 1$), we can conveniently label our states by ternary arrays of *step numbers* d_i,

$$d_i = (\Delta a_i \Delta \bar{c}_i)_2 = 2\Delta a_i + \Delta \bar{c}_i = 1 + 2\Delta a_i - \Delta c_i\,, \tag{44}$$

where the bar designates a binary complementation: $\bar{0} = 1, \quad \bar{1} = 0$. The step numbers d_i represent thus binary values of the rows of the AC tableau, and their possible values are listed in Table II. The corresponding orbital *occupancies (occupation numbers)* n_i, yielding the diagonal matrix representatives of weight generators,

$$\langle [\mathbf{m}'] | E_{ii} | [\mathbf{m}] \rangle = \delta([\mathbf{m}'], [\mathbf{m}]) n_i([\mathbf{m}]), \tag{45}$$

are then given by the digital sum of the AC tableau entries at each level (cf. Table I),

$$n_i \equiv n_i([\mathbf{m}]) = 2\Delta a_i + \Delta b_i = 1 + \Delta a_i - \Delta c_i = \Delta a_i + \Delta \bar{c}_i\,. \tag{46}$$

Table II.

Possible values of the step numbers d_i, Eq. (44), corresponding ΔABC tableau entries and occupation numbers n_i.

d_i	Δa_i	$\Delta \bar{c}_i$	Δc_i	Δb_i	ΔS_i	n_i
0	0	0	1	0	0	0
1	0	1	0	1	$\frac{1}{2}$	1
2	1	0	1	-1	$-\frac{1}{2}$	1
3	1	1	0	0	0	2

10. We can now easily construct the N-electron spin-adapted basis labeled by the ABC tableaux (or corresponding ΔAC or step number tableaux)* by starting with the highest weight $[a, b, c]$ given by Eq. (38) and subtracting at each level the four possible ABC rows (*i.e.*, $[0,0,1]$, $[0,1,0]$, $[1,-1,1]$ and $[1,0,0]$, *cf.* Table II), while keeping only the permissible (lexical) ones with non-negative entries. We note that ΔAC labeling is also very closely related with the Weyl tableaux labeling [73]. The total number of spin-adapted states for each irrep $\Gamma(a, b, c)$ is given by the dimension formula [10]

$$\dim\ \Gamma(a,b,c) = \frac{b+1}{n+1} \binom{n+1}{a} \binom{n+1}{c}, \tag{47}$$

where $\binom{m}{k} = m!/(m-k)!k!$ are the usual binomial coefficients. At each level i, starting with the highest weight $[a, b, c,]$ for $i = n$, we generate a number of permissible rows. Collecting all identical rows at each level we construct the so-called distinct row table (DRT), which efficiently represents even very large bases [12, 13]. This table lists all the distinct rows $[a_i b_i c_i]$ at each level i as well as up to the four chaining indices $j_{ik}(i = 0,1,2,3)$ corresponding to the four possible step numbers of Table II. It is also convenient to list chaining indices for reverse basis generation (starting with the lowest weight) and the dimensions associated with distinct rows and corresponding $U(i)$ subgroups. Using the DRT table, one can easily determine the lexical index for each ABC tableau and *vice versa*. A convenient graphical representation of the DRT was introduced by Shavitt [12, 13] by representing each distinct row $[a_i, b_i, c_i]$ by a vertex at the i-th level, starting with the highest weight $[a, b, c,]$ at the top and preserving an equidistant vertical level spacing for $i < n$ levels. The allowed step numbers are then represented by edges connecting corresponding vertices with the negative slope of the edge representing the step number. Introducing the bottom vertex corresponding to the 0-th level with the weight $[0,0,0]$ one obtains a two-rooted graph whose various paths connecting the top and the bottom vertices (roots) correspond to the individual spin-adapted states (for details see Shavitt [31] and [12, 13, 16, 24-30]).

The Shavitt graph representing pictorially the global structure of the N-electron n-level spin-adapted basis is in fact an expanded form of the Yamanouchi-Kotani branching diagram [15]. A similar DRT representation, which arises naturally in the UGA description, can also be exploited in other approaches to the spin-adaption problem, as has been shown for the Young-Yamanouchi states of the symmetric group approach by Duch and Karwowski [25], who introduced configuration occupation number graphs. A very general exploitation of the graphical representation of this type was recently given by Duch [75].

11. A crucial problem for any implementation of UGA scheme is an efficient evaluation of generator matrix elements and of their products in the Gel'fand- Tsetlin basis. This problem received a considerable attention in the past and, although very efficient algorithms are presently available, the improvements are still being

*A comparison of various labeling schemes for $U(n)$ canonical N-electron spin-adapted states may be found in Fig. 9 of [74].

sought. In view of Eq. (31) we can restrict our attention to the raising (or lowering) generators [cf. Table II and Eq. (45)].

Matrix elements of *elementary generators* $E_{i,i+1}$ are given by a simple formula as 0,1 or the square-root of a ratio of two consecutive integers (*cf.* Eqs. (31, 32) of [10] or Eq. (5.7) and Fig. 13 of [11] or Eq. (37) of [76]). Any column or row of an elementary generator matrix representative has at most two non-vanishing entries. The explicit formula mentioned above [10] determines a given column or row of the matrix representative for all the elementary raising (lowering) generators simultaneously.

The *non-elementary generator* matrix elements can be efficiently determined iteratively using the commutation relations, Eq. (33), when all the states are considered , as in the full CI calculations. However, for truncated basis (which must be invariably used for any but the simple models) the intermediate states which are needed in evaluating the commutator in Eq. (33) may be missing. Only when the basis truncation is carried out using the harmonic level excitation diagram (HLED) scheme [76] we are guaranteed that all the necessary intermediate states are at hand. However, this may be inconvenient in general applications. Moreover, one needs the next n levels in HLED when computing matrix elements of generator products. Thus, an independent and direct algorithm for the evaluation of generator matrix elements is highly desirable.

Shavitt was first to notice that the elementary generator formulas can be factored and extended this factorization to the general case using mathematical induction [13]. It was soon realized that this factorization is most easily obtained from the SU(2) representation of these matrix elements, particularly when exploiting the graphical methods of spin algebras [66,67]. The resulting angular momentum diagram for single generator matrix elements is immediately seen to separate into the product of $6j$-symbols, providing the segment values W_r, so that

$$\langle[\mathbf{m}']|E_{ij}|[\mathbf{m}]\rangle = \prod_{r\in\Omega} W_r\,, \tag{48}$$

where the range Ω is the set

$$\Omega \equiv (E_{ij}) = \{k\,|\min(i,j) \leq k \leq \max(i,j)\}. \tag{49}$$

The segment values W_r are given by simple expressions [14, 77], and have been evaluated for various phase conventions. Very much the same technique can also be exploited for the symmetric group based approaches [78, 79]. More importantly, the same technique can also be exploited for the matrix elements of *generator products*, in which case we find [14]

$$\langle[\mathbf{m}'']\,|E_{ij}E_{i'j'}|[\mathbf{m}]\rangle = \prod_{k\epsilon\Xi} W_k \sum_{X=0}^{1} \prod_{k'\epsilon\Xi'} W_{k'}(X), \tag{50}$$

where

$$\begin{aligned} \Xi &= \Omega \,\Delta\, \Omega' &&= (\Omega \cup \Omega') \setminus (\Omega \cap \Omega'), \\ \Xi' &= \Omega \cap \Omega'\,, \end{aligned} \tag{51}$$

designate the non-overlap and overlap generator product ranges and $\Omega \equiv \Omega(E_{ij})$, $\Omega' = \Omega(E_{i'j'})$ are defined by Eq. (49). All the necessary segment values have been tabulated in [14] and alternative derivations have also been published [77, 80]. Yet another derivation of these segment formulas is afforded by the Green-Gould approach [33] and will be described in the following section.

12. Knowing the matrix representatives for the orbital U(n) generators and their products, it is straightforward to obtain the matrix representative of the Hamiltonian, Eq. (34). The actual computational procedure which accomplishes this step will very much depend on the chosen strategy for the implementation of UGA. Let us only mention that most diagonalization algorithms for large sparse matrices are based on Krylov sequences and require the evaluation of the action of the Hamiltonian $\hat{H}$ on the trial vector $\|C_K\|$, with K labeling the spin-adapted states, *i.e.*,

$$\sigma_I = \sum_J H_{IJ} C_J. \tag{52}$$

The crucial term in this basic iteration step arises from the generator product term in $\hat{H}$, Eq. (34), leading to

$$\sigma'_I = \sum_{KJ} \sum_{i,j,k,\ell} (ij|k\ell) \gamma_{ij}^{IK} \gamma_{k\ell}^{KJ} C_J, \tag{53}$$

where we have used Mulliken's notation for two-electron integrals, $(ij|k\ell) = \langle ik|\hat{v}|j\ell\rangle$, and where we have designated the generator matrix elements in the spin-adapted basis $\{|k\rangle\}$ as

$$\gamma_{ij}^{KL} = \langle K|E_{ij}|L\rangle. \tag{54}$$

The computational strategy in evaluating the right hand side of (53) will depend on whether we choose an integral driven [17,20], loop driven [18], shape driven [19], internal interaction block driven [21] or other algorithm for UGA implementation. These aspects of UGA will be treated in detail by Shavitt's contribution to this series [31].

4. Green-Gould Representation Theory. An efficient and versatile evaluation of UGA generator matrix elements in the chosen spin-adapted basis represents a key ingredient for its successful implementation, particularly when large scale computations are intended. Consequently, this problem received considerable attention (*cf.*, *e.g.*, [16]) and different derivations have been investigated. Perhaps the most straightforward technique for the derivation of required segment values is based on graphical methods of spin algebras [66, 67], exploiting the relationship between the U(n) and U(2) or SU(2) irreps mentioned earlier [14]. Recently, another very general method, which can provide equivalent results and allows an extension to more general situations, has been exploited by Gould and Chandler [34]. This method follows Gould's general representation theory for unitary and orthogonal groups, which is in turn based on Green's polynomial identities for the generators of semi-simple Lie groups [33]. Gould and Chandler have given an alternative derivation of factorized expressions for matrix elements of UGA generators [34], Eq. (48), and of products of such generators, Eq. (50), although the actual evaluation and tabulation of segment values (which can be quite laborious) was not given. More importantly, however, they have also shown how to determine the matrix elements of U(2n) spinorbital generators among the various U(n) irrep bases [35], which will be required when considering the spin-orbit or other spin-dependent effects. Very recently this method was also employed to handle the system partitioning [36] exploiting bases adapted to the chain

$$\text{(55)} \qquad U(n) \supset U(n_1) \times U(n_2) \times \cdots \times U(n_k)\ ; \ \sum_{i=1}^{k} n_i = n\ .$$

Note that the hole-partile formalism [15] ($k = 2$) or the partitioning into core, active and virtual orbitals used in general multi-reference approaches ($k = 3$) represent special cases of such a partitioning. More general partitioning of this type will then be required for various "group-function" approaches*, such as atom-in-molecule, diatomics-in-molecule and other schemes [81]. We now briefly outline this approach for the simplest possible case in order to indicate its essential features.

1. We define the $n \times n$ matrix $\mathbf{E}$, whose elements are the U(n) generators E_{ij},

$$\text{(56)} \qquad \mathbf{E} \equiv \mathbf{E}^{(n)} := [E_{ij}]_{n\times n}, \quad (i,j = 1,\ldots,n).$$

As indicated, we shall drop the superscript (n) when no confusion can arise. As before, we shall not distinguish the U(n) generators and their representatives, unless necessary for better understanding. Moreover, we shall imply the summation convention over the repeated indices.

We first observe that the powers of $\mathbf{E}$ are uniquely defined since

$$\text{(57)} \qquad (\mathbf{E}^m)_{ij} = E_{ik}(\mathbf{E}^{m-1})_{kj} = (\mathbf{E}^{m-1})_{ik}E_{kj}.$$

It is then easily verified (using mathematical induction) that the U(n) commutation relations (22) can be generalized for an arbitrary power of $\mathbf{E}$ or, in fact, for any finite polynomial in $\mathbf{E}$, $p(\mathbf{E})$, obtaining

$$\text{(58)} \qquad [E_{ij}, p(\mathbf{E})_{k\ell}] = \delta_{jk}p(\mathbf{E})_{i\ell} - \delta_{i\ell}p(\mathbf{E})_{kj}\,.$$

We also note that the traces of the various powers of $\mathbf{E}$ represent Casimir invariants for U(n), *i.e.*,

$$\text{(59)} \qquad I_k := \operatorname{tr}(\mathbf{E}^k)\,.$$

In particular, $I_1 = E_{ii} = N$ and $I_2 = E_{ij}E_{ji}$ (summation convention implied). It is easily seen that I_k commute with any generator E_{ij},

$$\text{(60)} \qquad [I_k, E_{ij}] = 0\,,$$

*Here "group" means a chemical functional group rather than an algebraic structure.

so that, in view of Schur's lemma, I_k are invariant on any U(n) irrep. It is more difficult to show that at most n of these invariants are independent. The property (60) also implies that I_k form a mutually commuting set of Hermitean invariants since

$$[I_k, I_\ell] = 0 \quad , \qquad I_k^\dagger = I_k \,. \tag{61}$$

Examining the action of I_k on the highest weight state of a given irrep Γ we can find the recurrence relations yielding the constant value $\chi_\Gamma(I_k)$ which I_k takes on Γ. In particular, for the UGA irrep $\Gamma(a,b,c)$ of $U(n)$ we get

$$\begin{aligned} \chi_\Gamma(I_1) &= n - c + 1 = N \,, \\ \chi_\Gamma((I_2) &= a(n+3-a) + (n-c)(1+c) \,. \end{aligned} \tag{62}$$

2. Green showed [33] that the following polynomial identities hold on any irrep $\Gamma(\mathbf{m}_n)$ of U(n),

$$\prod_{r=1}^{n} (\mathbf{E} - \epsilon_r) = 0 \qquad \text{where} \qquad \epsilon_r = m_{rn} + n - r, \tag{63}$$

and

$$\prod_{r=1}^{n} (\bar{\mathbf{E}} - \bar{\epsilon}_r) = 0 \qquad \text{where} \qquad \bar{\epsilon}_r = r \; - 1 - m_{rn} \; = n - 1 - \epsilon_r, \tag{63'}$$

$\bar{\mathbf{E}}$ designating a contragredient or a conjugate representation,*

$$(\bar{\mathbf{E}})_{ij} = \bar{E}_{ij} = \; -E_{ji} \,. \tag{64}$$

These polynomial identities are not necessarily the minimum polynomial identities satisfied by $\mathbf{E}$. It can be shown that $\epsilon_r(\bar{\epsilon}_r)$ represent constant values which $\mathbf{E}$ takes on the rep $\overline{\Gamma}_1 \times \Gamma(\mathbf{m}_n)$ ($\Gamma_1 \times \Gamma(\mathbf{m}_n)$) with Γ_1 and $\bar{\Gamma}_1$ designating the fundamental vector representation and its conjugate, $\Gamma_1(E_{ij}) = \mathbf{e}^{ij}$ and $\bar{\Gamma}_1(E_{ij}) = -\mathbf{e}^{ji}, \mathbf{e}^{ij}$ being the $n \times n$ matric units, Eq. (21). It thus suffices to retain in the products (63) and (63′) those irreps, which are contained in the Clebsch-Gordan reduction of $\bar{\Gamma}_1 \times \Gamma(\mathbf{m}_n)$ or $\Gamma_1 \times \Gamma(\mathbf{m}_n)$, *i.e.*,

$$\begin{aligned} \bar{\Gamma}_1 \times \Gamma(\mathbf{m}_n) &= \bigoplus_r \Gamma(\mathbf{m}_n - \boldsymbol{\Delta}_r) \quad , \\ \Gamma_1 \times \Gamma(\mathbf{m}_n) &= \bigoplus_r \Gamma(\mathbf{m}_n + \boldsymbol{\Delta}_r) \quad , \end{aligned} \tag{65}$$

where

$$\boldsymbol{\Delta}_r := (0, \ldots, 0, 1, 0, \ldots, 0), \tag{66}$$

*Recall that a conjugate of a rep $\Gamma : g \mapsto \Gamma(g)$ is defined as $\bar{\Gamma} : g \mapsto [\Gamma^\dagger(g)]^{-1}$.

with 1 in the r-th position. Since the Γ's on the right hand side of (65) must satisfy the lexicality conditions, Eqs. (25), (28), we see that for the UGA irrep $\Gamma(a,b,c)$ there are at most three irreps present with $r_1 = a+b$, $r_2 = a$, $r_3 = n$ and $r_1 = a+b+1$, $r_2 = a+1$, $r_3 = 1$, respectively. Thus, for each $\Gamma(a,b,c)$ the following identities hold

$$(67) \quad \prod_{i=1}^{3}(\mathbf{E}-\epsilon_i) = (\mathbf{E}-\epsilon_1)(\mathbf{E}-\epsilon_2)\mathbf{E} = 0, \text{ with } \epsilon_1 = 1+c, \quad \epsilon_2 = n+2-a;$$

and

$$(67') \quad \prod_{i=1}^{3}(\bar{\mathbf{E}}-\bar{\epsilon}_i) = 0 \quad \text{with} \quad \bar{\epsilon}_1 = n-c, \quad \bar{\epsilon}_2 = a-1, \quad \bar{\epsilon}_3 = -2.$$

3. With polynomial identities (63) or (67) we can associate mutually orthonormal projectors,

$$(68) \qquad P[r] = \prod_{k=1}^{n} \frac{\mathbf{E}-\epsilon_k}{\epsilon_r-\epsilon_k} \quad \text{and} \quad \bar{P}[r] = \prod_{\substack{k=1\\(k\neq r)}}^{n} \frac{\bar{\mathbf{E}}-\bar{\epsilon}_k}{\bar{\epsilon}_r-\bar{\epsilon}_k},$$

such that

$$(69) \qquad P[r]P[s] = \delta_{rs}P[r] \quad \text{and} \quad \bar{P}[r]\bar{P}[s] = \delta_{rs}\bar{P}[r],$$

providing the resolution of the identity,

$$(70) \qquad \sum_{r=1}^{n} P[r] = \sum_{r=1}^{n} \bar{P}[r] = \hat{1}\ ,$$

as follows from Lagrange's interpolation formula. Thus, for the $\Gamma(a,b,c)$ irrep we get the following projectors

$$(71) \quad P[1] = \frac{(\mathbf{E}-\epsilon_2)\mathbf{E}}{(\epsilon_1-\epsilon_2)\epsilon_2}, \quad P[2] = \frac{(\mathbf{E}-\epsilon_1)\mathbf{E}}{(\epsilon_2-\epsilon_1)\epsilon_2}, \quad P[3] = \frac{(\mathbf{E}-\epsilon_1)(\mathbf{E}-\epsilon_2)}{\epsilon_1\epsilon_2},$$

and similarly for $\bar{P}[r]$, $\quad r = 1,2,3$.

4. Recall that a vector operator Ψ on $\mathrm{U}(n)$ is defined as a collection of operators $\Psi_i \quad (i = 1,\ldots,n)$ that satisfy the relations

$$(72) \qquad [E_{ij}, \Psi_k] = \delta_{jk}\Psi_i \quad .$$

It can be shown that any such vector operator can be resolved into a sum of shift components when acting on $\Gamma(\mathbf{m}_n)$,

$$(73) \qquad \Psi_i = \sum_r \Psi[r]_i \quad ,$$

such that $\Psi[r]_i|v\rangle$ transforms according to $\Gamma(\mathbf{m}_n + \mathbf{\Delta}_r)$ when $|v\rangle$ transforms according to $\Gamma(\mathbf{m}_n)$, and that these shift components result by an application of the projectors $P[r]$, namely

$$\Psi[r]_i = P[r]_{ij}\,\Psi_j = \Psi_j \bar{P}\,[r]_{ji} \quad . \tag{74}$$

Restricting ourselves to at most two-column irreps needed in UGA, we find that

$$\Psi[r]_i \quad : \quad [a_i, b_i, c_i] \to [a_i, b_i, c_i] + \delta_r, \tag{75}$$

and

$$\Psi^\dagger[r]_i \quad : \quad [a_i, b_i, c_i] \to \ [a_i, b_i, c_i] - \delta_r \quad , \tag{75'}$$

where

$$\delta_1 = [0, 1, -1] \qquad \text{and} \qquad \delta_2 = [1, -1, 0] \quad . \tag{76}$$

5. To derive the elementary generator matrix elements we consider any two adjacent subgroups in the canonical chain (29), *i.e.*,

$$\mathrm{U}(m+1) \supset \mathrm{U}(m), \tag{77}$$

and observe that the generators $E_{i,m+1} =: \Theta_i$ constitute a vector operator $\mathbf{\Theta}$ of $\mathrm{U}(m)$, since

$$[E_{ij}, \Theta_k] = \delta_{jk}\ \Theta_i \tag{78}$$

holds, as follows from $\mathrm{U}(n)$ commutation relations, Eq. (22) for E_{ij} or Eq. (58). Resolving $\mathbf{\Theta}$ into the shift components we get the relations (75) and (75′) with Ψ replaced by Θ. Thus

$$E_{m,m+1}\ |\{\mathbf{P}\}\rangle = \sum_{r=1}^{2} \Theta[r]_m |\{\mathbf{P}\}\rangle = \sum_{r=1}^{2} N_m^r\ |\{\mathbf{P} + \delta_r^m\}\rangle, \tag{79}$$

where

$$|\{\mathbf{P}\}\rangle = \left| \begin{matrix} \mathbf{P}_{m+1} \\ \mathbf{P}_m \\ (\mathbf{P}) \end{matrix} \right\rangle \quad \text{and} \quad |\{\mathbf{P} + \delta_r^m\}\rangle = \left| \begin{matrix} \mathbf{P}_{m+1} \\ \mathbf{P}_m + \delta_r \\ (\mathbf{P}) \end{matrix} \right\rangle \tag{80}$$

are the basis vectors labeled by the ABC tableaux, so that $\mathbf{P}_i = [a_i, b_i, c_i]$ and $(\mathbf{P})$ designates the remainder of the tableau. The coefficients N_m^r give the desired matrix elements of $E_{m,m+1}$. Choosing the real phases we can write

$$N_m^r = \langle \Theta^\dagger[r]_m\, \Theta[r]_m \rangle^{\frac{1}{2}}, \tag{81}$$

with the mean value evaluated for the state $|\{\mathbf{P}\}\rangle$, Eq. (80).

6. Exploiting the Wigner-Eckart theorem [82] for U(n) we find that

$$\Theta^{\dagger}[r]_i\,\Theta[r]_j = R^r_m \bar{P}[r]_{ij}, \tag{82}$$

where R^r_m is a U(m) invariant whose eigenvalues determine the square of the reduced matrix element. In particular

$$\Theta^{\dagger}[r]_m\,\Theta[r]_m = R^r_m \bar{C}^r_m, \tag{83}$$

where

$$\bar{C}^r_m = \bar{P}[r]_{mm} \tag{84}$$

is a U($m-1$) invariant which is diagonal in the canonical Gel'fand-Tsetlin basis. It thus remains to evaluate the invariants R^r_m and C^r_m on $\Gamma(a,b,c)$. The latter can be found directly be expressing them through $(\mathbf{E}^k)_{mm}$ (no sum over m), which gives

$$\chi_\Gamma(\bar{C}^1_m) = \frac{\Delta c_m(2+b_m)}{(2+m-c_m)(1+b_{m-1})}, \quad \chi_\Gamma(\bar{C}^2_m) = \frac{\Delta a_m\, b_m}{(1+a_m)(1+b_{m-1})}. \tag{85}$$

The reduced coefficient related R^r_m are determined by first calculating the trace of Eq. (82),

$$\begin{aligned} \Gamma^r_m :&= R^r_m \operatorname{tr}(\bar{P}[r]) = \sum_{j=1}^{m} \Theta^{\dagger}[r]_j \Theta[r]_j \\ &= \sum_{k,\ell=1}^{m} E_{m+1,k}\, E_{\ell,m+1}\, \frac{\bar{E}_{\ell k} - \bar{\epsilon}_\ell \delta_{\ell k}}{\bar{\epsilon}_r - \bar{\epsilon}_\ell}, \end{aligned} \tag{86}$$

and

$$\bar{J}_r := \operatorname{tr}(\bar{P}[r]), \tag{87}$$

which gives $R^r_m = \Gamma^r_m/\bar{J}_r$.

The Γ-invariant is then expressed through the $(\mathbf{E}^{(m+1)})^k_{m+1,m+1}$ invariants for $k = 1, 2$ and 3, which eventually gives

$$\chi_\Gamma(\Gamma^1_m) = \frac{\Delta \bar{c}_{m+1}(2+b_m)c_m}{1+b_{m+1}}, \tag{88a}$$

and

$$\chi_\Gamma(\Gamma^2_m) = \frac{\Delta a_{m+1}(m+1-a_m)b_m}{1+b_{m+1}}. \tag{88b}$$

Finally, the J-invariants can be expressed in terms of the dimension formula (47),

(89)
$$\chi_\Gamma(\bar{J}_1) = \frac{\dim\ \Gamma(a,b+1,c-1)}{\dim\ \Gamma(a,b,c)} = \frac{(2+b_m)c_m}{(1+b_m)(m+2-c_m)},$$
$$\chi_\Gamma(\bar{J}_2) = \frac{\dim\ \Gamma(a+1,b-1,c)}{\dim\ \Gamma(a,b,c)} = \frac{b_m(m+1-a_m)}{(1+b_m)(1+a_m)}.$$

Thus

(90a)
$$\chi_\Gamma(R^1_m) = \frac{(m+1-c_m)\,(1+b_m)\,\Delta\bar{c}_{m+1}}{1+b_{m+1}},$$

and

(90b)
$$\chi_\Gamma(R^2_m) = \frac{(1+a_m)\,(1+b_m)\Delta a_{m+1}}{1+b_{m+1}},$$

and substituting into Eqs. (81), (83) for $(N^r_m)^2 = \chi_\Gamma(R^r_m \bar{C}^r_m)$ yields finally

(91)
$$\langle\{\mathbf{P}_m + \delta_r\}\,|E_{m,m+1}\,|\{\mathbf{P}\}\rangle = \left[\frac{(1+b_m)\,(b_m+4-2r)}{(1+b_{m-1})\,(1+b_{m+1})}\right]^{\frac{1}{2}} \quad , \quad (r=1,2)$$

In a similar way we obtain the matrix elements for the lowering generator $E_{m+1,m}$ which, of course, will be given by Eq. (31). This result is identical with Eqs. (31, 32) of [10].

Generalization of this result to non-elementary generators [34], spinorbital generators [35], explicit formulae for the fundamental $\mathrm{U}(n_1) : \mathrm{U}(n_1) \times \mathrm{U}(n_2)$, $(n = n_1 + n_2)$ Wigner coefficients, their symmetry properties and general segment level formalism in an arbitrary multishell composite basis [36] have been given. For example, in the simplest case of partitioning into the two subsystems (note also the similarity of this case with the hole-particle formalism [15]),

$$\mathrm{U}(n) \equiv \mathrm{U}(n_1+n_2) \supset \mathrm{U}(n_1) \times \mathrm{U}(n_2),$$

we can relabel the $\mathrm{U}(n)$ generators as follows

$$a_{ij} = E_{ij}, \qquad (i,j = 1,\ldots,n_1)$$
$$b_{\mu\nu} = E_{\mu+n_1,\nu+n_1}, \qquad (\mu,\nu = 1,\ldots,n_2)$$
$$\psi_{\mu i} = E_{\mu+n_1,i} \quad \text{and} \quad \psi_{i\mu} = E_{i,\mu+n_1}\ .$$

The $\mathrm{U}(n)$ commutation relations [cf. Eq. (22) or Sec. 3.7] can then be re-written in terms of the generators a_{ij}, $b_{\mu\nu}$, $\psi_{\mu i}$, $\psi_{i\mu}$. Clearly, $[a_{i\mu}, b_{\mu\nu}] = 0$ and we obtain standard commutation relations for a_{ij}, and for $b_{\mu\nu}$, representing generators for

$\mathrm{U}(n_1)$ and $\mathrm{U}(n_2)$, respectively, while for the "charge transfer" generators $\psi_{\mu i}, \psi_{i\mu}$ we have, for example,

$$[a_{ij},\ \psi_{k\sigma}] = \delta_{kj}\ \psi_{i\sigma} \qquad [b_{\mu\nu},\ \psi_{k\sigma}] = -\delta_{\mu\sigma}\,\psi_{k\nu} \quad ,$$
$$[a_{ij},\ \psi_{\sigma k}] = -\delta_{ik}\psi_{\sigma j}, \qquad [b_{\mu\nu},\ \psi_{\sigma k}] = \delta_{\nu\sigma}\,\psi_{\mu k} \quad .$$

Thus, we can again regard these charge transfer operators as $\mathrm{U}(n_1)\times\mathrm{U}(n_2)$ tensor operators, which transform as $[1,\dot{0}]\otimes\overline{[1,\dot{0}]}$ and $\overline{[1,\dot{0}]}\otimes[1,\dot{0}]$, respectively* (for details, see [36]). Currently, work is in progress for a general multi-reference orbital partitioning [83].

5. Clifford Algebra UGA. In this section we wish to point out the fundamental role of Clifford algebras for finite-dimensional quantum-chemical orbital models of many-electron systems, to elucidate the relationship between Clifford algebra matric units and UGA generators [7] and to indicate the possibilities offered by CAUGA in handling of the many-electron correlation problem [7-9, 32].

5.1 Basic relationships. *1.* Recall [84] that a Clifford algebra C_m is an associative algebra generated by the Clifford numbers α_i satisfying the relations

$$\{\alpha_i, \alpha_j\} = 2\delta_{ij} \quad , \quad (i, j = 1, \ldots, m) \tag{92}$$

with the anticommutator defined by Eq. (3). The monomials $\alpha_1^{\nu_1}\ \alpha_2^{\nu_2}\ \ldots\ \alpha_m^{\nu_m}$ with $\nu_i = 0$ or 1 form the basis for C_m, since in view of Eq. (92) $\alpha_i^2 = 1$, so that $\dim\ C_m = 2^m$.

2. The fermionic algebra F_{2n} generated by $X_I^\dagger,\ X_I,\ I = 1, \ldots, 2n$ is isomorphic with the Clifford algebra C_{4n}, since we can define

$$\alpha_I = X_I + X_I^\dagger \quad , \qquad \alpha_{I+2n} = i(X_I - X_I^\dagger). \tag{93}$$

It is easily verified that α_I's satisfy Eq. (92).

3. We can construct a matric basis for F_{2n} by defining

$$e_{[I]\,[J]} = X_{[I]}^\dagger\ e_0\ X_{[J]} \quad , \tag{94}$$

with $X_{[I]}^\dagger$, $X_{[J]}$ defined by Eq. (15) and e_0 designating a vacuum projector

$$e_0 = \bar{N}_1\bar{N}_2\ldots\bar{N}_{2n}, \tag{95}$$

with $\bar{N}_I$ defined by Eq. (6). Since

$$X_{[I]}e_0 = e_0\ X_{[I]}^\dagger = 0, \tag{96}$$

*The symbol $\dot{0}$ indicates a repetition as required, *i.e.*, $[1,\dot{0}] \equiv [1,0,0,\ldots,0]$ with appropriate number of $0'$s.

and

$$e_0 X_{[I]}\, X^{\dagger}_{[J]}\, e_0 = \delta_{[I],\,[J]} e_0 \quad , \tag{97}$$

we easily find that indeed

$$e_{[I]\,[J]}\, e_{[K]\,[L]} = \delta_{[J]\,[K]}\, e_{[I]\,[L]}. \tag{98}$$

4. The diagonal matric units provide a resolution of the identify into idempotent orthogonal projectors

$$e_{\{I\}} = e_{[I]\,[I]} \ , \tag{99}$$

since

$$e^2_{\{I\}} = e_{\{I\}} \qquad , \qquad e^{\dagger}_{[I]} = e_{[I]} \tag{100}$$

$$e_{\{I\}}\, e_{\{J\}} = \delta_{\{I\},\{J\}}\, e_{\{I\}}, \tag{100'}$$

and

$$\hat{I} \equiv e = \sum_{\{I\}} e_{\{I\}}. \tag{100''}$$

It can be easily verified that in fact

$$e_{\{I\}} = N_{I_1}\, N_{I_2} \ldots N_{I_r}\, \bar{N}_{I_{r+1}} \ldots \bar{N}_{I_{2n}} \ . \tag{101}$$

Since the occupation number operators N_I and their conjugates $\bar{N}_I$, Eq. (6), mutually commute, the order of the operators on the right hand side of Eq. (101) is immaterial. We emphasize this symmetric nature of the product (101) by enclosing the index set I in curly brackets [cf. Eq. (14)].

5. The vacuum projector e_0, Eq. (95), can also be written in the form

$$e_0 = |0\rangle\langle 0| \, , \tag{102}$$

so that

$$e_{[I]\,[J]} = X^{\dagger}_{[I]}\, |0\rangle\langle 0| X_{[J]} = |\,[I]\rangle\langle [J]|, \tag{103}$$

where

$$|[I]\rangle \;=\; |\,[I_1\ I_2\ \ldots\ I_r]\rangle \;\equiv \left| \begin{matrix} I_1 \\ I_2 \\ \vdots \\ I_r \end{matrix} \right\rangle = X^{\dagger}_{[I]}\, |0\rangle \tag{16''}$$

is an r-particle antisymmetric state, Eqs. (16),(16′), and $\langle [I] \mid$ is its dual. The antisymmetric nature of these states is indicated either by enclosing the index set I in square brackets [cf. Eq. (14)] or by a vertical arrangement of spinorbital labels implying the one-column Weyl tableau [9].

The matric units (94) or (103) can be classified into the particle-number conserving ($\text{card}[I] = \text{card}[J]$) and non-conserving ($\text{card}[I] \neq \text{card}[J]$) ones. The former ones are closed under multiplication and form a subalgebra for each cardinality. In the special case of $\text{card}[I] = \text{card}[J] = 1$, when we can simply set $[I] = I$ and $[J] = J$, we obtain

$$e_{IJ} = \mid I\rangle\langle J \mid \quad , \tag{104}$$

which multiply in the same way as the GL($2n$) matric units, *i.e.*,

$$e_{IJ} \quad e_{KL} \;=\; \delta_{JK}\, e_{IL} \quad . \tag{105}$$

6. Any associative algebra can be turned into a Lie algebra by defining the Lie product of A and B as a commutator $[A, B]$, Eq. (8). We thus get

$$[\, e_{[I]\,[J]},\; e_{[K]\,[L]}] = \delta_{\,[J],\,[K]}\; e_{[I]\,[L]} - \delta_{\,[I],\,[L]}\; e_{[K]\,[J]}, \tag{106}$$

providing the generators of U(2^{2n}) or, in the one-particle subalgebra case (104), of U($2n$).

5.2. Generators. *1.* In the second-quantized formalism one usually realizes the U($2n$) generators as

$$e^I_J = X^\dagger_I X_J. \tag{107}$$

Generalizing this definition as suggested by Eqs. (94) or (103) we can write

$$e^{[I]}_{[J]} = X^\dagger_{[I]}\; X_{[J]}. \tag{108}$$

The relationship of these two possible bases was investigated in detail recently [7]. It has been shown (see also [6]) that both bases are related as

$$e_{[I]\,[J]} = e^{[I]}_{[J]}\; f_{r'}\,,\ \text{card}\ [J] = r', \tag{109}$$

where $f_{r'}$ is a projector onto the at most r'-particle layers of the Fock space, *i.e.*,

$$f_{r'} = \prod_{t=r'+1}^{2n} \frac{\hat{N}\ - t}{r' - t} \tag{110}$$

or, equivalently,

$$f_{r'} = \sum_{t=0}^{r'} |t\rangle\, \langle t|\,, \tag{110'}$$

where

$$(110'') \qquad |t\rangle\langle t| = \sum_{[I]} |[I_1 I_2 \dots I_t]\rangle \, \langle [I_1 I_2 \dots I_t]| \, .$$

Explicitly, we can also write

$$(111) \qquad e^{[I]}_{[J]} = e_{[I]\,[J]} = \sum_{I_1}{}' \, e_{[II_1]\,[JI_1]} + \sum_{I_1<I_2}{}' \, e_{[II_1I_2]\,[JI_1I_2]} + \cdots \quad ,$$

where the prime indicates that the summation extends over all spinorbitals not included in $[I]$ or $[J]$. Moreover, the sets $[II_1]$, $[II_1I_2]$, etc. are not necessarily ordered [*i.e.*, generalizing the definition (94)], thus ensuring that all the phases in the expansion (111) are +1.

2. Writing the spinorbitals $|I\rangle$ as products of orbital and spin kets, $|I\rangle = |i\rangle|\sigma\rangle$, we define one-particle orbital generators Λ_{ij} as a partial trace of spinorbital generators (107),

$$(112) \qquad \Lambda_{ij} = \sum_{\sigma} e^{i\sigma}_{j\sigma} = X^{\dagger}_{i\alpha}X_{j\alpha} + X^{\dagger}_{i\beta}X_{j\beta},$$

as in UGA.* These operators again satisfy the commutation relations of the type (22) and the Hermitean property (24), forming the generators of the orbital group U(n). We also introduce corresponding "multi-particle"orbital generators of U(2^n). In order to avoid multi-particle indices, we shall label the Clifford algebra spanning set monomials (basis vectors) by the orbital occupation numbers $(m_1 m_2 \dots m_n)$ with $m_i = 0$ or 1. Interpreting the occupation numbers as binary integers [69] $(m_1 m_2 \dots m_n)_2$, we can label these basis states and corresponding generators using integer indices

$$(113) \qquad p \equiv p\{m_i\} = 2^n - (m_i m_2 \dots m_n)_2 .$$

Thus, the orbital analogues of the operators $e_{[I]\,[J]}$ will be designated as E_{pq} with p and q labeling the ordered orbital sets $[i]$ and $[j]$ according to the labeling convention (113). Similarly, the basis states will be designated as $|p)$. For example, in the three-orbital case we have:

(114)

$(m_1m_2m_3)$	:	000	100	010	001	110	101	011	111
$(m_1m_2m_3)_2$	:	0	4	2	1	6	5	3	7
$\mid[i]\rangle$	:	$\|0\rangle$	$\|1\rangle$	$\|2\rangle$	$\|3\rangle$	$\|[12]\rangle$	$\|[13]\rangle$	$\|[23]\rangle$	$\|[123]\rangle$
$\mid p)$	:	$\|8)$	$\|4)$	$\|6)$	$\|7)$	$\|2)$	$\|3)$	$\|5)$	$\|1)$

*In order to distinguish easily U(n) and U(2^n) generators we now designate the former ones by Λ_{ij} and the latter ones again by E_{ij}.

The corresponding operators $E_{pq} = |p)(q|$ provide then a fundamental (one box) defining representation for $U(2^n)$.

We note that one can also construct spin-free higher rank analogues of the operators (108), which proved to be useful in various many-body theory applications [7, 85, 86]. Recently, we have found these replacement (also called excitation or reduced density) operators [7] to be very useful in deriving the spin-adapted linear coupled-cluster equations in the multi-reference case [87].

3. In view of the embedding $U(2^n) \supset U(n)$, Eq. (1), we can express the $U(n)$ generators Λ_{ij} in terms of the $U(2^n)$ generators E_{pq} [8, 69]. Analogously to (111) we write

$$\Lambda_{ij} = \sum_{\{k\}}{}' \; |[i\{k\}]\rangle\langle[j\{k\}]| \quad . \tag{115}$$

where the sets $[i\{k\}] = [i \quad k_1 \quad k_2 \ldots k_r]$ and $[j\{k\}] = [j\, k_1\, k_2 \ldots k_r]$ are not necessarily ordered and the prime indicates that $i, j \notin \{k\}$. For example, in the above considered case of $U(8) \supset U(3)$ we get using (115) that

$$\begin{array}{lclclcl} \Lambda_{12} & = & |1\rangle\langle 2| & + & |[13]\rangle\langle[23]| & = & E_{46} + E_{35} \;, \\ \Lambda_{13} & = & |1\rangle\langle 3| & + & |[12]\rangle\langle[32]| & = & E_{47} - E_{25} \;, \text{etc.} \end{array} \tag{116}$$

Generally we get

$$\Lambda_{ij} = (-1)^{j-i+1} \sum_{p,q,r} (-1)^{S_2(q)} \quad E_{p+\tau,p+\tau'} \quad , (i \neq j) \tag{117}$$

where

$$\tau \equiv \tau_{ij}(r,q) \;, \qquad \tau' \equiv \tau_{ji}(q,r),$$

$$\tau_{k\ell}(u,v) = 2^{n-\ell}(2^{\ell-k+1}u + 2v + 1) \quad ,$$

with

$$1 \leq p \leq 2^{n-j} \quad , \quad 0 \leq q \leq 2^{j-i-1} - 1 \quad , \quad 0 \leq r \leq 2^{i-1} - 1 \quad ,$$

and $S_2(q)$ designating the digital sum of q_2 (the binary representation of q). Similarly, for the weight generators we get

$$\Lambda_{ii} = \sum_{p=1}^{2^n} m_i^p E_{pp} \,, \tag{118}$$

where m_i^p designates the occupation numbers in the p-th basis state [using the labeling convention (114)]. Interpreting the fundamental states as Slater determinants, the generator expressions (117) and (118) can also be regarded as an encoding of Slater rules for one-particle replacement operators. In the following we shall need expressions (117) only for the states with a given (fixed) particle number. It is easily seen that while the general representation of Λ_{ij} Eq. (117), will contain 2^{n-2} generators, only

$$\left[\binom{n-2}{a-1}+\binom{n-2}{c-1}\right]\frac{1}{1+\delta_{b,0}} \tag{119}$$

of them will be needed when considering the irrep $\Gamma(a,b,c)$, given by Eq. (38). For the basis states $|p)$ associated with a given fixed particle number we can set up a table, whose entries $\pm(i,j)$ indicate the generator Λ_{ij} transforming the state labeling the column into that labeling the row. Such a representation is possible since each E_{pq} is contained at most once in some Λ_{ij} and all the nonvanishing coefficients in the representation (117) equal ± 1. For example, considering U(4) 2-particle states, we immediately find the following $U(2^n)$ basis vectors to be relevant:

	$\|p)$	(m_1, m_2, m_3, m_4)
	4	1100
	6	1010
	7	1001
	10	0110
	11	0101
(120)	13	0011

so that the table mentioned above is

		4	6	7	10	11	13
	4	-	(2,3)	(2,4)	$-(1,3)$	$-(1,4)$	-
	6		-	(3,4)	(1,2)	-	$-(1,4)$
	7			-	-	(1,2)	(1,3)
	10				-	(3,4)	$-(2,4)$
	11					-	(2,3)
(121)	13						-

Thus, for example, $\Lambda_{24}\,|7) = |4)$, $\Lambda_{14}\,|11) = -|4)$, etc. (cf. Eqs. (59) and (67) of [8]).

5.3 States. *1.* Using the fundamental one-box spinor representation for $U(2^n)$ discussed above, we can construct general two-column irreps of $U(n)$ and represent the corresponding spin-adapted states as linear combinations of $U(2^n)$ two-box states. We note, however, that the $U(2^n) \downarrow U(n)$ subduction is not multiplicity free. In this respect, the subgroup chain (1) can provide additional labels, if desired. Moreover, the SO(2n) group, involving two-particle generators of the type $X_i^\dagger X_j^\dagger$ or $X_i X_j$, and the SO($2n+1$) group, which also involves single particle creators and annihilators, are suitable to handle the particle-number non-conserving operators [8].

It can be shown [37-39] that, generally, any p-column irrep of U(n) is contained at least once in the totally symmetric p-box irrep of U(2^n). We illustrate the structure of CAUGA carrier spaces on an example below.

2. Consider, for simplicity, the four orbital case ($n = 4$). To obtain possible U(4) irreps carried by the totally symmetric two-box irrep $[2\ \dot{0}] \equiv [2]$ of $U(2^4) = U(16)$, we consider the symmetrized power $([0]+[1]+[1^2]+\cdots+[1^4])^{\otimes 2}$. Applying Littlewood-Richardson rules we find that this $U(16)$ irrep of dimension 136 subduces to 19 components listed in Table III.

Table III.

Decomposition of the subduced rep $[2\dot{0}] \downarrow$ U(4) of U(16) into the irreducible components, listed by their associated particle number N. Their dimensions, highest weight states (HWS) and the number M, Eq. (122), of needed two box tableaux are also given.

N	U(4) Components	Dim	HWS	M
0	$[0]$	1	$[16\|16]$	1
1	$[1]$	4	$[8\|16]$	4
2	$[1^2], [2]$	$6, 10$	$[4\|16], [8\|8]$	$6, 10$
3	$2[1^3], [21]$	$4, 20$	$[2\|16], [4\|8]$	$4, 24$
4	$3[1^4], [21^2], [2^2]$	$1, 15, 20$	$[1\|16], [2\|8], [4\|4]$	$1, 16, 21$
5	$2[21^3], [2^2 1]$	$4, 20$	$[1\|8], [2\|4]$	$4, 24$
6	$[2^2 1^2], [2^3]$	$6, 10$	$[1\|4], [2\|2]$	$6, 10$
7	$[2^3 1]$	4	$[1\|2]$	4
8	$[2^4]$	1	$[1\|1]$	1

In this table we have arranged the resulting U(4) components by the particle number N they carry. We observe that only the low-dimensional high-spin ($S = 3/2$ and 2) irreps appear more than once in this decomposition (for $3 \le N \le 5$). The number of two-box states $M \equiv M(a,b,c)$, which are required to represent the states of a given irrep $\Gamma(a,b,c) \equiv [2^a 1^b 0^c]$, is easily seen to be given as $\dim([1^{a+b}] \otimes [1^a])$ and equals

$$M = \binom{n}{a}\binom{n}{c}, \qquad \text{when} \quad b \neq 0, \quad \text{and} \tag{122}$$
$$M = \tfrac{1}{2}\binom{n}{a}\left[\binom{n}{a} + 1\right], \qquad \text{when} \quad b = 0\,.$$

3. Designating the two-box states of $[2\ \dot{0}]$ irrep of U(2^n) as

$$[i|j] \quad \equiv \quad \left|\,\boxed{i\,|\,j}\,\right\rangle \quad , \tag{123}$$

we can represent any spin-adapted state associated with the $U(n)$ irrep $\Gamma(a,b,c)$ as a linear combination of states $[i|j]$. In particular, the highest weight state of $\Gamma(a,b,c)$ can be shown [8] to be represented by $[2^c|2^{b+c}]$ (*cf.* Table III for an example). The above given expressions (122) for the number of two-box states (123), which are required for the representation of $\Gamma(a,b,c)$ states, can also be interpreted as the number of possible binary representatives which we can form by permiting 0's and 1's in the HWS components in all possible ways.

As an example, consider $N=3$ doublet irrep [21], so that $a=b=1$ and $c=2$ when $n=4$. We thus need $M=\binom{4}{2}\binom{4}{1}=24$ two-box tableaux to represent 20 doublet states of the [21] irrep (*cf.* Table III).* The HWS in this case is [4|8] or, in a binary representation [1100|1000]. Thus, the $\binom{4}{2}=6$ possible integers labeling the first box are listed in (120), while the $\binom{4}{1}=4$ integers labeling the second box are clearly:

$$\begin{array}{rcl} 8 & : & 1000 \\ 12 & : & 0100 \\ 14 & : & 0010 \\ 15 & : & 0001 \end{array} \tag{124}$$

Consequently, all possible 2-box tableaux can be represented by a 6×4 array with rows labeled by indices (120) and columns by (124), (*cf.* also Figs. 7-9 of [8]). The action of Λ_{ij} generators within the first set of indices was given in table (121) and for the second set we find similarly

(125)

	8	12	14	15
8	-	(1,2)	(1,3)	(1,4)
12		-	(2,3)	(2,4)
14			-	(3,4)
15				-

These tables can also be represented by the following HLED-type diagrams,

(126)

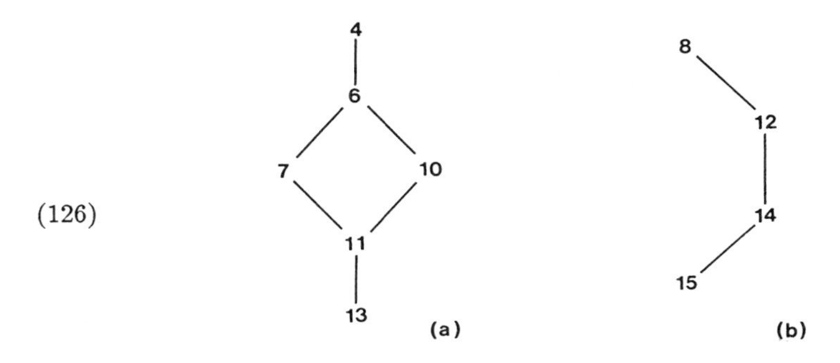

where only elementary lowering generator edges $\Lambda_{i,i-1}$ are indicated using the following convention

*Note that these 24 tableaux can represent 20 doublet states of [21] and 4 quartet states of $[1^3]$. However, for the latter, we prefer to use the tableaux associated with $[1^3] \otimes [0]$ rep with the HWS [2–16] (rather than $[2] \otimes [1]$ used above) as explained in Ref. [9].

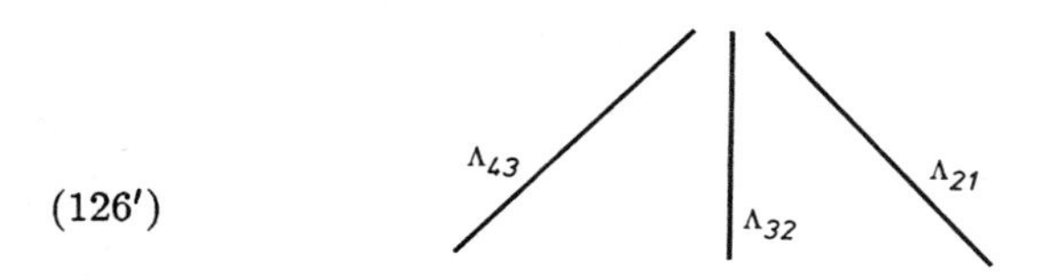

In the singlet case when $b = S = 0$, the HWS is $[2^c|2^c]$, so that both sets labeling rows and columns of the array just mentioned are identical. In view of the symmetry of the states (123), the part of the array below (or above) the diagonal need not be considered.

4. The matrix elements of E_{pq} generators within the two-box state basis (123) are very simple indeed: when acting on such states, the operator E_{pq} replaces the index q (if present) by the index p, and only if the double occupancy is created or destroyed, an extra factor of $\sqrt{2}$ arises. We reiterate that double occupancies will only ocur when $b = S = 0$. Defining the unnormalized states $(i|j)$ as

$$(i|j) = \sqrt{1+\delta_{ij}} \quad [i|j] \quad , \tag{127}$$

we can write the action of E_{pq} as follows

$$E_{pq}(i|j) = \delta_{qi}(p|j) + \delta_{qj}(i|p)\,. \tag{128}$$

This rule and the representation (117) determine the action of Λ_{ij} generators as summarized in tables (121) and (125) for the special case considered. This action is also easily determined directly [7] using the binary representation of two-box states when we realize that Λ_{ij} interchanges 0 and 1 in positions i and j in order to give a nonzero result. The sign is determined by the number of 1′s between the i-th and j-th positions. For example

$$\begin{aligned} \Lambda_{23}[6|14] &= \Lambda_{23}[1\underset{\sim}{0}\underset{\sim}{1}0|0\underset{\sim}{0}\underset{\sim}{1}0] \\ &= [1\underset{\sim}{1}\underset{\sim}{0}0|0010] + [1010|0\underset{\sim}{1}\underset{\sim}{0}0] \\ &= [4|14| + [6|12]\,, \end{aligned} \tag{129}$$

or

$$\begin{aligned} \Lambda_{13}[10|14] &= \Lambda_{13}[\underset{\sim}{0}1\underset{\sim}{1}0|\underset{\sim}{0}0\underset{\sim}{1}0] \\ &= -[\underset{\sim}{1}1\underset{\sim}{0}0|0010] + [0110|\underset{\sim}{1}0\underset{\sim}{0}0] \\ &= -[4|14] + [10|8]\,, \end{aligned} \tag{130}$$

as can be easily verified using tables (121) and (125).

5. In contrast to UGA, which exploits the Gel'fand-Tsetlin canonical basis, CAUGA does not require any particular coupling scheme, although the most natural basis involves the valence-bond Weyl-Rumer states [8]. The simplest version of

CAUGA would result if we use directly the orthonormal, though not spin-adapted, basis of two-box states (123). Such an approach is equivalent to the exploitation of a simple Slater determinant basis as implemented at the full CI level in the vectorized version by Knowles and Handy [23] (except for the singlet S=0 case where the additional symmetry was pointed out but not used).

The spin-adapted CAUGA states can be either obtained by a canonical construction [8, 9] method or using the U(n) Clebsch-Gordan coefficients or S_N outer product reduction coefficients [9]. The former method is best illustrated by exploiting the HLED scheme [76], which can also be useful in standard UGA [24]. We start from the HWS (or, similarly, the lowest weight state) representing the zeroth level of our diagram, and apply to it all elementary lowering (raising) generators (starting with $\Lambda_{n,n-1}$), obtaining the set of first level states. Repeating this procedure we generate the whole basis (unless we terminate the procedure at some level). In CAUGA, this procedure will generate non-orthogonal Weyl-Rumer states [28], whose Schmidt orthogonalization gives Gel'fand-Tsetlin basis. For example, in case of $\Gamma(1,1,2)$ irrep of U(4) we obtain the HLED shown in Fig. 1. Clearly, this HLED represents a "direct" product of single-box graphs (126). In order to distinguish easier the subgraphs (126a) and (126b), the edges of the latter one are dashed. Note that the four groups of three two-box tableaux (marked A-D in Fig. 1), having the same occupation numbers, represent two doublets with three open shells (the third linear combination representing the quartet state) and appear on the same co-diagonal in the "expanded"array of relevant two box states, which are associated with the $[1^2] \otimes [1]$ representation, as shown in Fig. 2. Note also that the relevant Gel'fand-Tsetlin or valence bond Weyl-Rumer states have always the same structure in terms of two-box tableaux belonging to each group as indicated in Fig. 1. Clearly, any other coupling scheme, yielding another set of spin-adapted states (*i.e.*, another linear combinations of the three two-box tableaux states in our case) can be used when desired.

6. It has been shown recently [9] that a general system partitioning with respect to the chain (55) can be exploited in the CAUGA formalism. Two distinct ways of constructing the partitioned bases have been outlined [9]. Such a partitioning will generally result in a substantial factorization of the highly-dimensional irreps into a considerable number of much smaller spaces (*cf.* examples in [9]). These partitioned sub-problems can be given a simple interpretation, depending on the nature of partitioning, and can be helpful in implementing various physical or computational criteria to achieve the much needed truncation of many-electron bases. Further development in this regard is highly desirable.

7. The CAUGA scheme is also amenable to a general point group symmetry adaptation [31] to the chain $U(n_i) \supset G \supset G(s)$, where n_i is the dimension defining a pure configuration and $G(s)$ designates the canonical chain supplying a unique labeling, following the procedure of Chen *et al.* [88].

Currently, we are also exploring an exploitation of CAUGA in the symmetry-adapted multi-reference coupled-cluster approach, based on the cluster ansatz due to Jeziorski and Monkhorst [89]. The possibility to express the configurations spanning an excitation manifold through generator products acting on a few reference configurations,* which can also be mutually inter-related in this way , and the possibility to calculate easily the action of arbitrary generator products, seem to be very promising in this regard [87].

We must also mention the possibility to exploit parastatistics [91] in the formulation of the many-body problem [38, 39], with para-Fermi operators creating and annihilating spin-averaged parafermions of order 2, which satisfy generalized Pauli principle. It was shown [38] that the Green-Gould approach outlined in Sec. 4, is equally suitable for this formulation, yielding particularly simple expressions for the particle number non-conserving $SO(2n+1)$ generators.

It is our belief that a further exploitation of Clifford algebras can provide very simple, straightforward and transparent formulation of the many-fermion problem, which will yield not only an automatically spin-adapted, but also computationally versatile and efficient formalism for both shell-model and coupled-cluster approaches to the correlation problem. There remain numerous possibilities to exploit a rich internal structure of CAUGA bases and $U(n)$ operators indicated above, and to implement them in both molecular orbital and valence-bond type approaches. An essential role of $U(2^n)$ group in the general formulation of the independent particle model has also been recognized [92], and would deserve further investigation.

Acknowledgements. The author wishes to express his sincere gratitude to the Workshop organizer Prof. D. G. Truhlar and to the Institute for Mathematics and its Applications, in particular to Profs. H. Weinberger and W. Miller, Jr., University of Minnesota, for their kind hospitality, stimulating atmosphere and the opportunity to take part in the Workshop on Atomic and Molecular Structure and Dynamics. Continued support by NSERC is also gratefully acknowledged.

---

*Similarly as in Matsen's generator bases [30,90], but avoiding non-orothogonality and overcompleteness.

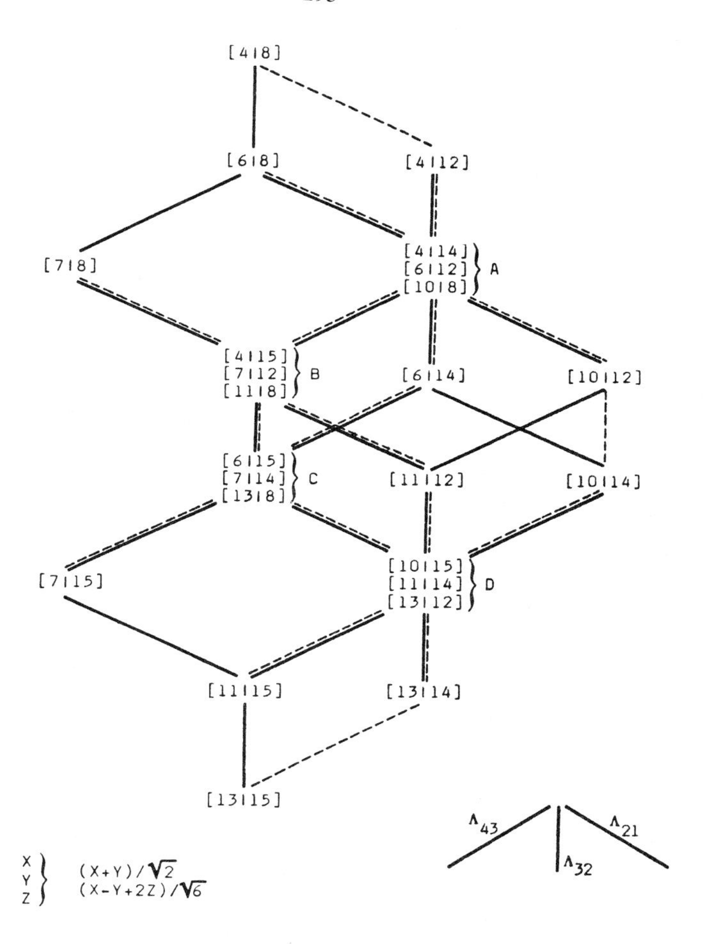

Fig. 1. HLED scheme for the U(4) irrep $\Gamma(1,1,2)$ with the HWS [4|8]. The triples of two-box tableaux A-D indicate either the canonical Gel'fand-Tsetlin states shown in the lower left hand side corder, or the Rumer states $(X+Y)\ /\ \sqrt{2}$ and $(X+Z)\ /\ \sqrt{2}$, or any other pair of linearly independent doublets. The edges represent the elementary lowering generators according to the convention (126′) shown in the lower right hand side corner. Full and dashed lines correspond to subdiagrams (126a) and (126b), respectively.

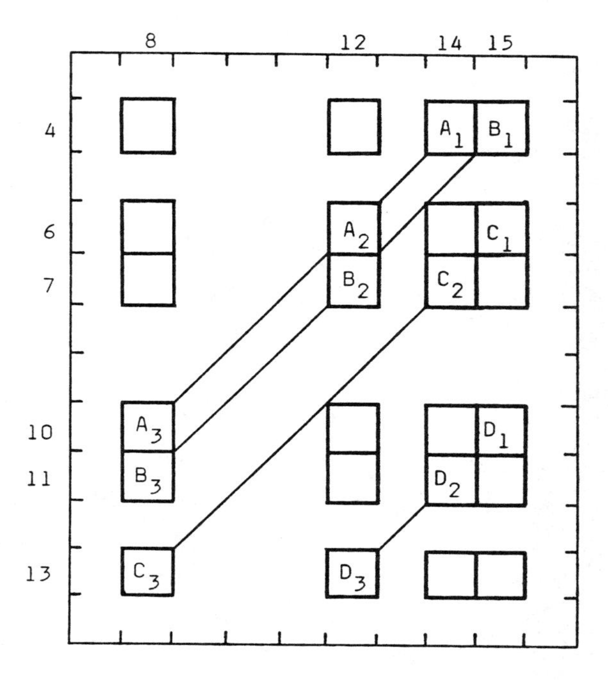

Fig. 2. A schematic representation of the structure of the two-box basis states for the $\Gamma(1,1,2)$ irrep of U(4) associated with the product rep $[1^2] \otimes [1]$. The pertinent $U(2^n)$ single box basis states are given by (120) and (124) in this case. Note that the triples of two-box states labeled A-D in Fig. 1 appear on the same co-diagonal when the 6×4 array of relevant two-box states is represented on a two-dimensional rectangular grid.

REFERENCES

[1] S. GOSHEN (GOLDSTEIN) AND H. J. LIPKIN, Annals of Phys., 6 (1959), p. 301;
A.O. BARUT AND A. BÖHM, Phys. Rev. B, 139 (1965), p. 1107.

[2] N. MUKUNDA, L. O'RAIFEARTAIGH AND E. C. G. SUDARSHAN, Phys. Rev. Lett., 15 (1965), p. 1041.

[3] A. O. BARUT, A. BÖHM AND Y. NE'EMAN (eds.), *Dynamical Groups and Spectrum Generating Algebras*, World Scientific, Singapore, 1987, in press.

[4] *Cf., e.g.*, B. G. ADAMS, J. CIZEK AND J. PALDUS, Advan. Quantum Chem., in press; to be reprinted in [3].

[5] *Cf.* P. E. S. WORMER, Ph.D. Thesis, University of Nijmegen, The Netherlands, 1975.

[6] M. MOSHINSKY AND C. QUESNE, J. Math. Phys., 11 (1970), p. 1631.

[7] J. PALDUS AND B. JEZIORSKI, Theoret. Chim. Acta, in press.

[8] J. PALDUS AND C. R. SARMA, J. Chem. Phys., 83 (1985), p. 5135.

[9] J. PALDUS, M. J. GAO AND J. Q. CHEN, Phys. Rev. A, 35 (1987), p. 3197.

[10] J. PALDUS, J. Chem. Phys., 61 (1974), p. 5321.

[11] J. PALDUS, in *Theoretical Chemistry: Advances and Perspectives*, Vol. 2 (H. Eyring and D.J. Henderson, eds.), Academic, New York, 1976, pp. 131–190.

[12] I. SHAVITT, Int. J. Quantum Chem., Symp., 11 (1977), p. 131.

[13] I. SHAVITT, Int. J. Quantum Chem., Symp., 12 (1978), p. 5.

[14] J. PALDUS AND M. J. BOYLE, Phys. Scr., 21 (1980), p. 295.

[15] J. PALDUS AND M. J. BOYLE, Phys. Rev. A, 22 (1980), p. 2299;
M. J. BOYLE AND J. PALDUS, Phys. Rev. A, 22 (1980) p. 2316.

[16] J. HINZE (ED.), *The unitary group for the evaluation of electronic energy matrix elements* (Lecture Notes in Chemistry, Vol. 22), Springer, Berlin, 1981.

[17] P. E. M. SIEGBAHN, J. Chem. Phys., 70 (1979), p. 5391; 72 (1980), p. 1647.

[18] B. R. BROOKS AND H. F. SCHAEFER, J. Chem. Phys., 70 (1979), p. 5092.

[19] P. SAXE, D. J. FOX, H. F. SCHAEFER AND N. C. HANDY, J. Chem. Phys., 77 (1982), p. 5584.

[20] H. LISCHKA, R. SHEPARD, F. BROWN AND I. SHAVITT, Int. J. Quantum Chem., Symp., 15 (1981), p. 91.

[21] V. R. SAUNDERS AND J. H. VAN LENTHE, Mol. Phys., 48 (1983), p. 923.

[22] P. E. M. SIEGBAHN, Chem. Phys. Lett., 109 (1984), p. 417.

[23] P. J. KNOWLES AND N. C. HANDY, Chem. Phys. Lett., 111 (1984), p. 315.

[24] M. S. ROBB AND U. NIAZI, Comp. Phys. Rep., 1 (1984), p. 127.

[25] W. DUCH AND J. KARWOWSKY, Comp. Phys. Rep., 2 (1985), p. 93.

[26] J. PALDUS, in *Symmetries in Science II* (B. Gruber and R. Lenczewski, eds.), Plenum, New York, 1986, pp. 429–446.

[27] I. SHAVITT, in *New Horizons in Quantum Chemistry* (P.-O. Löwdin and B. Pullman, eds.), Reidel, Dordrecht, The Netherlands, 1983, pp. 279–293.

[28] R. PAUNCZ, *Spin Eigenfunctions: Construction and Use*, Plenum, New York, 1979, Chap. 9.

[29] S. WILSON, *Electron Correlation in Molecules*, Clarendon, Oxford, 1984, Chap. 5.

[30] F. A. MATSEN AND R. PAUNCZ, *The Unitary Group in Quantum Chemistry*, Elsevier, Amsterdam, 1986.

[31] I. SHAVITT, this volume.

[32] M. J. GAO, J. Q. CHEN AND J. PALDUS, Int. J. Quantum Chem., 32 (1987), p. 133.

[33] H. S. GREEN, J. Math. Phys., 12 (1971), p. 2106;
M. D. GOULD, J. Math. Phys., 21 (1980), p. 444; 22 (1981), p. 15.

[34] M. D. GOULD AND G. S. CHANDLER, Int. J. Quantum Chem., 25 (1984), pp. 553, 603, 1089; 27 (1985), p. 878 (E).

[35] M. D. GOULD AND G. S. CHANDLER, Int. J. Quantum Chem., 26 (1984), p. 44.

[36] M. D. GOULD AND J. PALDUS, Int. J. Quantum Chem., 30 (1986), p. 327;
M. D. GOULD, *ibid.*, 30 (1986), p. 364.

[37] C. R. SARMA AND J. PALDUS, J. Math. Phys., 26 (1985), p. 1140.

[38] M. D. GOULD AND J. PALDUS, Phys. Rev. A, 34 (1986), p. 804.

[39] M. D. GOULD AND J. PALDUS, J. Math. Phys., in press.

[40] P.-O. LÖWDIN, Phys. Rev., 97 (1955), pp. 1474, 1490, 1509;
Advan. Chem. Phys., 2 (1959), p. 207.

[41] R. MCWEENY AND B. T. SUTCLIFFE, *Methods of Molecular Quantum Mechanics*, Academic, New York, 1969.

[42] F. A. Berezin, *The Method of Second Quantization*, Academic, New York, 1966; for an elementary introduction see: J. Paldus and J. Cizek, Advan. Quantum Chem., 9 (1975), p. 105.
[43] *Cf.*, *e.g.*, I. Shavitt, in *Modern Theoretical Chemistry, Vol. 3, Methods of Electronic Structure Theory*, (H.F. Schaefer III, ed.), Plenum, New York, 1977, pp. 189–276.
[44] B. O. Roos, Chem. Phys. Lett., 15 (1972), p. 153;
B. O. Roos and P. E. M. Siegbahn, in *Modern Theoretical Chemistry*, Vol. 3, *Methods of Electronic Structure Theory*, (H.F. Schaefer III, ed.), Plenum, New York, 1977, pp. 277–318.
[45] R. J. Buenker and S. Peyerimhoff, Theor. Chim. Acta, 39 (1975), p. 217;
R. J. Buenker, S. Peyerimhoff and W. Butscher, Mol. Phys., 35 (1978), p. 771.
[46] D. J. Thouless, *The Quantum Mechanics of Many-Body Systems*, Academic, New York, 1961.
[47] Ref. [29], Chap. 4.
[48] S. Wilson, K. Jankowski and J. Paldus,, Int. J. Quantum Chem., 23 (1983), p. 1781.
[49] J. Cizek, J. Chem. Phys., 45 (1966), p. 4256;
Advan. Chem. Phys., 14 (1969), p. 35.
[50] J. Hubbard, Proc. Roy. Soc. (London), A240 (1957), p. 539;
A243 (1958), p. 336;
A244 (1958), p. 199.
[51] J. Paldus, J. Cizek and I. Shavitt, Phys. Rev. A, 5 (1972), p. 50.
[52] J. Paldus, in *New Horizons in Quantum Chemistry*, (P.-O. Löwdin and B. Pullman, eds.), Reidel, Dordrecht, The Netherlands, 1983, pp. 31–60.
[53] *Cf.*, *e.g.*, J. Paldus, P. E. S. Wormer, F. Visser and A. van der Avoird, J. Chem. Phys., 76 (1982), p. 2458.
[54] W. D. Laidig, P. Saxe and R. J. Bartlett, J. Chem. Phys., 86 (1987), p. 887.
[55] Ref. [28], Chaps. 1-9.
[56] W. I. Salmon and K. Ruedenberg, J. Chem. Phys., 57 (1972), p. 2776.
[57] P. E. S. Wormer and J. Paldus, Int. J. Quantum Chem., 16 (1979), p. 1307;
18 (1980), p. 843;
J. Paldus and P. E. S. Wormer, *ibid.*, 16 (1979), p. 1321.
[58] M. Kotani, A. Amemiya, E. Ishiguro and T. Kimura, *Tables of Molecular Integrals*, 2nd ed., Maruzen, Tokyo, 1963, Chap. 1 and references therein.
[59] R. Serber, Phys. Rev., 45 (1934), p. 461;
J. Chem. Phys., 2 (1934), p. 697.
[60] C. M. Reeves, Commun. ACM, 9 (1966), p. 276;
B. T. Sutcliffe, J. Chem. Phys., 45 (1966), p. 235.
[61] F. A. Matsen, Advan. Quantum Chem., 1 (1964), p. 59.
[62] M. Moshinsky, *Group Theory and the Many-Body Problem*, Gordon and Breach, New York, 1968, in *Many-Body Problems and Other Selected Topics in Theoretical Physics* (M. Moshinsky, T.A. Brody and G. Jacob, eds.) Gordon and Breach, New York, 1966, p. 289.
[63] M. Moshinsky and T. H. Seligman, Ann. Phys. (N.Y.), 66 (1971), p. 311.
[64] For an overview see L.C. Biedenharn and J.D. Louck, *Angular Momentum in Quantum Physics, Theory and Application*, Addison-Wesley, Reading, Mass., 1981, Chap. 5.
[65] J. Schwinger, *On Angular Momentum*, U.S. Atomic Energy Commission Report NYO-3071, 1952 (unpublished) [reprinted in *Quantum Theory of Angular Momentum*, (L.C. Biedenharn and H. van Dam, eds.), Academic, New York, 1965, pp. 229–279].
[66] A. P. Jucys, I. B. Levinson, and V. V. Vanagas, *Mathematical Apparatus of the Theory of Angular Momentum*, Institute of Physics and Mathematics of the Academy of Sciences of the Lithuanian S.S.R., Mintis, Vilnius, (in Russian) [English translations: Israel Program for Scientific Translations, Jerusalem, 1962; Gordon and Breech, New York, 1964].
[67] E. El Baz and B. Castel, *Graphical Methods of Spin Algebras in Atomic, Nuclear and Particle Physics*, Dekker, New York, 1972.
[68] M. D. Gould, private communication.
[69] R. S. Nikam and C. R. Sarma, J. Math. Phys., 25 (1984), p. 1199.
[70] I. M. Gel'fand and M. L. Tsetlin, Dokl. Acad. Nauk SSSR, 71, (1950), pp. 825, 1070.
[71] G.de B. Robinson, *Representation Theory of the Symmetric Group*, University of Toronto Press, Toronto, 1961.
[72] G. E. Baird and L. C. Biedenharn, J. Math. Phys., 4 (1963), p. 1449;
5 (1964), p. 1723, 1730;
6 (1965), p. 1847.

[73] J. Paldus, Phys. Rev. A, 14 (1976), p. 1620.
[74] J. Paldus, Ref. [16], pp. 1–50.
[75] W. Duch, *GRMS or Graphical Representation of Model Spaces*, Lecture Notes in Chemistry, Vol. 42, Springer, Berlin, 1986.
[76] J. Paldus, in *Electrons in Finite and Infinite Structures*, (P. Phariseau and L. Scheire, eds.), Plenum, New York, 1977, p. 411.
[77] P.W. Payne, Int. J. Quantum Chem., 22 (1982), p. 1085.
[78] J.-F. Gouyet, R. Schranner and T. H. Seligman, J. Phys. A, 8 (1975), p. 258.
[79] G. W. F. Drake and M. Schlesinger, Phys. Rev. A, 15 (1977), p. 1990.
[80] I. Shavitt, in Ref. [16], p. 50.
[81] *Cf., e.g.,* R. McWeeny, Proc. Roy. Soc. (London), A253 (1959), p. 242, or Sec. 7.3, p. 208 of Ref. [29].
[82] See, *e.g.*, Ref. [64];
and L. C. Biedenharn and J. D. Louck, *The Racah-Wigner Algebra in Quantum Theory*, Addison-Wesley, Reading, Mass., 1981;
and L.C. Biedenharn, A. Giovannini and J.D. Louck, J. Math. Phys., 8 (1967), p. 691.
[83] M.D. Gould and G.S. Chandler, private communication.
[84] E.J. Cartan, *The Theory of Spinors*, MIT Press, Cambridge, 1967, p. 85;
see also C.C. Chevalley, *The Algebraic Theory of Spinors*, Columbia University Press, Morningside Heights, New York, 1954;
H. Boerner, *Representations of Groups*, 2nd ed., North-Holland, Amsterdam, 1970;
A. Ramakrishnan, *L-Matrix Theory or the Grammar of Dirac Matrices*, Tata McGraw Hill, Bombay, India, 1972.
[85] W. Kutzelnigg, J. Chem. Phys., 82 (1985), p. 4166, and references therein.
[86] J. Hinze and J. T. Broad, in Ref. [16], pp. 332-344.
[87] B. Jeziorski and J. Paldus, unpublished results.
[88] J. Q. Chen, F. Wang, M. J. Gao and Z. R. Yu, Scientia Sinica, 9 (1980), p. 116.
[89] B. Jeziorski and H. J. Monkhorst, Phys. Rev. A, 24 (1984), p. 1668.
[90] F. A. Matsen, Int. J. Quantum Chem., Symp., 18 (1984), p. 43.
[91] Y. Ohnuki and S. Kamefuchi, *Quantum Field Theory and Parastatistics*, Springer, New York, 1982.
[92] H. Fukutome, Progr. Theoret. Phys., 65 (1981), p. 809;
H. Fukutome and S. Nishiyama, *ibid.*, 72 (1984), p. 239 and references therein.

UNITARY GROUP APPROACH TO CONFIGURATION INTERACTION CALCULATIONS OF THE ELECTRONIC STRUCTURE OF ATOMS AND MOLECULES

ISAIAH SHAVITT*

Abstract. The application of the unitary group approach (UGA) in its graphical form (GUGA) to the calculation of electronic wave functions of molecules, particularly by the configuration interaction (CI) method, is reviewed. The discussion covers the various representations of the Gel'fand-Tsetlin basis, including the distinct row table (DRT) and its graphical representation, and the evaluation of matrix elements of the unitary group generators and of the electronic Hamiltonian. The overall organization of multireference CI calculations based on the graphical unitary group approach, including considerations relating to efficient operation on vector computers, are described.

Key words. unitary group, configuration interaction, electronic structure

AMS(MOS) subject classifications. 81G50, 22E70

1. INTRODUCTION

Electronic structure theory for atoms and molecules has been discussed in this Workshop by Almlöf [1]. The mathematical basis of the application of the unitary group approach (UGA) to electronic structure theory has been discussed by Paldus [2]. The present contribution will focus on the application of this approach to large-scale electronic structure calculations, primarily in the context of the configuration interaction (CI) method. It will emphasize the "graphical" form of the unitary group approach (GUGA), in which a graphical representation of the Gel'fand-Tsetlin basis and of generator matrix elements is used to help in the derivations and to expose various structural aspects of the calculations.

---

* Department of Chemistry, Ohio State University, Columbus, Ohio 43210.
This work was supported by the National Science Foundation under Grant CHE-8219408.

The principal goal of the development presented here is the design of efficient procedures for multireference CI calculations in the direct CI mode. In general, the CI method [3] obtains an approximate solution of the electronic Schrödinger equation as an expansion, in terms of a suitable basis, in a subspace of the N-electron Hilbert space (see Section 2). The expansion coefficients are determined variationally, by minimization of the expectation value of the quantum mechanical electronic Hamiltonian operator in that subspace, requiring the solution of a matrix eigenvalue problem for the matrix representation of that operator. The N-electron basis functions, also called *configurations* (or *configuration state functions*) are expressed in terms of a set of one-electron spatial functions called *orbitals* (Section 2). One of the objectives of the unitary group approach is to provide a convenient and systematic way for the construction of a suitable N-electron basis from the one-electron basis.

The term *multireference CI* refers to a particular procedure for specification of the subspace for the wave function expansion, in which this subspace is spanned by a set of *reference configurations* (an optimized, relatively small set of configurations) plus all other configurations obtained from the former by promotion of one or two electrons from orbitals occupied in the reference configurations to unoccupied (excited) orbitals.

In *conventional CI* [3] the matrix representation of the electronic Hamiltonian is computed explicitly, in some systematic order, and saved for use in the iterative solution of the eigenvalue problem. Each element of this (usually sparse) matrix is computed as a linear combination of (zero or more) one-electron and two-electron integrals of corresponding components of the Hamiltonian operator over the orbitals. The unitary group approach can be used to derive the formulas for these linear combinations (Section 3).

Current CI calculations often use expansions containing $O(10^6)$ or more terms. The conventional CI approach is limited by the size of the Hamiltonian matrix that can be stored and handled in the computer (even in external disk or tape files), as well as by difficulties in the design of efficient (vectorizable) procedures for the calculation of the matrix elements. The number of orbital integrals can also be quite large, of $O(n^4)$ for n orbitals (with $n \approx O(10^2)$, typically), so neither the Hamiltonian matrix nor the list of integrals can usually be stored entirely within the central memory of the computer. Sequential calculation of the Hamiltonian matrix elements requires essentially random access to the integrals, while sequential processing of the integrals produces essentially random sequences of matrix element contributions, causing serious difficulties for the organization of CI computations.

The *direct CI* approach [4] overcomes these difficulties by avoiding the explicit computation and storage of the Hamiltonian matrix. Instead, in each iteration, the integrals are processed in a systematic sequence, producing contributions to Hamiltonian matrix elements which are utilized immediately in the iterative procedure for the solution of the eigenvalue problem and discarded. A key requirement for this approach is an efficent procedure for the identification of the matrix element (in terms of its row and column indices) to which each contribution belongs. This is another aspect of the problem to which the unitary group techniques, particularly in the graphical formulation presented here, make an important contribution (Section 2). Because of this characteristic, as well as because it facilitates the calculation of the individual matrix element contributions, the graphical unitary group approach has made possible the generalization of the direct CI method from the limited number of cases previously possible [4] to arbitrary electronic states and more general CI expansion types.

The iterative methods used for finding the appropriate eigenvalue and eigenvector of the Hamiltonian matrix take advantage of the diagonal dominance of this matrix and of the availability of a good starting approximation (due to the optimization of the reference configurations). They employ either a variational approach, based on the minimization of the Rayleigh quotient, or a perturbation approach, using the diagonal of the matrix as a zero-order operator, to obtain successive corrections to the eigenvector estimate [3]. The most common approach [5] expands the desired eigenvector in the space spanned by the initial eigenvector approximation and the successively computed and orthogonalized correction vectors, requiring the solution of a small projected eigenvalue problem in each iteration. In all cases, the rate-determining step in the calculation is the evaluation of the vector obtained by multiplying the Hamiltonian matrix by the current eigenvector approximation. It is the optimization of this step which is the primary aim of the application of the unitary group approach to the CI method.

The graphical unitary group approach and its application to the configuration interaction method will be presented here in three parts:

(1) Representations of the Gel'fand-Tsetlin basis, covering various tableaux representations, step vectors, and the distinct row table and its graphical representation (Section 2).

(2) Evaluation of matrix elements of the electronic Hamiltonian and of the generators of the unitary group (Section 3).

(3) Overall organization of configuration interaction calculations based on the graphical unitary group approach (Section 4).

Relevant results from the previous lectures [1, 2] will be summarized, where appropriate, without derivation or detailed attribution. Further details can also be found in several papers and reviews [3, 6–27].

2. REPRESENTATIONS OF THE GEL'FAND-TSETLIN BASIS

2.1. Hilbert spaces and the unitary group. The electronic wave function of an N-electron system, such as a molecule, is typically computed as an expansion in a subspace of a finite-dimensional N-electron Hilbert space. The N-electron Hilbert space is constructed as the antisymmetric component of the N-th rank tensor power of a finite-dimensional one-electron Hilbert space. The one-electron Hilbert space, in turn, is a tensor product of two smaller one-electron spaces, one on the domain of the three spatial coordinates of the electron, spanned by a basis of n spatial orbitals, and the other on the domain of the two-valued spin coordinate, spanned by a basis of two one-electron spin functions (commonly denoted α and β). In what follows it will be assumed that the chosen one-electron spatial basis is orthonormal.

As discussed by Paldus in this Workshop [2] (see also [8, 22]), the unitary group U(n) can be used to organize and label an orthonormal, spin-adapted basis for the N-electron Hilbert space. Wave function expansions for electronic states of definite N (number of electrons) and

S (total spin quantum number) built upon an (orthonormal) one-electron spatial basis of n orbitals can be related to particular irreducible representations (irreps) of U(n). Specifically, the fixed-S subspace of that Hilbert space provides a carrier space for such an irrep, and a convenient basis for that irrep is obtained by adaptation to the subgroup chain

$$U(n) \supset U(n-1) \supset \ldots \supset U(2) \supset U(1), \tag{1}$$

providing the desired orthonormal spin-adapted N-electron basis functions for the wave function expansion, together with a convenient labeling scheme for these basis functions. This basis is known as the *Gel'fand-Tsetlin basis*, and the individual basis functions are referred to as *Gel'fand states*. They are described directly in terms of the n orthonormal spatial basis orbitals, without explicit reference to the spin functions. In fact, the component of the total spin state (i.e., the eigenvalue of the S_z operator) is left unspecified, and can be assigned separately, when necessary, in terms of an irrep of U(2) conjugate to the above U(n) irrep (resulting from the decomposition $U(2n) \supset U(n) \times U(2)$ [2]).

2.2. The Gel'fand tableau representation. Each irrep $\Gamma\{\mathbf{m}_n\}$ of U(n) is uniquely labeled by its *highest weight vector*

$$\mathbf{m}_n = (m_{1n}\ \ m_{2n}\ \ \cdots\ \ m_{nn}), \tag{2}$$

with non-increasing integer components. The only irreps relevant for electronic states are those for which

$$m_{in} = 0,\ 1,\ \text{or}\ 2 \tag{3}$$

(related to the fact that electrons have two spin states, so that no more than two electrons can occupy any spatial orbital). Furthermore, for an N-electron state with total spin quantum number S, the highest weight vector must satisfy

$$\sum_i m_{in} = N, \tag{4}$$

$$\sum_i \delta_{1,m_{in}} = 2S \tag{5}$$

(where the sum of Kronecker deltas in (5) counts the number of 1's in $\mathbf{m}_n$). The individual basis functions of the Gel'fand-Tsetlin basis for $\Gamma\{\mathbf{m}_n\}$ are uniquely specified by listing the highest weight vectors $\mathbf{m}_j$ of the irreps of the U(j) subgroups (j = n-1, n-2, ..., 1) obtained by subduction in the chain (1). (The uniqueness of this labeling is a consequence of the simple reducibility of the subduced representations $\Gamma\{\mathbf{m}_j\}\downarrow U(j-1)$.) These weight vectors are collected in a triangular array,

(6)
$$[m] = \begin{bmatrix} m_{1n} \quad m_{2n} \quad m_{3n} \quad \cdots\cdots \quad m_{nn} \\ m_{1,n-1} \; m_{2,n-1} \; \cdots \; m_{n-1,n-1} \\ \cdots\cdots\cdots\cdots\cdots\cdots \\ \cdots\cdots\cdots\cdots\cdots \\ \cdots\cdots\cdots\cdots \\ \cdots\cdots\cdots \\ m_{12} \; m_{22} \\ m_{11} \end{bmatrix}$$

known as a *Gel'fand tableau*. Its components are subject to the betweenness condition

(7)
$$m_{ij} \geq m_{i,j-1} \geq m_{i+1,j},$$

which is a consequence of the subduction properties of the chain. Gel'fand tableaux which satisfy condition (3) (and can therefore be used in the description of electronic wave functions) are called *electronic Gel'fand tableaux*.

2.3. The Paldus tableau and step vector representations. Paldus noted [6] that condition (3) for electronic Gel'fand tableaux allows a simpler representation of the weight vectors and the Gel'fand tableaux in terms of non-negative integer triples (a_j b_j c_j), giving the number of 2's, 1's, and 0's, respectively, in the weight vector $\mathbf{m}_j$. Each basis vector of the Gel'fand-Tsetlin basis can therefore be represented by the *Paldus tableau* (referred to as an *ABC tableau* by Paldus)

$$(8) \qquad [a \;\; b \;\; c] = \begin{bmatrix} a_n & b_n & c_n \\ a_{n-1} & b_{n-1} & c_{n-1} \\ \cdots & \cdots & \cdots \\ \cdots & \cdots & \cdots \\ a_1 & b_1 & c_1 \\ 0 & 0 & 0 \end{bmatrix},$$

in which a bottom row, $a_0 = b_0 = c_0 = 0$, has been added for later convenience. Obviously,

$$a_j + b_j + c_j = j,$$

making one of the columns redundant, but it is convenient to keep all three columns anyway. The top row of this tableau, which identifies the irrep of U(n), is usually written (a b c), without subscripts.

In terms of the Paldus tableau, the conditions (4, 5) take the form

$$(9) \qquad 2a + b = N,$$

$$(10) \qquad b = 2S.$$

The relationship between the Paldus tableau and the corresponding N-electron basis function (Gel'fand state) can be described in terms of the construction of the Gel'fand state by the successive spin-coupled addition of orbitals, $j = 1, 2, \ldots, n$, beginning with an empty function (corresponding to the bottom row (0 0 0)), and resulting at each level j (j is called the *level index*) in an intermediate function having N_j electrons and adapted to total spin quantum number S_j according to

$$(11) \qquad 2a_j + b_j = N_j,$$

$$(12) \qquad b_j = 2S_j.$$

The betweenness conditions (7) restrict the relationship of each pair of successive rows (levels) of the tableau to one of at most four types. These types are best described in terms of the differences

(13) $\Delta z_j = z_j - z_{j-1}$ $(z = a, b, c, N, S, \quad j = 1, 2, \ldots, n)$.

Each level j is identified with an orbital of the one-electron basis and

(14) $$n_j = \Delta N_j = 0, 1, \text{ or } 2$$

is the electron occupancy of orbital j in the Gel'fand state. The four possibilities, and their interpretation in terms of the electron occupancies and spin coupling of the corresponding Gel'fand state, are listed in Table 1.

Also shown in Table 1 is the *step number*

(15) $$d_j = 3\Delta a_j + \Delta b_j = 0, 1, 2, \text{ or } 3,$$

which can be used to label the four cases. The *step vector*

(16) $$\mathbf{d} = (d_n \ d_{n-1} \ \ldots \ d_1)$$

provides another unique representation of the Gel'fand state, and satisfies

(17) $$\sum_j d_j = 3a + b = \frac{3}{2}N - S$$

(as well as other conditions to ensure that a_j, b_j, c_j are nonnegative at every level). Figure 1 illustrates the various representations of the same Gel'fand state.

TABLE 1
The four possible relationships between successive rows of a Paldus tableau, and their interpretation in terms of orbital occupancy and spin coupling.

d_j	Δa_j	Δb_j	Δc_j	$n_j = \Delta N_j$	ΔS_j
0	0	0	1	0	0
1	0	1	0	1	1/2
2	1	-1	1	1	-1/2
3	1	0	0	2	0

$$
\begin{array}{c}
7\\ 6\\ 5\\ 4\\ 3\\ 2\\ 1\\ 0
\end{array}
\qquad
\begin{bmatrix}
2 & & 2 & & 2 & & 1 & & 1 & & 0 & & 0\\
& 2 & & 2 & & 1 & & 1 & & 0 & & 0 & \\
& & 2 & & 2 & & 1 & & 0 & & 0 & & \\
& & & 2 & & 1 & & 1 & & 0 & & & \\
& & & & 2 & & 1 & & 0 & & & & \\
& & & & & 2 & & 1 & & & & & \\
& & & & & & 2 & & & & & &
\end{bmatrix}
\qquad
\begin{bmatrix}
3 & 2 & 2\\
2 & 2 & 2\\
2 & 1 & 2\\
1 & 2 & 1\\
1 & 1 & 1\\
1 & 1 & 0\\
1 & 0 & 0\\
0 & 0 & 0
\end{bmatrix}
\qquad
\begin{bmatrix}
3\\ 1\\ 2\\ 1\\ 0\\ 1\\ 3
\end{bmatrix}
\qquad
\begin{array}{|c|c|}
\hline
1 & 1\\ \hline
2 & 5\\ \hline
4 & 7\\ \hline
6 & \multicolumn{1}{c}{}\\ \cline{1-1}
7 & \multicolumn{1}{c}{}\\ \cline{1-1}
\end{array}
$$

(a) (b) (c) (d)

$$
(1/\sqrt{6})\left(|1\ \bar{1}\ \bar{2}\ 4\ 5\ 6\ 7\ \bar{7}| + |1\ \bar{1}\ 2\ \bar{4}\ 5\ 6\ 7\ \bar{7}| - 2|1\ \bar{1}\ 2\ 4\ \bar{5}\ 6\ 7\ \bar{7}|\right)
$$

(e)

FIG. 1. *Five representations of a Gel'fand state for* $N = 8$, $S = 1$, $n = 7$: *(a) Gel'fand tableau, (b) Paldus tableau, (c) step vector, (d) Weyl tableau, (e) Slater determinants. The first column identifies the level indices* j *for (a–c).*

2.4. The Weyl tableau representation. Another common representation of the Gel'fand states, closely related to the Young tableau representation used in the symmetric group approach, is provided by the *Weyl tableau.* An electronic Weyl tableau for an N-electron state with total spin S consists of N boxes arranged in two columns, with $\frac{1}{2}N + S = a + b$ boxes in the first column and $\frac{1}{2}N - S = a$ boxes in the second column (Fig. 1(d)). Orbital labels (the integers 1, 2, ..., n) are placed in the boxes, one to a box, subject to the condition that the numbers increase down each column and are non-decreasing in each row from left to right.

The correspondence between the Weyl tableau notation and the other representations of the Gel'fand states is best described in terms of the step vector representation. To obtain the Weyl tableau corresponding to a given step vector, the step numbers are examined in the order $d_1, d_2, \ldots, d_n$. If $d_j = 0$, the integer j is not used in the tableau. If $d_j = 1$ or 2, the integer j is placed in the first available box (from the top) in the first or second column, respectively. If $d_j = 3$, the integer j is placed in the first available box in each of the two columns. Thus each orbital label appears in the tableau a number of times equal to the electron occupancy of the corresponding orbital.

Prescriptions for the explicit expression of the electronic Gel'fand states in terms of Slater determinants (antisymmetrized products of spinorbitals) have been given, based on either the Weyl tableau notation [17] or the Paldus tableau and step vector notations [23].

2.5. The distinct row table. Examination of the Gel'fand or Paldus tableaux representing a non-trivial Gel'fand-Tsetlin basis shows that various subtableaux recur identically in many members of the basis. Failure to take advantage of this fact in applications leads to extensive redundancies in the calculations. In fact, it is the clear demonstration of these redundancies which is one of the principal merits of the tableau approach. It is clear that all the tableaux of the basis are constructed from a relatively small repertory of distinct rows, just as the many words of a language are constructed from a small number of letters constituting its alphabet. It has therefore been proposed [18] that the N-electron basis be represented compactly in terms of a *distinct row table* (DRT), listing all the distinct rows and their interconnections that occur in its Paldus tableaux.

An example of a DRT for a full CI expansion (i.e., an expansion which includes all N-electron basis functions appropriate, in terms of spin and symmetry properties, to the electronic state under investigation which can be constructed from the given orbital basis) for the case $n = 6$, $N = 5$, $S = \frac{1}{2}$ is shown in Table 2 [18, 23] (assuming no spatial symmetry restrictions). Each row $(a_k \ b_k \ c_k)$ in this table is assigned a *level index* $j = a_k + b_k + c_k$, which determines the level it may occupy in any Paldus tableau, and a unique *running index* k. A set of (downward) *chaining indices* k_{dk} is provided to show which rows may occur at successive levels of the tableaux. Specifically, k_{dk} identifies the running index of the row at level j-1 which is related to the row k at level j by step number d. Chaining indices which lead to nonexistant rows (e.g., negative a, b, or c) are omitted, and can be assumed to have the value zero. Upward chaining indices can be defined in a similar manner, and may be useful in some applications of the DRT representation.

A set of *counting indices* y_{dk} and x_k are included in the DRT to allow the easy determination of a unique *lexical index* $m = 1, 2, \ldots, m_{max}$, assigned to the Gel'fand states in a *lexical order*. The ket notation $|m\rangle$ will be used to refer to the Gel'fand state with lexical

TABLE 2

The distinct row table for n = 6, N = 5, S = $\frac{1}{2}$ *(the* y_{0k} *column has been omitted, since all its entries are zero).*

j	k	a_k	b_k	c_k	k_{0k}	k_{1k}	k_{2k}	k_{3k}	y_{1k}	y_{2k}	y_{3k}	x_k
6	1	2	1	3	2	3	4	5	75	125	170	210
5	2	2	1	2	6	7	8	9	20	40	55	75
	3	2	0	3	7		9	10	20	20	40	50
	4	1	2	2	8	9	11	12	15	35	39	45
	5	1	1	3	9	10	12	13	20	30	36	40
4	6	2	1	1	14	15	16	17	3	9	12	20
	7	2	0	2	15		17	18	6	6	14	20
	8	1	2	1	16	17	19	20	3	11	12	15
	9	1	1	2	17	18	20	21	8	14	17	20
	10	1	0	3	18		21	22	6	6	9	10
	11	0	3	1	19	20			1	4	4	4
	12	0	2	2	20	21			3	6	6	6
	13	0	1	3	21	22			3	4	4	4
3	14	2	1	0		23		24	0	1	1	3
	15	2	0	1	23		24	25	1	1	3	6
	16	1	2	0		24		26	0	2	2	3
	17	1	1	1	24	25	26	27	2	5	6	8
	18	1	0	2	25		27	28	3	3	5	6
	19	0	3	0		26			0	1	1	1
	20	0	2	1	26	27			1	3	3	3
	21	0	1	2	27	28			2	3	3	3
	22	0	0	3	28				1	1	1	1
2	23	2	0	0				29	0	0	0	1
	24	1	1	0		29		30	0	1	1	2
	25	1	0	1	29		30	31	1	1	2	3
	26	0	2	0		30			0	1	1	1
	27	0	1	1	30	31			1	2	2	2
	28	0	0	2	31				1	1	1	1
1	29	1	0	0				32	0	0	0	1
	30	0	1	0		32			0	1	1	1
	31	0	0	1	32				1	1	1	1
0	32	0	0	0								1

index m. The lexical order can be defined in terms of either of the Gel'fand or Paldus tableau or step vector representations. If the various rows of the Gel'fand or Paldus tableau of a state are arranged in succession as a single vector, beginning with the top row, then the lexical order can be determined by comparing the integers in corresponding positions in this vector, from left to right, for the different

Gel'fand states. One state precedes another in the lexical order if the former has the greater integer in the first position in which their respective vectors differ (the former state is said to have a higher *weight* than the latter). Equivalently, if the components of the step vectors of the different states are compared in order, beginning with the highest component d_n, one state precedes another if the former has the smaller integer in the first position in which their respective vectors differ.

A lexical order can be defined in a similar manner for subtableaux made up of rows j, j-1, ..., 0 (i.e., valid tablaux for U(j)). The counting index x_k is defined to equal the number of different subtableaux having the distinct row k as their top row at level j that can be constructed in the DRT. The counting index y_{dk} (d = 0, 1, 2, or 3) is defined to equal the number of such subtableaux having row k as their top row at level j which lexically precede all subtableaux with the same row at level j and with row k_{dk} at level j-1. Assigning x = 1 to the bottom row (0 0 0), all the counting indices can be computed recursively, in order of increasing level index, from the equations

$$y_{0k} = 0,$$

(18)
$$y_{dk} = y_{d-1,k} + x_{k_{d-1,k}} \qquad (d = 1, 2, 3, 4),$$

$$x_k = y_{4k},$$

where it is understood that $x_{k_{d-1,k}} = 0$ for any missing chaining index $k_{d-1,k}$. The lexical index m corresponding to step vector **d** can then be calculated from

(19)
$$m(\mathbf{d}) = 1 + \sum_{j=1}^{n} y_{d_j k_j},$$

where k_j is the running index of the distinct row belonging to that state at level j. The x value for the top row gives the total number m_{max} of Gel'fand states contained in the DRT. The DRT in the example of Table 2 describes 210 Gel'fand states in terms of 32 distinct rows. The compactness of the DRT representation is demonstrated by the example of a complete (full CI) expansion for N = 10, S = 0, n = 30, which consists of 4,035,556,161 Gel'fand states, and is generated by a DRT of 511 rows.

A more intuitive explanation of the counting scheme will be given in the next subsection, in terms of the graphical representation of the DRT. A set of counting indices for reverse lexical order can be defined in a similar manner in terms of the upward chaining indices, with the recursive calculation begun at the top level [23].

Bases for subsets of the appropriate N-electron Hilbert space (i.e., for partial CI expansions) can be constructed by omitting selected chaining indices (and rows) from the DRT. A contiguous set of lexical indices for the retained Gel'fand states can still be computed in this case from the same equations (18, 19). While this procedure does not provide complete flexibility in the selection of the retained states, it can be used to generate most of the common types of CI expansions (e.g., all single and double excitations from various reference spaces).

2.6. The graphical representation. The relationships of the various component of the DRT can be visualized conveniently in terms of a two-rooted, hierarchical, directed graph [18]. In this graph, referred to as the *distinct row graph* (also called *Shavitt graph*), each row of the DRT is represented by a *vertex*, labeled by the corresponding running index k, and each chaining index k_{dk} is represented by an *arc* (directed edge), directed from the vertex with label k_{dk} to that with label k. The hierarchy of the vertices is then in order of increasing level indices j of the corresponding rows of the DRT. The unique bottom and top rows of the DRT are represented by the source and sink roots, also called the *tail* and the *head* of the graph, respectively. The vertices and arcs of the graph are assigned *weights* equal to the values of the corresponding x_k and y_{dk} counting indices, respectively.

It is convenient to arrange the vertices of the distinct row graph on a regular grid, such that the vertical and horizontal positions of a vertex correspond, respectively, to the level index j and to the a and b values of the corresponding row of the DRT. The level indices are arranged to increase upwards, as in the Gel'fand and Paldus tableau notations and in the step vectors. The horizontal grid is arranged according to increasing value, from right to left, of the quantity $a\alpha + b$, where the scale parameter α ($\alpha > b_{max}$) is chosen so that the vertices for any given value of a are clearly separated horizontally

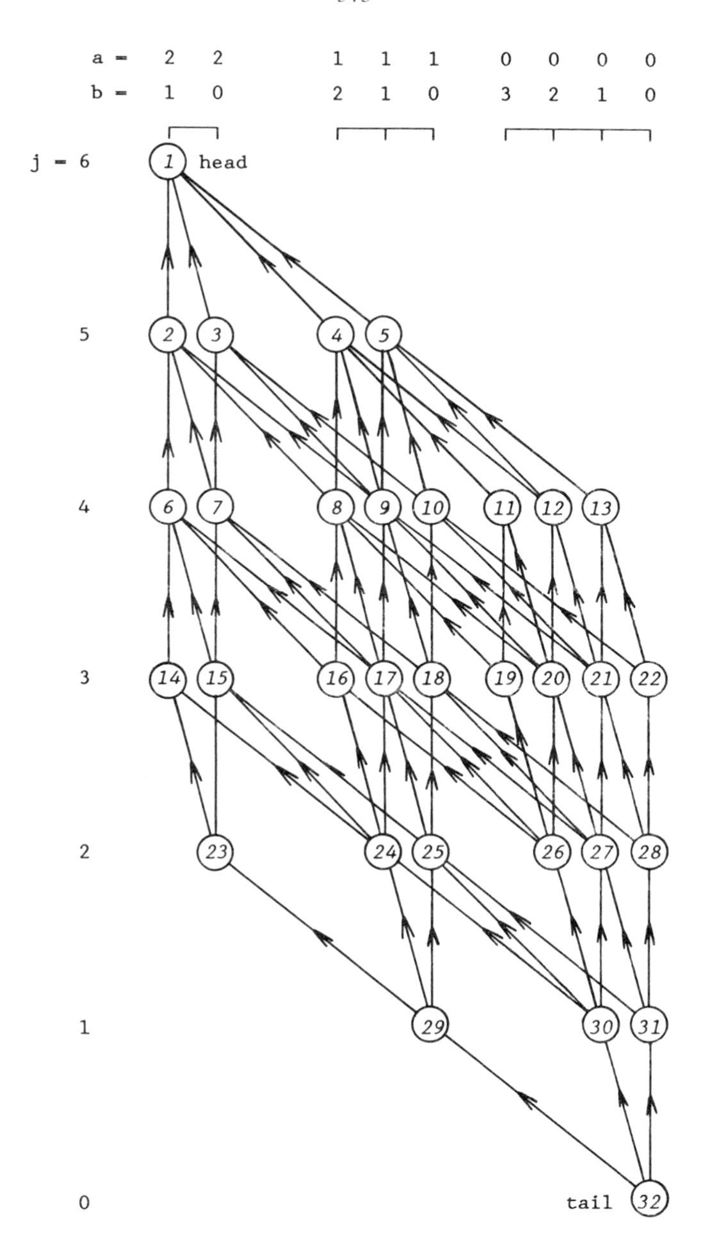

FIG. 2. *The graph describing the distinct row table of Table 2. The* a *and* b *values of the vertices are shown at the top, and the level indices* j *are shown on the left. Vertex labels* k *are circled.*

from those for any other value. (The value of c is implicitly determined for each vertex from a and b and the level index j.) With this arrangement, different step number values are represented on the graph by lines of different slope. In particular, vertical lines correspond to zero step numbers (empty orbitals), while increasing step number values are represented by lines with increasing deviation from the vertical. Each level of the graph, *comprising the vertices of that level plus the arcs connected to them from below*, represents one of the orbitals of the one-electron basis. The graph representing the DRT of Table 2 is shown in Fig. 2, with each vertex labeled by its running index k.

A one-to-one correspondence exists between the Gel'fand states represented in the DRT and the *directed walks* (paths) on the graph leading from its tail to its head and following the directions of the arcs. If the weight of any walk is defined as the sum of the weights of the arcs it traverses, it is seen from (19) that the lexical index of each of the Gel'fand states is equal to the weight of the corresponding walk, plus one. One walk precedes another in lexical order if, at the highest level at which they differ, the former is to the left of the latter. As in the case of the Paldus tableau, each walk describes the construction of the corresponding Gel'fand state in terms of the successive spin-coupled addition of the orbitals, in order of their label values (level indices j), beginning at the vacuum state (represented by the tail of the graph). At each level the corresponding arc describes (through its slope) the role of that orbital in the Gel'fand state, and the corresponding vertex specifies (through its a and b values) the number of electrons N_j and spin quantum number S_j reached at that point.

The distinct row graph can now be used to provide a clear explanation of the counting scheme and the assignment of lexical indices. The vertex weights (the x_k indices) obviously count the number of directed walks connecting each vertex with the tail. The weight of a vertex is simply obtained as the sum of the weights of the (at most four) vertices directly connected with it from below. The evaluation of this sum is equivalent to the recursion process in (18). The intermediate sums in (18) are the arc weights. The arc weight y_{dk} ($d = 0, 1, 2, 3$) represents the number of directed walks connecting vertex k with the tail which, at level $k-1$, pass to the left of vertex

k_{dk} (and thus lexically precede all such walks passing through k_{dk}). With this interpretation of the arc weights, it is easy to verify that the lexical indices of the Gel'fand states are indeed given as the weights of the corresponding walks, plus one, in agreement with (19). The easy determination of the lexical index of a Gel'fand state from the set of arc weights is of key importance for the feasibility of multireference direct CI calculations, as noted in the Introduction.

An alternative counting scheme, based on upward counting indices [18] and connections to the head of the graph, can similarly be defined, and results in a different ordering of the Gel'fand states. This "reverse lexical order" is useful in some implementations of multireference CI calculations (Section 4).

As will be shown in the following sections, the graphical representation provides important benefits for the implementation and execution of large-scale electronic wave function calculations. It provides a clear visualization of the structure of the N-electron basis, leading to better insight into the structure of the required calculations [28–33]. The graphcial approach proved helpful in the design of effective algorithms for matrix element evaluation (Section 3) and has been very helpful in devising the overall strategy and organization of the computations (Section 4). In fact, the distinct row graph can be said to provide a template on which the calculations are carried out. With proper organization, the rate-determining steps of the calculations can be made to flow smoothly and efficiently, with little need for conditional branches and with a minimum of redundant computations, and can be well-adapted for efficient operation on vector-oriented computers.

The graphical representation of N-electron bases has also been adapted to other schemes based on the symmetric group or other approaches [34–36]. Generalizations of the Paldus tableau and the DRT and graphical approach for particles having more than two spin states have been made by Kent and Schlesinger [37, 38]. Various other generalizations of the unitary group approach, including particle-hole based schemes, other coupling schemes, and the use of Clifford algebra, have been discussed in Paldus's contribution [2], where detailed references are provided.

3. EVALUATION OF MATRIX ELEMENTS

3.1. Generators of the unitary group. A realization of the infinitesimal *generators* of the unitary group U(n) on the many-electron Hilbert spaces can be achieved in the form [2, 8]

$$E_{ij} = \sum_{\sigma=\alpha,\beta} X_{i\sigma}^{\dagger} X_{j\sigma} \qquad (i, j = 1, 2, \ldots, n), \tag{20}$$

where $X_{i\sigma}^{\dagger}$ and $X_{i\sigma}$ are the Fermion (antisymmetric) creation and annihilation operators, respectively, for an electron in spatial orbital i with spin σ. The anticommutation properties of the Fermion operators [2] lead to the commutation relation

$$[E_{ij}, E_{k\ell}] = \delta_{kj} E_{i\ell} - \delta_{i\ell} E_{kj} \tag{21}$$

for the operators defined in (20), which is the commutation property required for the unitary group generators, justifying the identification of those operators as generators of U(n). An equivalent expression of these generators [27] is in terms of the dyadic (ket-bra) operators,

$$E_{ij} = \sum_{\mu=1}^{N} |i(\mu)\rangle\langle j(\mu)|, \tag{22}$$

where $i(\mu)$ is spatial orbital i as a function of the coordinates of electron μ. The generator E_{ij} is classified as a *raising generator*, *weight genearator*, or *lowering generator*, depending on whether $i < j$, $i = j$, or $i > j$, respectively. These names reflect the effect of the generator on the weight (the sequence of weight vectors $\mathbf{m}_j$ which determines the lexical order) of the state on which it operates [8].

As seen from (20), when a generator E_{ij} operates on a Gel'fand state $|m\rangle$, it has the effect of moving an electron from orbital j to orbital i. In general, the result is a linear combination of zero or more Gel'fand states belonging to the same U(n) irrep as $|m\rangle$. If $i < j$ (raising generator), then all the resulting states $|m'\rangle$ are related to the original state according to

(23) $$z'_\ell = z_\ell \quad (z = a, b, c, N, S, \quad \ell < i \text{ or } \ell \geq j),$$

(24) $$d'_\ell = d_\ell \quad (\ell < i \text{ or } \ell > j),$$

(25) $$N'_\ell = N_\ell + 1 \qquad (i \leq \ell < j),$$

(26) $$S'_\ell = S_\ell \pm \frac{1}{2} \qquad (i \leq \ell < j),$$

(27) $$b'_\ell = b_\ell \pm 1 \qquad (i \leq \ell < j).$$

The fact that the difference $b'_\ell - b_\ell$ in (27) must be odd is obvious from (11) and (25). The restriction to unit difference, and thus also the proof of (26), will be demonstrated later. Finally,

(28) $$m' < m,$$

so that the matrix representation of the raising generator in the lexically-ordered Gel'fand-Tsetlin basis is strictly upper triangular. Similar relationships can be obtained for lowering generators, which are the adjoints of the raising generators,

(29) $$E^\dagger_{ij} = E_{ji},$$

and have strictly lower triangular representations. The weight generators simply return the occupancy of the corresponding orbital in the Gel'fand state on which they operate,

(30) $$E_{jj}|m\rangle = n_j|m\rangle,$$

and thus have diagonal representations.

3.2. The unitary group form of the Hamiltonian. The spin-independent Hamiltonian of an N-electron system,

(31) $$H = \sum_{\mu=1}^{N} h(\mu) + \frac{1}{2}\sum_{\mu\neq\nu} v(\mu,\nu),$$

which is a symmetric sum (over the electrons) of one-electron and two-electron operators, can be recast into a "second-quantized" form involving sums over the orbitals and spin functions of a one-electron basis,

$$(32)\qquad H = \sum_{i,j} \sum_{\sigma} h_{ij} X^{\dagger}_{i\sigma} X_{j\sigma} + \frac{1}{2} \sum_{i,j,k,\ell} \sum_{\sigma,\tau} [ij;k\ell] X^{\dagger}_{i\sigma} X^{\dagger}_{k\tau} X_{\ell\tau} X_{j\sigma},$$

in which the number of electrons no longer appears. This form is equivalent to (31) within all many-electron Hilbert spaces constructed from the given one-electron basis. The coefficients of the operator products in (32) are one-electron and two-electron integrals over the spatial orbitals,

$$(33)\qquad h_{ij} = \langle i|h|j\rangle,$$

$$(34)\qquad [ij;k\ell] = \langle i(1)k(2)|v(1,2)|j(1)\ell(2)\rangle.$$

Using (20), and defining the two-body operator [23]

$$(35)\qquad e_{ij,k\ell} = E_{ij}E_{kl} - \delta_{kj}E_{i\ell} = e_{k\ell,ij}$$

(equal to the normal product $N[E_{ij}E_{k\ell}]$, relative to the physical vacuum, of the Fermion operators in the generator product), the second-quantized Hamiltonian can be expressed as a bilinear form in the unitary group generators,

$$(36)\qquad H = \sum_{i,j} h_{ij}E_{ij} + \frac{1}{2} \sum_{i,j,k,\ell} [ij;k\ell] e_{ij,k\ell}.$$

Determination of the energy and expansion coefficients of an electronic wave function by CI or similar methods requires the calculation of the matrix elements of the Hamiltonian in the N-electron basis. Using the Gel'fand-Tsetlin basis and the unitary group Hamiltonian (36), a matrix element between two Gel'fand states with lexical indices m and m' can be written in the form

$$(37)\qquad \langle m'|H|m\rangle = \sum_{i,j} h_{ij} \langle m'|E_{ij}|m\rangle + \frac{1}{2} \sum_{i,j,k,\ell} [ij;k\ell] \langle m'|e_{ij,k\ell}|m\rangle.$$

The matrix elements of the unitary group generators E_{ij} and the two-body operators $e_{ij,k\ell}$ in the Gel'fand-Tsetlin basis are, therefore, the coefficients (also called "coupling coefficients") of the one-electron and two-electron integrals in the Hamiltonian matrix element expression. These generator matrix elements are thus the key link between the orbital basis and the N-electron orthonormal, spin-adapted Gel'fand-Tsetlin basis. This link is the principal formal problem of the CI method, and the unitary group approach, through (37), provides a very effective solution. Once practical procedure are available for determining the matrix elements of these unitary group operators, it becomes possible to design appropriate algorithms for the determination of Hamiltonian matrix element formulas and for the efficient calculation of their numerical values.

3.3. Matrix elements of unitary group generators. Matrix elements of weight generators E_{jj} are particularly simple to obtain, based on (30),

(38) $$\langle m'|E_{jj}|m\rangle = \delta_{m'm} n_j,$$

where n_j is the occupancy of orbital j in $|m\rangle$. Furthermore, because of the adjoint property (29) of raising and lowering generators, their matrix elements (which are all real) are related by

(39) $$\langle m'|E_{ij}|m\rangle = \langle m|E_{ji}|m'\rangle.$$

Thus it is sufficient to consider the determination of matrix elements of raising generators.

From the earlier discussion it is obvious that a raising generator matrix element $\langle m'|E_{ij}|m\rangle$, $i < j$, can be nonzero only if the conditions (23–28) are fulfilled. In terms of the graphical representation of the two Gel'fand states, the corresponding requirement is that the walks describing these states on the graph must coincide from the tail to level i-1 and from the vertex at level j to the head. Between these levels the two walks form a *loop*, as shown in Fig. 3, with the arcs belonging to state $|m'\rangle$ being entirely to the left (i.e., lexically preceding) those of state $|m\rangle$. It is also found that the value of the matrix element is independent of the *lower walk* and *upper walk*, connecting the loop to the tail and head, respectively. This value depends only on the *shape* of the loop and on the b values of its vertices, and will be called the *loop value*.

For later convenience, each loops is divided into *loop segments*, each consisting of the vertices of the loop at a particular level plus

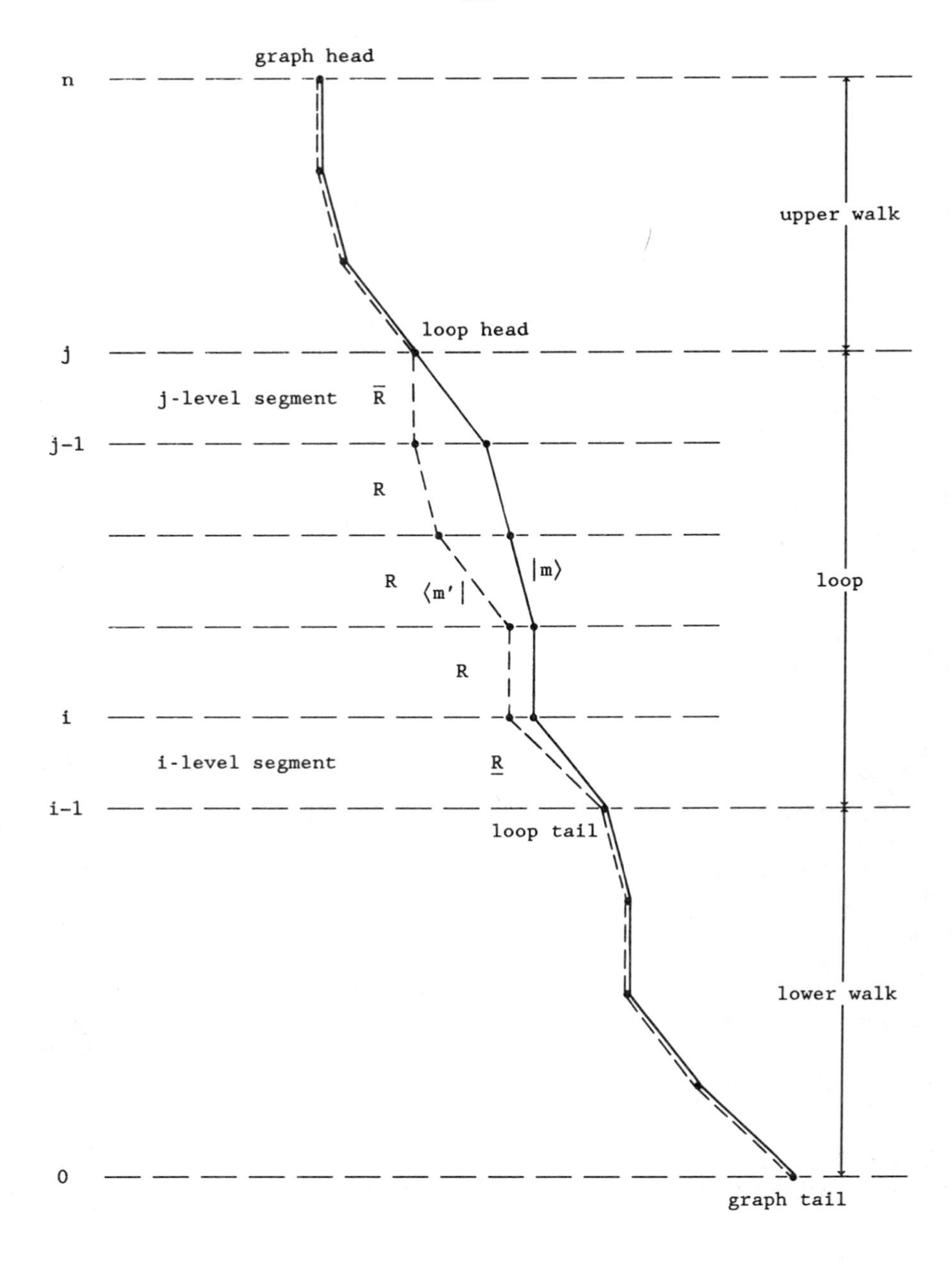

FIG. 3. *Graphical representation of a nonzero matrix element* $\langle m'|E_{ij}|m\rangle$ *of a unitary group raising generator [19]. The walks belonging to the states* $|m\rangle$ *and* $|m'\rangle$ *are shown as full and dashed lines, respectively. Level indices are shown on the left.*

the arcs attached to them from below (see Fig. 3). The number of segments is the *loop length* ($j-i+1=5$ in Fig. 3). The *range* of a generator E_{ij} is defined as the set of levels consisting of levels i and j and all levels between them, and will be denoted by the symbol $\{i:j\}$. The *length* of the range is $\max(i,j) - \min(i,j) + 1$, and is equal to the length of the loops representing matrix elements of that generator.

The history of derivations of unitary group generator matrix element formulas has been reviewed by Paldus [22]. The most powerful approach is based on the graphical techniques of spin algebra (see, e.g., [14, 22]). Here we shall present derivations entirely within the unitary group approach, using the graphical representation of the Gel'fand states [19, 23]. The starting point for the derivations are the simple formulas obtained for *elementary generators* $E_{j,j\pm1}$. These formulas [6–8] are a special case of the otherwise unwieldy general matrix element formulas of Gel'fand and Tsetlin [39] (see [8]), and can also be obtained by other approaches. A graphical representation of the matrix elements of the elementary raising generators [19] is shown in Fig. 4. Only the loop part of the matrix element is shown, since the matrix element value (loop value) is independent of the lower and upper walks. The eight loop types shown are the only types which produce nonzero values. The loop values are given in terms of the b value of the *loop head*.

For matrix elements of nonelementary raising generators, Paldus originally proposed [6] the use of the recursion formula

$$E_{ij} = [E_{i,j-1}, E_{j-1,j}], \tag{40}$$

which is a special case of the commutation relation (21). The matrix elements are obtained in the form

$$\langle m'|E_{ij}|m\rangle = \sum_{m''} \Big(\langle m'|E_{i,j-1}|m''\rangle\langle m''|E_{j-1,j}|m\rangle - \langle m'|E_{j-1,j}|m''\rangle\langle m''|E_{i,j-1}|m\rangle\Big). \tag{41}$$

However, the direct implementation of this method is difficult to incorporate into a general CI treatment, especially when (as is usually the case) not all the intermediate states $|m''\rangle$ appearing in the sum are included in the CI expansion.

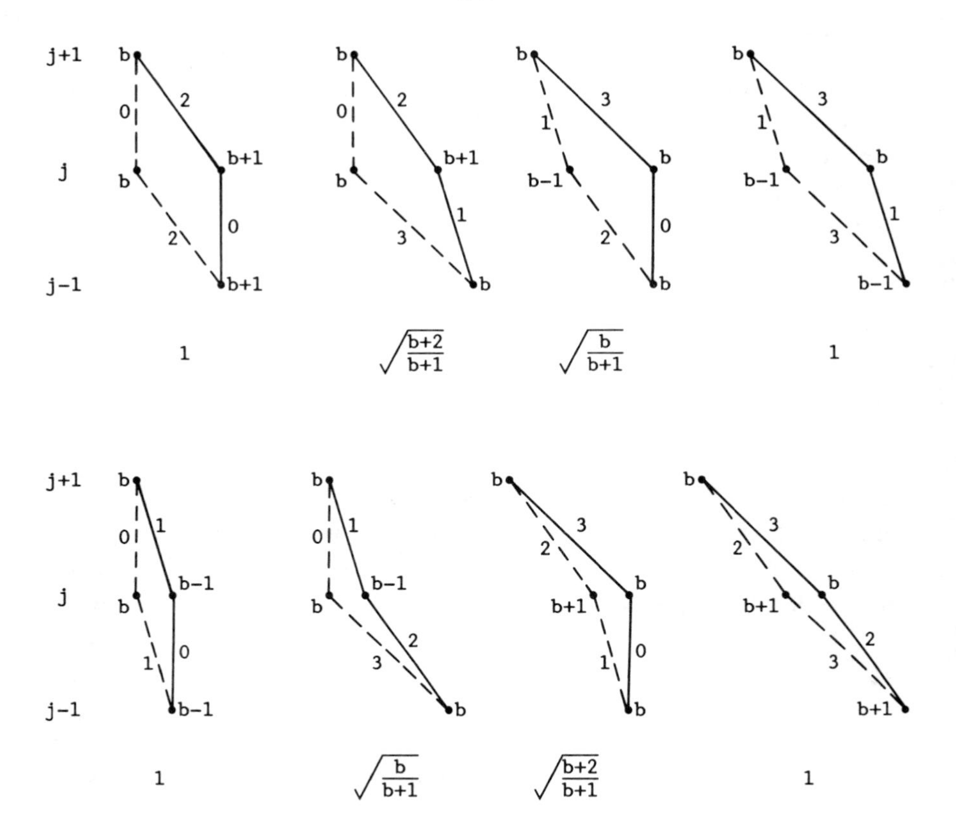

FIG. 4. The eight elementary loops which give nonzero matrix elements $\langle m'|E_{j,j+1}|m\rangle$ for elementary raising generators [19]. Vertex b values and arc step numbers d are shown, and the loop values are given below the loops in terms of the b value of the loop head. The level indices are shown on the left.

A formally equivalent, but practically much more convenient procedure for the evaluation of generator matrix elements can be derived using the graphical unitary group representation [19]. Starting with the conjecture that the loop value can be factored into segment values, independent contributions from the individual loop segments, it can be verified that the conjecture holds and all the required segment values can be obtained. The derivation will be inductive, starting from the elementary loops, and showing that if the conjecture holds for loops up to some length, then it is true for successively longer loops. The resulting formula will take the form [19, 23]

(42) $$\langle m'|E_{ij}|m\rangle = \prod_{k=i}^{j} W(Q_k;\ d'_k d_k,\ \Delta_k,\ b_k),$$

where Q_k is the segment type symbol, as further described below, and

(43) $$\Delta_k = b_k - b'_k.$$

Each segment consists of two arcs, the ket line, belonging to the ket state $|m\rangle$ and having step number d_k (described by a full line in all the figures) and the bra line, belonging to $\langle m'|$ and having step number d'_k (described by a dashed line). The segment value W will be expressed in terms of b_k, the b value of the upper vertex on the ket line.

Segment type symbols R and L will be used to denote segments of raising and lowering generators, respectively. A bar over or under the symbol will indicate a top segment (the top segment of a loop) or a bottom segment, respectively.

It can be verified by inspection that the factorization conjecture holds for all the elementary loops of Fig. 4. In fact, the choice of segment values which gives the correct values for these loops is not unique. One convenient choice is shown in Fig. 5. The eight segments in that figure can be classified into two groups (A, B, E, F and C, D, G, H, using the labels in Fig. 5), depending upon the Δ_k value at their open end. Since upper segments ($\bar{R}$) of one of these groups can only be connected with bottom segments ($\underline{R}$) of the same group, the segment values can be chosen independently for the two groups. Two arbitrary functions, f(b) and g(b), can be chosen, one for each group; if the upper segment values of a group are multiplied by the corresponding function and the bottom segment values divided by it (after adjusting the b values in the arguments of f(b) or g(b) to match), all the loop values remain unchanged. These results are summarized in Table 3. Unless stated otherwise, the choice f(b) = g(b) = 1 will be assumed from here on.

With (42) confirmed for all loops of length $j - i + 1 = 2$, the inductive proof for longer loops will be carried out graphically [19]

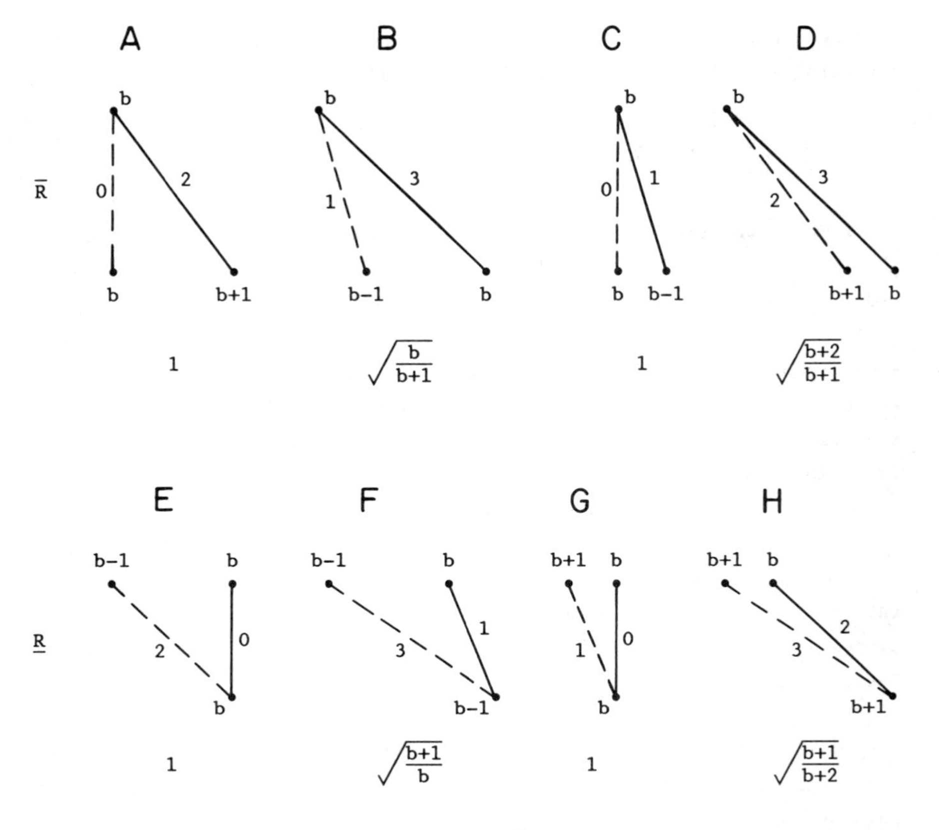

FIG. 5. *The four top segments* $\overline{R}$ (A–D) *and the four bottom segments* $\underline{R}$ (E–F) *obtained by factorization of the elementary loops [19]. Vertex* b *values and arc step numbers* d *are shown, and the segment values for* $f(b) = g(b) = 1$ *are given below the segments.*

(Fig. 6), employing a more general version of the recursion (40),

$$E_{ij} = [E_{ik}, E_{kj}] \qquad (i < k < j), \tag{44}$$

and using the corresponding summation over intermediate states analogous to (41),

$$\langle m'|E_{ij}|m\rangle = \sum_{m''} \Big(\langle m'|E_{ik}|m''\rangle \langle m''|E_{kj}|m\rangle - \langle m'|E_{kj}|m''\rangle \langle m''|E_{ik}|m\rangle \Big). \tag{45}$$

TABLE 3

All nonzero segment values for top and bottom raising generator segments [19]. The segment values are given in terms of the b *value of the top vertex on the ket lines. The arbitrary functions* f(b) *and* g(b) *are included. The labels are those shown in Fig. 5.*

Label	Segment value	Label	Segment value
	Top segments $\overline{R}$		Bottom segments $\underline{R}$
A	$f(b+1)$	E	$\frac{1}{f(b)}$
B	$f(b)\sqrt{\frac{b}{b+1}}$	F	$\frac{1}{f(b)}\sqrt{\frac{b+1}{b}}$
C	$g(b-1)$	G	$\frac{1}{g(b)}$
D	$g(b)\sqrt{\frac{b+2}{b+1}}$	H	$\frac{1}{g(b)}\sqrt{\frac{b+1}{b+2}}$

Assuming that the factorization (42) is correct for all loop of length $k-i+1$ (such as for E_{ik}) and for all loops of length $j-k+1$ (E_{kj}), it will be shown that it is also correct for all loops of length $j-i+1$, i.e., for E_{ij}.

From the conditions (23–24) and Fig. 3, it is clear that, in order to contribute to the sum, the walk describing intermediate state $|m''\rangle$ in the first term of (45) must coincide with the walk for state $|m'\rangle$ from the vertex at level k upwards, and with the walk for state $|m\rangle$ from the vertex at level k–1 downwards. The k-level arc on this walk must therefore connect the vertex at level k–1 on the $|m\rangle$ walk with the vertex at level k on the $|m'\rangle$ walk. If this connection corresponds to a valid arc, there will be exactly one nonzero contribution to the sum (45) from the first term; otherwise, there will be none. A similar result, with the roles of $|m\rangle$ and $|m'\rangle$ reversed, is obtained for the second term. This process is independent of the details of the parts of the original loops above and below level k, and is illustrated in Fig. 6. It is also seen from Fig. 6 that the new loop segments created at level k through the interconnections (dotted lines) between the two walks are top and bottom segments, and their contribution to the two factors in each term of (45) is therefore known (since it has been assumed that the factorization is correct for the shorter loops). The segment value for the original k-level segment (called a *middle segment*) can therefore be determined as at most a difference of two products of top and bottom loop segment values. This process is shown graphically for one such middle segment in Fig. 6, and the results for all middle segments with nonzero values are given in Table 4 and Fig. 7. The proof of the factorization equation (42) is thus complete. It is also seen that condition (26, 27) for nonzero raising generator matrix elements is a consequence of the requirement for at least one valid interconnection between the two walks at each level.

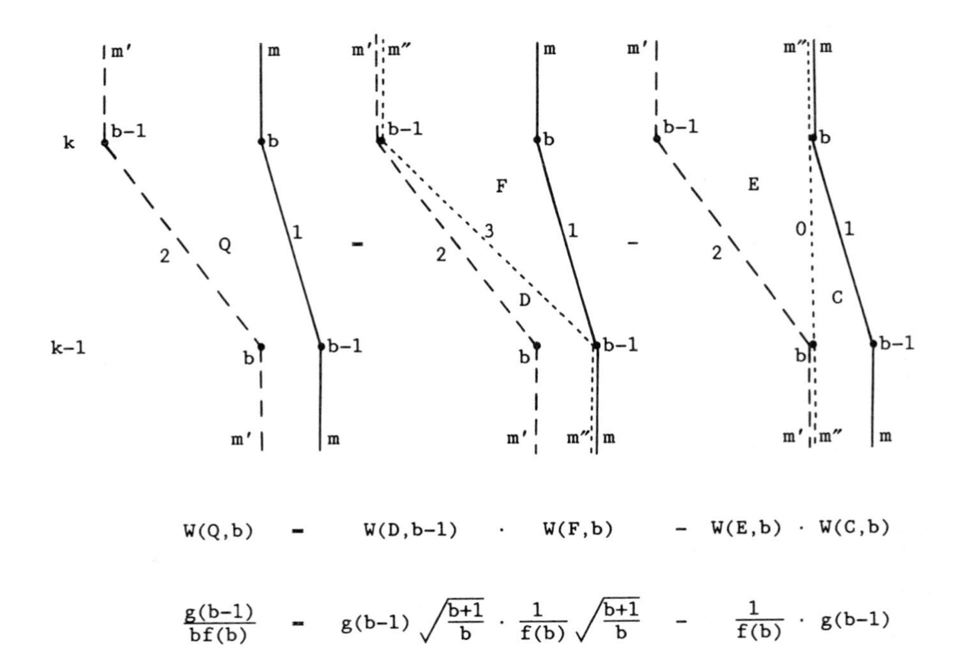

FIG. 6. *Example of the derivation of the value of a middle segment of a raising generator [19] (see (45)). The intermediate states are indicated by dotted lines, and the top and bottom segments created by the insertion of the intermediate state are indicated by the corresponding segment labels (see Tables 3, 4). Vertex* b *values and arc step numbers* d *are shown. Level indices are given on the left.*

TABLE 4

All nonzero segment values for middle raising generator segments R *and their derivation by the process of Fig. 6 [19]. The segment values are given in terms of the* b *value of the top vertex on the ket lines. The arbitrary functions* f(b) *and* g(b) *are included. An abbreviated notation* W(Q,b) *is used for segment values, showing the label for* Q *and the* b *value. The labels in the first column and in the derivations are those in Figs. 5, 7.*

Label	Segment value	Derivation
I	1	W(A,b-1)W(E,b)
J	$\frac{f(b-1)}{f(b)} \frac{\sqrt{(b-1)(b+1)}}{b}$	W(B,b-1)W(F,b)
K	$-\frac{f(b+1)}{f(b)}$	-W(E,b)W(A,b)
L	-1	-W(F,b)W(B,b)
M	1	W(C,b+1)W(G,b)
N	$-\frac{g(b-1)}{g(b)}$	-W(G,b)W((C,b)
O	$\frac{g(b+1)}{g(b)} \frac{\sqrt{(b+1)(b+3)}}{b+2}$	W(D,b+1)W(H,b)
P	-1	-W(H,b)W(D,b)
Q	$\frac{g(b-1)}{bf(b)}$	W(D,b-1)W(F,b) - W(E,b)W(C,b)
R	$-\frac{f(b+1)}{(b+2)g(b)}$	W(B,b+1)W(H,b) - W(G,b)W(A,b)

The segment values of lowering generator segments are numerically equal to those of the raising generator segments obtained by interchanging bra and ket lines, but their formal expressions are different because their values are expressed in terms of the b value at the top of the new ket line. Complete segment value tables for raising, lowering, and weight generators can be found in Ref. [23] (see also Ref. [14]).

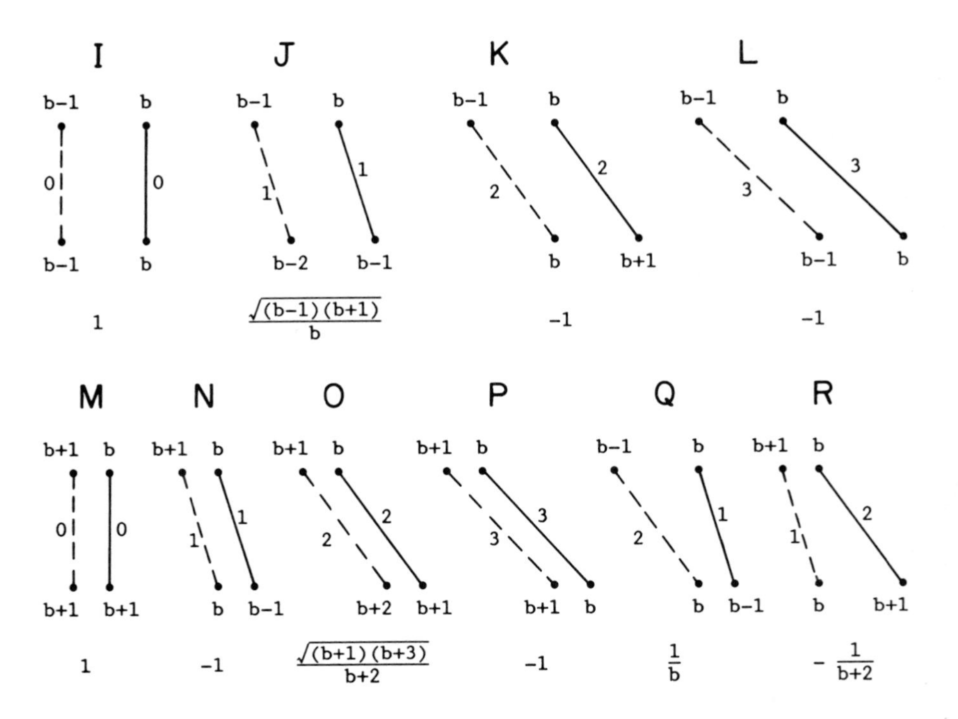

FIG. 7. *The ten middle segments* R *with nonzero segment values [19]. Vertex* b *values and arc step numbers* d *are shown, and the segment values for* f(b) = g(b) = 1 *are given below the segments.*

3.4. Matrix elements of two-body operators. Considering first the generator product $E_{ij}E_{k\ell}$, rather than the two-body operator $e_{ij,k\ell}$ (35), the matrix elements of this product can be computed, in principle, by a summation over intermediate states,

$$\langle m'|E_{ij}E_{k\ell}|m\rangle = \sum_{m''} \langle m'|E_{ij}|m''\rangle\langle m''|E_{k\ell}|m\rangle. \tag{46}$$

This process is illustrated in the graphical representation in Fig. 8, where the dotted lines represent intermediate states. The first example, Fig. 8(a), represents a product of two generators with non-overlapping ranges. In this case there is just one intermediate state, which coincides with $|m\rangle$ in the $(i:j)$ range and with $|m'\rangle$ in the $(k:\ell)$ range. The rules for nonzero matrix elements in this case are a simple extension of the rules (23–27) for single generators, and the value of

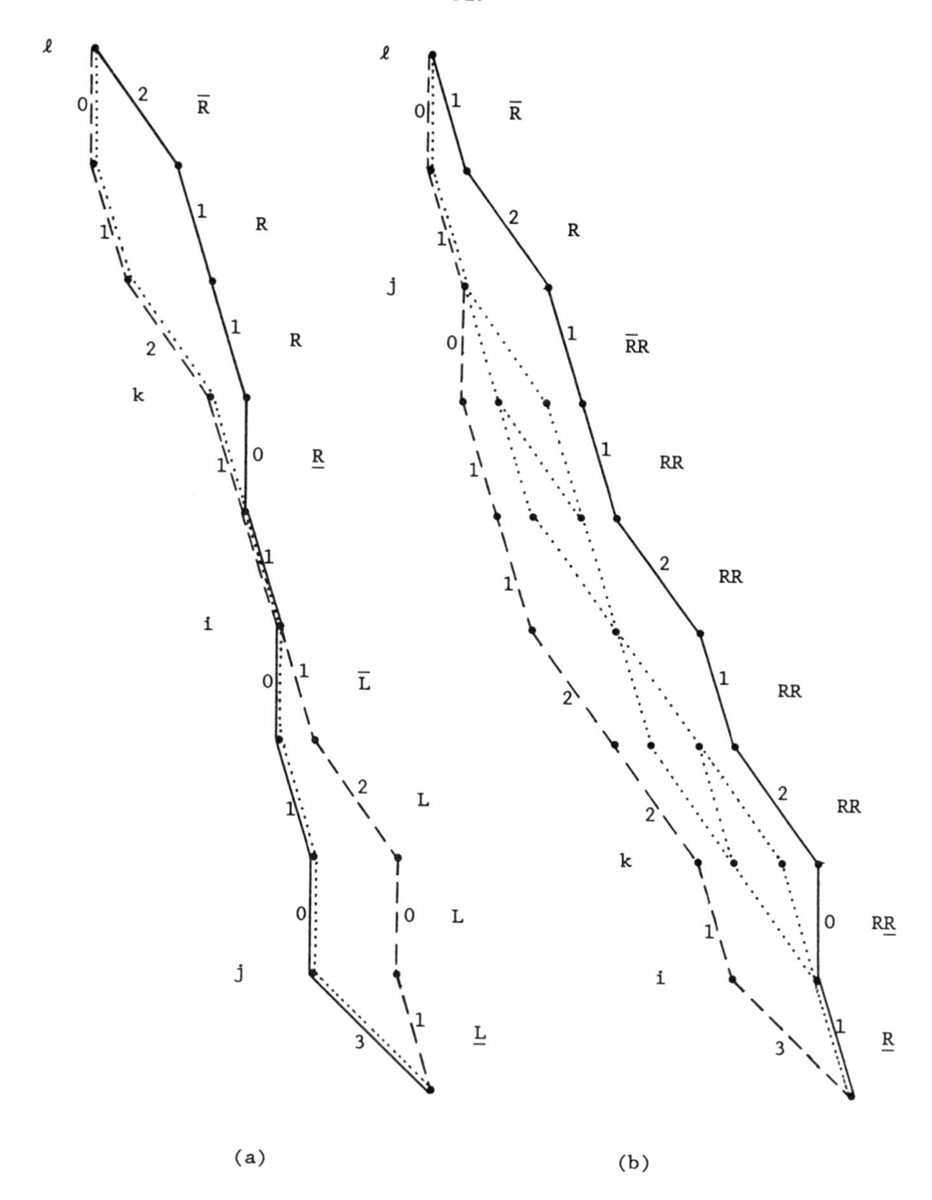

FIG. 8. *Two examples of loops representing matrix elements of the generator product* $E_{ij}E_{k\ell}$. *(a) A product of one lowering and one raising generator with non-overlapping ranges* $(j < i < k < \ell)$. *(b) A product of two raising generators with overlapping ranges* $(i < k < j < \ell)$. *The intermediate arcs (dotted lines) give rise to nine intermediate states in (b). Arc step numbers, segment type symbols, and relevant level indices are shown [23].*

the matrix element is given simply by the product of the values of the two loops in its graphical representation.

The second example, Fig. 8(b), involves overlapping generator ranges, and has contributions from several intermediate states $|m''\rangle$ (nine in the example shown). It was shown [19] that instead of explicit term by term summation, the sum in (46) can be evaluated recursively, level by level, by a simple extension of the factorization procedure for single-generator matrix elements. The conditions (23–27) for each of the two factors in (46) result in there being at most two intermediate vertices, with $b'' = b \pm 1$ (and, simultaneously, $b'' = b' \pm 1$, so that $\Delta = b - b' = 0, \pm 2$) at every level in the interior of the overlap range. (Of the four possible interconnections between two pairs of intermediate vertices at successive levels, no more than three can be valid intermediate arcs.) A set of values can be assigned to each *two-body segment* which can make nonzero contributions to the sum in (46), one value for each valid intermediate arc in that segment. The value associated with a particular intermediate arc is simply the product of the values of the two one-body segments into which the two-body segment is decomposed by that arc (both factors being expressed in terms of a common b value, corresponding to the upper vertex of the ket line). The values for all intermediate arcs for a two-body segment can be arranged in a $p \times q$ matrix, where p and q (each equal to 1 or 2) are the numbers of valid intermediate vertices at the top and bottom, respectively, of the segment. The complete matrix element can still be computed as a product, Eq. (42), over individual segment contributions, except that the contributions of the two-body segments can take the form of 1×2, 2×2, or 2×1 matrices (with at most three nonzero elements in the 2×2 case). The order of the multiplication is such that matrices belonging to higher-level segments appear to the left of those belonging to lower-level segments. Segment type symbols consisting of two letters (RR, RL, etc., reflecting the raising or lowering classification of each of the two factors in the generator product) with possible bars over or under the letters (see Figs. 8, 9) are assigned to the two-body segments in an obvious extension of the one-body segment notation. Examples of several two-body segments and their values are given in Fig. 9.

The computation of the $\delta_{kj}E_{i\ell}$ contribution in the evaluation of matrix elements of the two-body operator (35) can be avoided in most cases by choosing between the two equivalent forms $e_{ij,k\ell}$ and $e_{k\ell,ij}$.

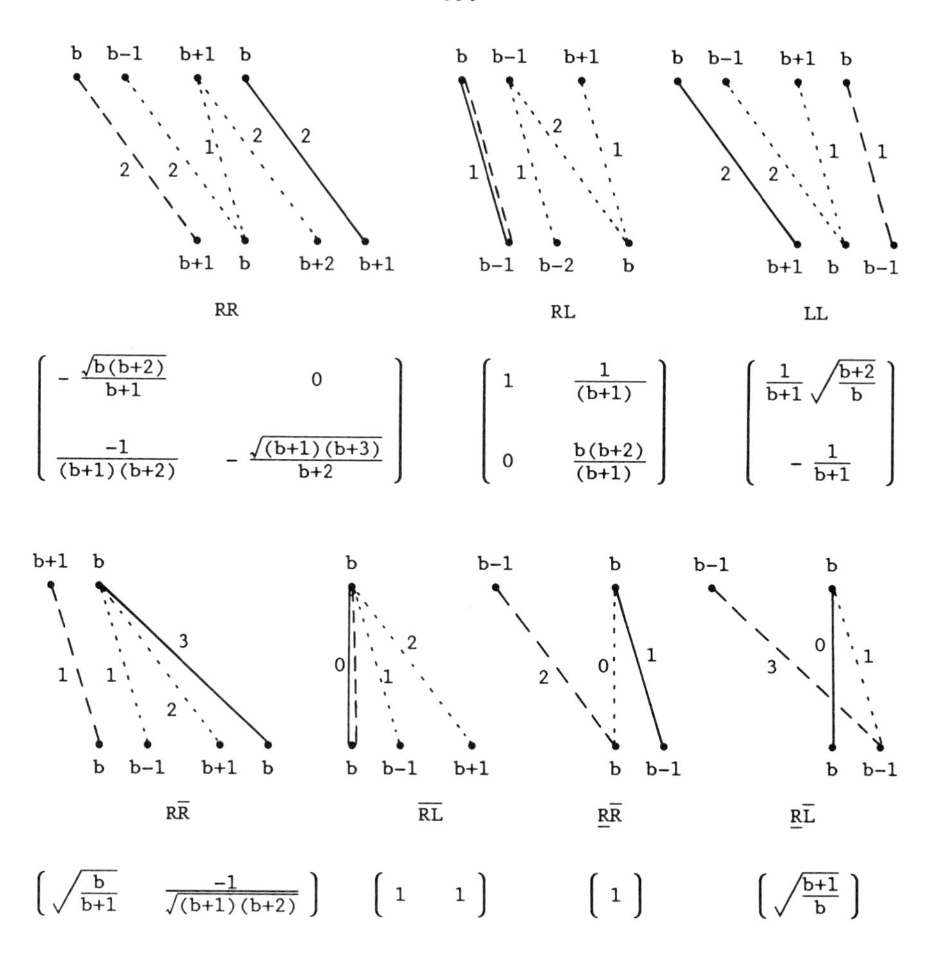

FIG. 9. *Examples of segments for the product of two generators, using the procedure of Ref. [19]. Intermediate arcs are shown as dotted lines. Vertex* b *values and arc step numbers are shown. Segment type symbols and segment-value matrices are given below the segments.*

This factorization scheme for two-body operators (introduced in [19], but first written explicitly as a matrix product by Kent and Schlesinger [40]) is quite workable. It has been derived entirely within the unitary group approach, making use of the graphical representation. However, an alternative and usually more efficient procedure has been derived by the graphical techniques of spin algebra by Drake and Schlesinger [16] and by Paldus and Boyle [14]. It obtains the overlap range contribution to the matrix element as a sum of two products of scalar segment values, the complete matrix element being given in the form

$$(47) \quad \langle m'|E_{ij}E_{k\ell}|m\rangle = \left\{\prod_{p\in S_1} W(Q_p;d'_pd_p,\Delta_p,b_p)\right\}\sum_{x=0,1}\left\{\prod_{p\in S_2} W_x(Q_p;d'_pd_p,\Delta_p,b_p)\right\}$$

where $S_2 = (i:j)\cap(k:\ell)$ is the overlap range and $S_1 = (i:j)\cup(k:\ell) - S_2$. The factors in the first product (S_1) are ordinary one-body segment values. Each of the two-body segments in S_2 has associated with it two values, W_0 and W_1, corresponding to singlet and triplet intermediate coupling, respectively, of the two electrons (see [14, 23] for further details).

Once the possibility of such an alternative factorization is known, it is possible to derive the relevant W_0 and W_1 segment values, and thus confirm (47), entirely within the graphical unitary group formalism. The derivation begins with the elementary generator product $E_{j,j+1}E_{j,j+1}$, which moves two electrons from orbital j+1 to orbital j. The only loop type describing nonzero matrix elements of this operator is of the form shown in Fig. 10(a). Using the matrix factorization method described above, its value is obtaines as

$$(48) \qquad \begin{pmatrix} \sqrt{\frac{b}{b+1}} & \sqrt{\frac{b+2}{b+1}} \end{pmatrix} \begin{pmatrix} \sqrt{\frac{b}{b+1}} \\ \sqrt{\frac{b+2}{b+1}} \end{pmatrix} = 2.$$

Since two electrons in the same orbital must be coupled singlet, the triplet contributions W_1 are set to zero and the value 2 is assigned to the product of the W_0 values of the two segments of this loop. As in the case of one-body segment values, there is some arbitrariness in how this value is to be factored between the two segments. A simple choice is to assign the value $W_0 = \sqrt{2}$ to each of the two segments ($\overline{RR}$ and $\underline{RR}$).

Considering next the loop section in Fig. 10(b), in which the top segment is the same as that in Fig. 10(a), the product of the two relevant segment-value matrices yields zero, so that it is necessary to assign $W_0 = 0$ to the lower segment in this case. In fact, it is easily

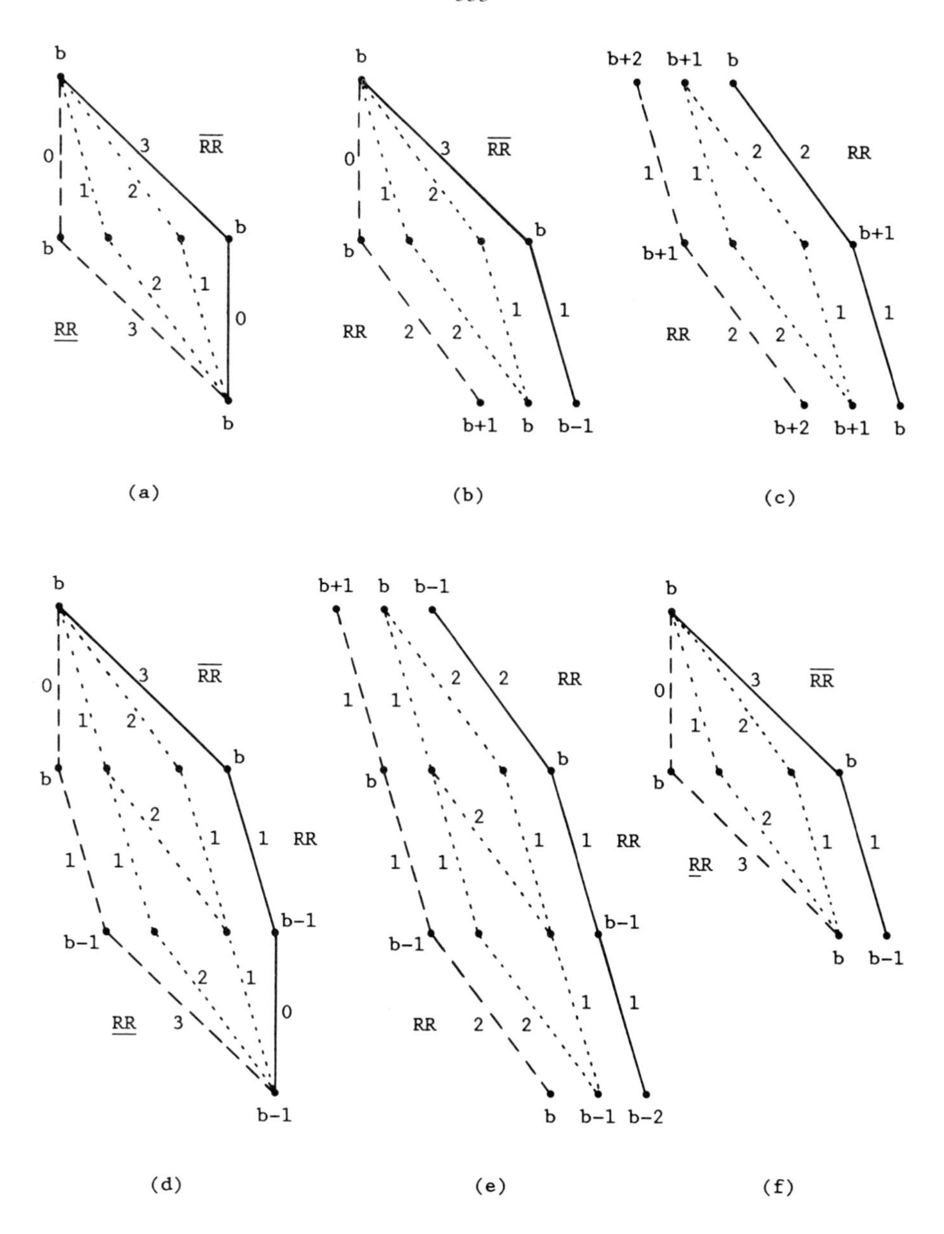

FIG. 10. *Examples of loop sections used for the derivation of two-body segment values for the factorization scheme (47) (see text). Each of these loop sections has no more than one intermediate vertex at its top and bottom vertex levels, and thus has a scalar value which can be determined by the matrix factorization procedure. Vertex* b *values, arc step numbers, and segment type symbols are shown.*

found that $W_0 = 0$ for all segments in which $\Delta = b - b' = \pm 2$ at either the top or the bottom of the segment (since $\Delta = \pm 2$ requires triplet coupling). The W_1 values for segments of this type can be found by considering the loop section in Fig. 10(c). Again, the contribution of this section can be evaluated as a product of the relevant segment-value matrices, and factored in an arbitrary way between the W_1 values of the two segments (since it has been established that the singlet-coupled contribution is zero).

The W_0 and W_1 values for other two-body segments can be obtained by attaching them to segments whose corresponding values have already been established and using the matrix factorization scheme to determine the correct value for the loop section. The loop sections must be chosen so that they have no more than one intermediate vertex at the top and bottom, so that the matrix product is a scalar. Examples are shown in Fig. 10(d–f). Similar procedures are used for generator products involving one or two lowering generators.

A formal proof of the equivalence of the evaluation of two-body loop values as a product of segment-value matrices and by the method of Eq. (47) can be achieved by a matrix transformation procedure which leads to a diagonalization of the segment-value matrices in the interior of the overlap range. Considering the case of a product of two raising generators $E_{ij}E_{k\ell}$ $(i<j,\ k<\ell)$ with overlapping ranges, and defining $p = \max(i,k)$, $q = \min(j,\ell)$ $(p<q)$, so that $\{p:q\}$ is the overlap range, the product of segment-value matrices in that range is

$$W_{\{p:q\}} = \mathbf{W}(Q_q)\mathbf{W}(Q_{q-1})\mathbf{W}(Q_{q-2})\ldots\mathbf{W}(Q_p), \tag{49}$$

where the other arguments of the $\mathbf{W}$ matrices have been suppressed for simplicity. Defining the matrices

$$\mathbf{U}(0,b) = \mathbf{V}(0,b) = \frac{1}{\sqrt{2(b+1)}}\begin{pmatrix} \sqrt{b} & \sqrt{b+2} \\ \sqrt{b+2} & -\sqrt{b} \end{pmatrix},$$

$$\mathbf{U}(\pm 2,b) = \begin{pmatrix} 0 & 1 \end{pmatrix}, \qquad \mathbf{V}(\pm 2,b) = \begin{pmatrix} 0 \\ 1 \end{pmatrix}, \tag{50}$$

which satisfy

$$\text{(51)} \qquad \mathbf{U}(0,b)\mathbf{V}(0,b) = \begin{pmatrix} 1 & 0 \\ 0 & 1 \end{pmatrix}, \qquad \mathbf{U}(\pm 2,b)\mathbf{V}(\pm 2,b) = 1,$$

the product $\mathbf{U}(\Delta,b)\mathbf{V}(\Delta,b)$ (where $\Delta = b - b' = 0, \pm 2$) can be inserted between the factors in (49) in the form

$$\text{(52)} \qquad \begin{aligned} W_{(p:q)} = {} & [\mathbf{W}(Q_q)\mathbf{U}(\Delta_{q-1},b_{q-1})][\mathbf{V}(\Delta_{q-1},b_{q-1})\mathbf{W}(Q_{q-1})\mathbf{U}(\Delta_{q-2},b_{q-2})]\ldots \\ & \ldots[\mathbf{V}(\Delta_{p+1},b_{p+1})\mathbf{W}(Q_{p+1})\mathbf{U}(\Delta_p,b_p)][\mathbf{V}(\Delta_p,b_p)\mathbf{W}(Q_p)]. \end{aligned}$$

It is easily verified that the matrix product within each square bracket in the interior of the overlap range takes the diagonal form

$$\text{(53)} \quad \mathbf{V}(\Delta_r,b_r)\mathbf{W}(Q_r)\mathbf{U}(\Delta_{r-1},b_{r-1}) = \begin{pmatrix} W_0(Q_r) & 0 \\ 0 & W_1(Q_r) \end{pmatrix} \qquad (p+1 \le r \le q-1)$$

for all valid RR segment shapes. The transformed matrices obtained for the top and bottom of the overlap range ($\overline{R}R$, $R\overline{R}$, $\overline{RR}$ and $\underline{R}R$, $R\underline{R}$, $\underline{RR}$ segment types) have the forms

$$\begin{pmatrix} W_0(Q_q) & W_1(Q_q) \end{pmatrix} \quad \text{and} \quad \begin{pmatrix} W_0(Q_p) \\ W_1(Q_p) \end{pmatrix},$$

respectively. Several examples of the derivation of the transformed segment-value matrices are shown in Fig. 11. An equivalent derivation holds for the product of two lowering generators. The procedure for raising-lowering (or lowering-raising) products is basically similar, though slightly more involved. In all cases, the result is a product of a 1×2 matrix, zero or more diagonal 2×2 matrices, and a 2×1 matrix, in that order, which is expressible as a sum of two products of scalar quantities, as given in Eq. (47).

Obviously, the $\mathbf{U}(\Delta,b)$ and $\mathbf{V}(\Delta,b)$ matrices may be multiplied and divided, respectively, by an arbitrary function of Δ and b without affecting the calculated loop values, thus providing the indeterminacy mentioned earlier in the choice of the individual segment values W_x. In fact, this arbitrary function can be chosen differently for RR, for LL, and for RL and LR generator products.

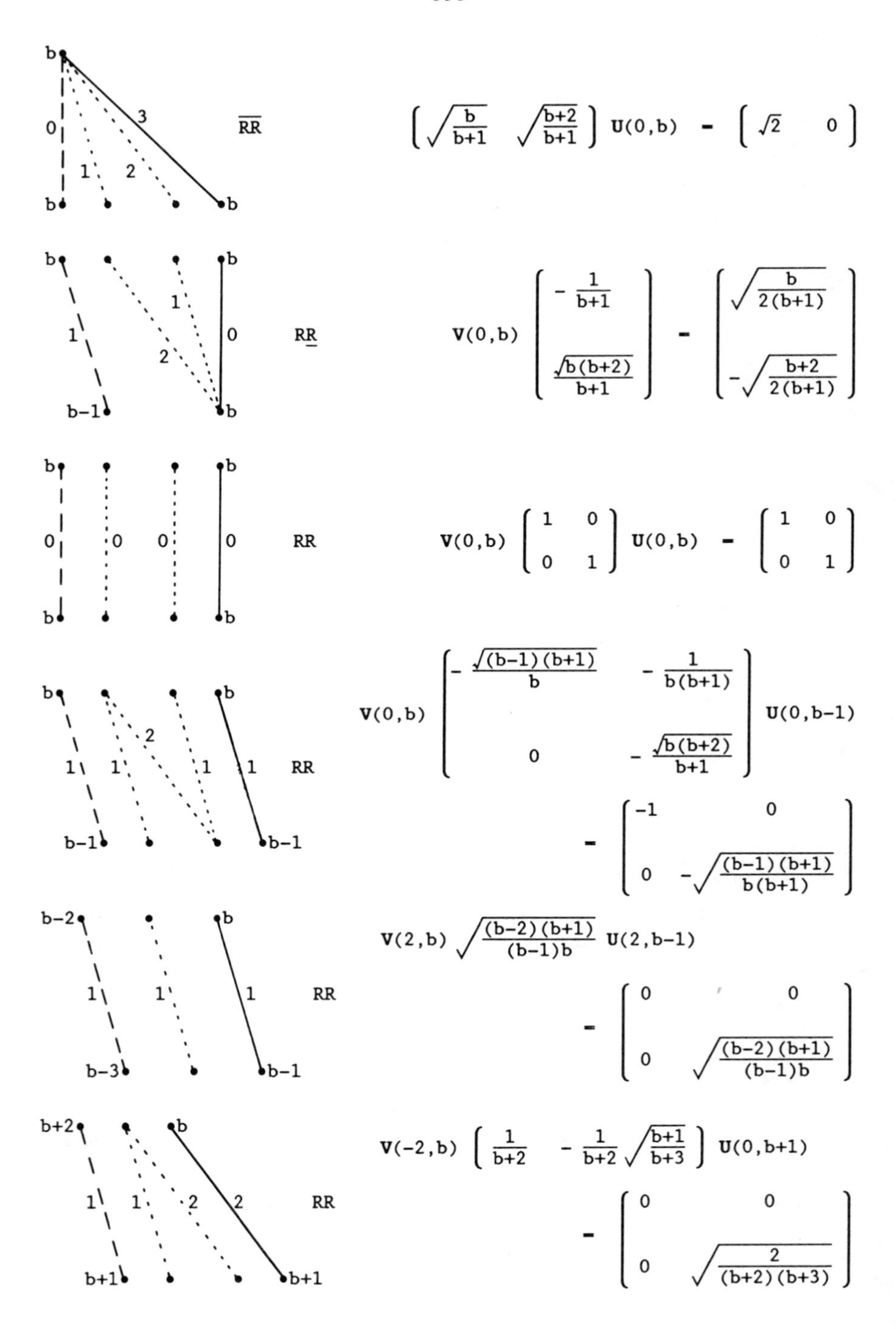

FIG. 11. *Examples of transformation of two-body segment value matrices.*

This improved approach does not require any reference to intermediate states, and makes it possible to incorporate the $\delta_{kj}E_{i\ell}$ contributions in the two-body segment values, so that these values produce the matrix elements of $e_{ij,k\ell}$ directly (the only affected segment types are $\overline{R}\underline{R}$, $\underline{L}\overline{L}$, $\overline{RL}$, and $\underline{LR}$, which simply become equal in value to the corresponding segments obtained by reversing the order of the two generator factors). Complete tables of segment values have been given in Ref. [23]. Similar tables, derived by graphical spin algebra techniques, have been given in [14] (see also [41]).

4. ORGANIZATION OF CONFIGURATION INTERACTION CALCULATIONS

4.1 Organization of the orbital space. Full CI calculations, which utilize the complete antisymmetric N-electron Hilbert space generated by the one-electron spatial orbital basis (Sections 1, 2) are generally impractical except for inadequately small basis sets or very small molecules. It is therefore necessary to identify a suitable subspace of the N-electron space in which a satisfactory approximate solution can be found [1, 3]. The usual procedure is to define the desired N-electron subspace in terms of a set of one or more reference configurations which are expected to make the major contribution to the desired wave function. The subspace for the wave function expansion is generated by adding to these reference configurations other configurations obtained by promoting ("exciting") a limited number of electrons (typically no more than two) from orbitals occupied in the reference configurations to other orbitals.

The reference configurations are usually constructed from a relatively small subset of the orbital basis, and the resulting reference wave function is usually optimized by self-consistent field (SCF) or multiconfigurational SCF (MCSCF) calculations (for single-reference and multireference expansions, respectively). This process results in an optimized partitioning of the one-electron orbital space into an *internal* orbital subspace, spanned by orbitals occupied in one or more of the reference configurations, and its complement, the *external* orbital subspace, spanned by the remaining ("excited" or "virtual") orbitals.

The internal subspace is further partitioned, in general, into subspaces spanned by *inactive* (or doubly-occupied) and *active* orbitals. The inactive orbitals are those which are doubly (i.e., fully) occupied in all the reference configurations, while the active orbitals (usually valence orbitals involved in the chemical or physical processes being studied) vary in their occupancy among the different reference configurations.

There often is an additional class of "frozen" orbitals, usually describing inner electron shells, which are left doubly occupied in all the configurations of the CI expansion (not just the reference configurations). Since it is possible to eliminate explicit consideration of these orbitals in the CI calculations by, in effect, folding their contributions into the one-electron terms of the Hamiltonian, they will be ignored in the remainder of this discussion. Similarly, some high-energy virtual orbitals may be discarded from the orbital space with little effect on the results.

4.2. Organization of the N-electron basis. The simplest structure of the DRT (and of its graphical representation) for a multireference CI expansion is obtained if the external orbitals are placed at the beginning of the ordered orbital list, so that they appear at the lowest levels of the distinct row graph. Any electrons occupying these orbitals in a configuration state function (Gel'fand state) are therefore spin-coupled among themselves before being coupled to the rest of the function. An example of the distinct row graph describing a CI expansion containing a set of reference configurations and all single and double excitations relative to them is shown in Fig. 12 (see also [23]). The reference configurations are represented in this graph by walks which are confined to the arcs drawn as heavy lines in the figure.

When high-quality basis sets are used, the vast majority of the orbitals in a CI expansion belong to the external space. The simple structure of the external orbital part of the graph is clearly demonstrated in Fig. 12. This structure is quite general for expansions limited to single and double excitations (this limitation being the most common). There are at most four vertices at each level in this part of the graph. The rightmost vertex ($a = b = 0$) at level j is traversed by all paths describing Gel'fand states in which no electrons occupy any of

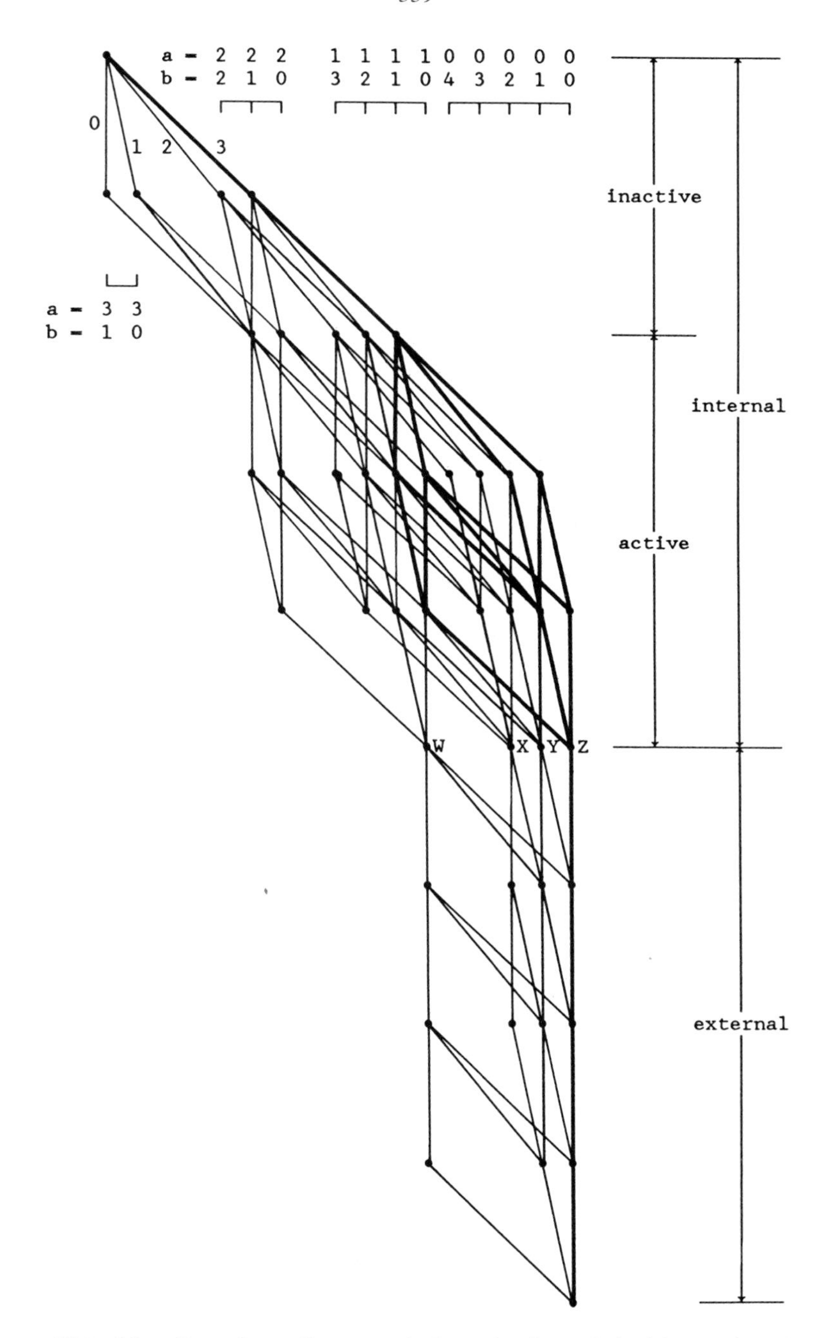

FIG. 12. *Example of a graph for single-and-double-excitation CI with eight reference configurations (heavy lines). Vertex* a *and* b *values are shown at the top, and the orbital subspaces are shown on the right (typically, the external part would be much longer than shown).*

the first j orbitals. The next vertex $(a = 0, b = 1)$ corresponds to a single electron in the subspace composed of the first j orbitals. The two leftmost vertices $(a = 0, b = 2$ and $a = 1, b = 0)$ correspond to two electrons distributed among the first j orbitals, with triplet $(S = 1)$ or singlet $(S = 0)$ spin coupling, respectively. Obviously, the reference configurations, which have no electrons in the external orbitals, are restricted to the rightmost vertical path in the external space.

A fairly simple structure is also obtained for the part of the graph representing the inactive (doubly-occupied) orbital subspace, particularly if the corresponding levels are placed at the top of the graph (as has been done in Fig. 12). The number of vertices at each level may be as large as six in this case, depending on the overall spin state. Again, the reference configurations are restricted to the rightmost path in this space. The most complex part of the graph is therefore limited to the typically small number of levels describing the active orbitals.

The simple and general structure of parts of the graph, and particularly of its longest part which represents the external orbital subspace, can be exploited very effectively in the organization of the CI calculations [23, 24]. All the detailed analysis required for matrix element formula determination and lexical index identification can be confined to the internal space, while the contributions of the external space, which take the same form in all calculations, can be built into the structure of the computer programs in an efficient and vectorizable way. Important elements in this organization are the numbering scheme chosen for the Gel'fand states and the treatment of spatial symmetry.

In choosing the numbering scheme it is convenient, in order to bring together blocks of Gel'fand states which require similar treatment, to deviate from the original lexical ordering. Specifically, it is convenient to bring together all states which share a common *boundary vertex*, the vertex at the boundary between the internal and external spaces. It is also convenient to have all the reference configurations near the beginning of the ordered list of Gel'fand states. To this end, the order chosen in the current implementation [24] combines reverse lexical ordering in the internal space with ordinary lexical ordering in the external space.

The *ordinal index* of a Gel'fand state, i.e., its position in the ordered list of states (replacing the lexical index when nonlexical ordering is used), is determined, first, by the *internal path* (the part of the walk representing the state within the internal space), and then, among all states which share a common internal path, by the *external path*. The internal paths are ordered first by their boundary vertex, from right to left, so that the paths having the rightmost boundary vertex (labeled Z in Fig. 12), and thus including all the reference configurations, appear first, and then by reverse lexical ordering among all paths sharing a common boundary index. (An additional modification in the internal space ordering in this implementation, relating to the step-number order, will not be discussed here.) An *index vector* is generated by the computer program, associating with each internal path the ordinal index of the first of the set of Gel'fand states sharing this internal path. The index vector entry can be set to a distinctive value, such as -1, to flag any internal path which is to be excluded, for any reason, from the CI expansion.

An important aspect of the implementation is the treatment of spatial symmetry. This treatment is currently limited to symmetries represented by Abelian point groups. Provided the orbitals are symmetry adapted to the point group, each internal path is also symmetry adapted, and the irrep to which it belongs can easily be identified. Thus, in additional to the index vector described above, each internal path also has associated with it an irrep number entry in a *symmetry vector*. The external orbitals are arranged in symmetry blocks, and the external paths that are to be combined with any given internal path in the CI expansion are limited to those paths which will give the desired overall symmetry of the electronic state under investigation. This limitation is easily incorporated in the assignment of the index vector entries, so that all included Gel'fand states are numbered consecutively. Gel'fand states described by walks passing through the rightmost boundary vertex have their symmetry determined entirely by the internal path. Internal paths of this type which do not have the irrep number entry appropriate to the electronic state are flagged (in the index vector) to be excluded from the expansion, and their exclusion is taken into account in the assignment of index vector entries to the remaining internal paths.

The index vector and symmetry vector each contain one entry for each internal path in the graph (whether this path is to be included or excluded in the CI expansion). These vectors are typically much shorter than the length of the CI expansion, and thus the computer resource requirements for their manipulation and storage are usually modest.

4.3. Direct CI iterations. The principal step in the iterations leading to the solution of the matrix eigenvalue problem for the CI expansion coefficients and energy [5] is the multiplication of the Hamiltonian matrix **H** by the current eigenvector estimate **v**,

$$\mathbf{w} = \mathbf{H}\mathbf{v}. \tag{54}$$

In the direct CI approach [4], as discussed in the Introduction, this multiplication is carried out without explicit construction of **H** (except for the main diagonal, which is computed once and stored). Instead, the one-electron and two-electron integrals (33, 34) are considered in some convenient systematic sequence, and the nonzero contributions of each integral to the elements of **H** (see (37)) are determined and used directly in the multiplication (54). The hermitian (real symmetric) nature of **H** is utilized in this treatment, so that corresponding off-diagonal contributions are computed once and used twice, accounting for the upper and lower triangles of the matrix. Other equalities and relationships corresponding to index permutations in the operators (35) and integrals (34) are also utilized. This process is repeated in each iteration.

In typical CI calculations, well over 80% of the Hamiltonian matrix elements are zero. An approach in which each matrix element is considered in a systematic sequence, its formula determined and checked for nonzero value, and evaluated, may spend too much time identifying and discarding zero elements. The key to a successful implementation is that nonzero terms should be generated directly, without having to consider and discard zero terms. An additional difficulty of the conventional CI approach is that the Hamiltonian matrix is often too large to store. The direct CI approach avoids both of these problems. It can be made quite efficient because most nonzero elements of **H** have only one or two integral contributions.

The direct treatment of nonzero contributions is easily achieved in the graphical unitary group approach through the process of *loop construction* [23]. Using the graph as a template, nonzero-valued loops are constructed on the graph by combining nonzero-valued loop segments in an incremental tree-like iterative process, proceeding from one loop to the next by replacing its bottom segment (or top segment, when reverse lexical order is used) by the next compatible segment, if there is one, or going back through the levels until a level is found at which segment replacement can lead to new loops. During this process partial segment value products, partial point group irrep products, and partial arc weight sums are accumulated and saved, to reduce the amount of computation required to obtain the relevant information for each completed loop.

Each completed loop can be used with all possible connections to the head and tail of the graph to produce matrix element contributions. The different contributions generated from a single loop are all equal, and the identity of the corresponding Gel'fand states can be determined from the walk weights. Because of the constant walk weight difference between the bra and ket lines for all such contributions from a single loop, all the corresponding matrix elements lie on the same codiagonal of the matrix. With forward lexical indexing, all the connections to the graph tail, for a fixed connection to the head, represent a contiguous sequence of such matrix elements on that codiagonal, facilitating vectorized computer processing. With the mixed ordering scheme described in Section 4.2 (lexical in the external space, reverse lexical in the internal space), the determination and processing of all the connections to the graph head is also simplified, in most cases, compared to a strict lexical ordering approach.

This clear demonstration of the structural features of direct CI calculations achieved by the graphical approach is extremely useful in the design of the computational procedures to eliminate redundant processing, to streamline the operation of the computer programs, and to enhance vectorization.

As discussed in Section 4.2, explicit loop construction for each calculation can be limited to the internal part of the graph. Specifically, internal parts of loops are constructed, using the graph as a template, and all relevant information about them is listed in a *formula file*. This formula file is generated once and used in each iteration, in conjunction with the correspondingly ordered integrals file, to compute the matrix element contributions, and carry out the multiplication (54), by extending the incomplete internal loops into the external space in all ways compatible with the spatial symmetry of the electronic state under investigation. Because of the simple structure of the external space, there is a limited number of types, or shapes, of loop extensions which need be considered, and the processing of each shape (or set of related shapes) is carried out by a specific, individually designed section of the computer program for maximum efficiency. Examples of external loop extensions are shown in Fig. 13. A complete list of external loop extension types was given in an unpublished report [42], and formed the basis for the current computer implementation of the method [24]. This approach was later rediscovered and called the *shape driven* approach by Saxe *et al.* [30].

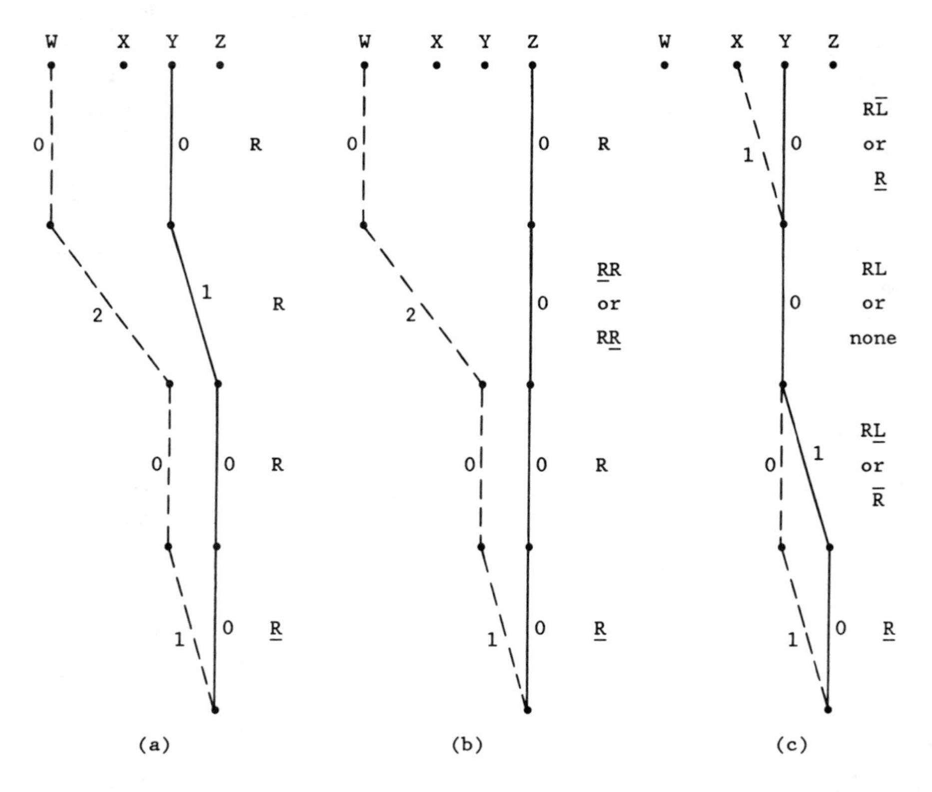

FIG. 13. *Examples of external loop extensions for: (a) one external index, (b) two external indices, and (c) three external indices. Boundary vertices are given at the top, and segment type symbols and arc step numbers are shown.*

The internal loop sections in the formula file are generated and arranged in four sets, according to the number (0, 1, 2, or 3) of external indices (indices referring to external orbitals) in the corresponding operators and integrals (34, 35). The simple case of four external indices requires no precomputed loop data, and is handled entirely by appropriate program sections in the iterative process. Within each set, the loop sections are arranged according to the specific internal indices involved, and this arrangement is reflected in the ordering of the integrals file, which is organized in blocks of integrals with common internal indices. (This discussion has ignored the relatively trivial case of the one-electron contributions, which are handled in an analogous manner as part of the treatment of appropriate two-electron terms.)

For each set of internal indices, the internal loop sections are arranged according to the different combinations of boundary vertices involved (W, X, Y, or Z, Fig. 12), since the details of loop extension depend strongly on the boundary vertices of the internal loop section. The loop sections are easily generated in this order because of the reverse lexical order used in the internal space. The formula file is independent of the external space and, therefore, on the size of the orbital basis. This file can therefore be reused when the basis set size is changed.

The integrals file takes advantage of spatial symmetry blocking (restricted to Abelian point groups), and this aspect is easily taken into account in the external space processing, since the process of loop extension is designed to omit walks with incompatible symmetries, which would correspond to missing zero integrals.

4.4. **Summary.** The approach and procedures described here have been implemented in the COLUMBUS system of computer programs for MCSCF and multireference CI calculations [24], and have been adapted for efficient operation on vector-oriented computers by reorganizing most external loop processing as matrix-matrix and matrix-vector multiplications [43]. Multireference CI calculations using more than a million configurations are being carried out routinely with these programs on current supercomputers and array processors. In most cases, the computation time has been found to vary almost linearly with the CI expansion length.

It is clear that the graphical unitary group approach has been of great value in the derivation of algorithms for general direct CI calculations in a spin-adapted basis and in guiding the design of the computer implementation. In addition to providing an effective formalism for the unified treatment of permutational antisymmetry and spin symmetry, the approach has provided considerable insight into the structure of the required calculations. One of the remaining shortcomings is the limitation of the treatment of spatial symmetry in the present formalism to Abelian point groups. A more general symmetry treatment would require deviation from the strict Gel'fand tableau spin coupling scheme, perhaps using the techniques developed by Gould and Paldus [44] (see also [2]). Procedures for the evaluation of matrix elements of spin-dependent operators have been developed by Gould and Chandler [45].

The insight into the structure of CI calculations provided by the graphical unitary group approach has influenced the design of other CI programs, and has helped spark a general increase in the power of CI treatments in recent years. The approach has also been applied to MCSCF calculations [46] and to the electron propagator method [25, 47, 48]. Attempts to apply it to quasidegenerate perturbation theory [49] and coupled cluster methods are continuing.

REFERENCES

[1] J. Almlöf, Electronic structure theory (in this volume).

[2] J. Paldus, Lie algebraic approach to many-electron correlation problem (in this volume).

[3] I. Shavitt, The method of configuration interaction, in Methods of Electronic Structure Theory (edited by H. F. Schaefer III), Plenum Press, New York, 1977, pp. 189-275.

[4] B. O. Roos and P. E. M. Siegbahn, The direct configuration interaction method from molecular integrals, in Methods of Electronic Structure Theory (edited by H. F. Schaefer III), Plenum Press, New York, 1977, pp. 277-318.

[5] E. R. Davidson, The iterative calculation of a few of the lowest eigenvalues and corresponding eigenvectors of large real-symmetric matrices, J. Comput. Phys. 17 (1975), pp. 87–94.

[6] J. Paldus, Group theoretical approach to the configuration interaction and perturbation theory calculations for atomic and molecular systems, J. Chem. Phys. 61 (1974), pp. 5321–5330.

[7] J. Paldus, *A pattern calculus for the unitary group approach to the electronic correlation problem*, Int. J. Quantum Chem., Symp. No. 9 (1975), pp. 165–174.

[8] J. Paldus, *Many-electron correlation problem. A group theoretical approach*, in *Theoretical Chemistry. Advances and Perspectives* (edited by H. Eyring and D. Henderson), Vol. 2, Academic Press, New York, 1976, pp. 131-290.

[9] J. Paldus, *Unitary-group approach to the many-electron correlation problem: Relation of Gelfand and Weyl tableau formulations*, Phys. Rev. A 14 (1976), pp. 1620–1625.

[10] J. Paldus, *Unitary group approach to the many-body correlation problem*, in *Electrons in Finite and Infinite Structures* (edited by P. Phariseau and L. Scheire), Plenum Press, New York, 1977, pp. 411–429.

[11] J. Paldus, *Unitary group approach to molecular electronic structure*, in *Group Theoretical Methods in Physics* (edited by W. Beiglböck, A. Böhm, and E. Takasugi), Springer-Verlag, Berlin, 1979, pp. 51–65.

[12] P. E. S. Wormer and J. Paldus, *Configuration interaction matrix elements. I. An algebraic approach to the relationship between unitary group generators and permutations*, Int. J. Quantum Chem. 16 (1979), pp. 1307–1319.

[13] J. Paldus and P. E. S. Wormer, *Configuration interaction matrix elements. II. Graphical approach to the relationship between unitary group generators and permutations*, Int. J. Quantum Chem. 16 (1979), pp. 1321–1335.

[14] J. Paldus and M. J. Boyle, *Unitary group approach to the many-electron correlation problem via graphical methods of spin algebras*, Phys. Scr. 21 (1980), pp. 295–311.

[15] J.-F. Gouyet, R. Schranner, and T. H. Seligman, *Spin recoupling and n-electron matrix elements*, J. Phys. A 8 (1975), pp. 285–298.

[16] G. W. F. Drake and M. Schlesinger, *Vector-coupling approach to orbital and spin-dependent tableau matrix elements in the theory of complex spectra*, Phys. Rev. A 15 (1977) 1990–1999.

[17] W. G. Harter and C. W. Patterson, *A Unitary Calculus for Electronic Orbitals (Lecture Notes in Physics 49)*, Springer-Verlag, Berlin, 1976.

[18] I. Shavitt, *Graph theoretical concepts for the unitary group approach to the many-electron correlation problem*, Int. J. Quantum Chem., Symp. No. 11 (1977), pp. 131–148.

[19] I. Shavitt, *Matrix element evaluation in the unitary group approach to the electron correlation problem*, Int. J. Quantum Chem., Symp. No. 12 (1978), pp. 5–32.

[20] I. Shavitt, *The utilization of Abelian point group symmetry in the graphical unitary group approach to the calculation of correlated electronic wavefunctions*, Chem. Phys. Lett. 63 (1979), pp. 421–427.

[21] J. Hinze, editor, *The Unitary Group for the Evaluation of Electronic Energy Matrix Elements (Lecture Notes in Chemistry 22)*, Springer-Verlag, Berlin, 1981.

[22] J. Paldus, *Unitary group approach to many-electron correlation problem*, in [21], pp. 1–50.

[23] I. Shavitt, *The graphical unitary group approach and its application to direct configuration interaction calculations*, in [21], pp. 51–99.

[24] H. Lischka, R. Shepard, F. B. Brown, and I. Shavitt, *New implementation of the graphical unitary group approach for multireference direct configuration interaction calculations*, Int. J. Quantum Chem., Symp. No. 15 (1981), pp. 91–100.

[25] G. Born and I. Shavitt, *A unitary group formulation of open-shell electron propagator theory*, J. Chem. Phys. 76 (1982), pp. 558–567.

[26] I. Shavitt, *The unitary group and the electron correlation problem*, in *New Horizons in Quantum Chemistry* (edited by P.-O. Löwdin and B. Pullman), Reidel, Dordrecht, 1983, pp. 279–293.

[27] F. A. Matsen and R. Pauncz, *The Unitary Group in Quantum Chemistry*, Elsevier, Amsterdam, 1986.

[28] B. R. Brooks and H. F. Schaefer III, *The graphical unitary group approach to the electron correlation problem. Methods and preliminary applications*, J. Chem. Phys. 70 (1979), pp. 5092–5106.

[29] B. R. Brooks, W. D. Laidig, P. Saxe, N. C. Handy, and H. F. Schaefer III, *The loop-driven graphical unitary group approach: A powerful method for the variational description of electron correlation*, Phys. Scr. 21 (1980).

[30] P. Saxe, D. J. Fox, H. F. Schaefer III, and N. C. Handy, *The shape-driven graphical unitary group approach to the electron correlation problem. Application to the ethylene molecule*, J. Chem. Phys. 77 (1982), pp. 5584–5592.

[31] P. E. M. Siegbahn, *Generalization of the direct CI method based on the graphical unitary group approach. I. Single replacements from a complete CI root function of any spin, first order wave functions*, J. Chem. Phys. 70 (1979), pp. 5391–5397.

[32] P. E. M. Siegbahn, *Generalization of the direct CI method based on the graphical unitary group approach. II. Single and double replacements from any set of reference configurations*, J. Chem. Phys. 72 (1980), pp. 1647–1656.

[33] P. E. M. Siegbahn, *factorization of the direct CI coupling coefficients into internal and external parts*, in [21], pp. 119–135.

[34] W. Duch and J. Karwowski, *Symmetric group graphical approach to the configuration interaction method*, in [21], pp. 260–271.

[35] W. Duch and J. Karwowski, *Symmetric group graphical approach to the configuration interaction method*, Int. J. Quantum Chem. 22 (1982), pp. 783–824.

[36] W. Duch, *GRMS or Graphical Representation of Model Spaces*, Vol. 1, Springer-Verlag, Berlin, 1986.

[37] R. D. Kent and M. Schlesinger, *Graph theoretical approach to the evaluation of matrix elements of one-body operators in SU(n)*, J. Chem. Phys. 84 (1986), pp. 1583–1589.

[38] R. D. Kent and M. Schlesinger, *Graphical unitary group approach to arbitrary spin representations*, Int. J. Quantum Chem. 30 (1986), pp. 737–750.

[39] I. M. Gel'fand and M. L. Tsetlin, *Finite dimensional representation of the group of unimodular matrices*, Dokl. Akad. Nauk SSSR 71 (1950), pp. 825–828 (in Russian).

[40] R. D. Kent and M. Schlesinger, *Two-body-operator matrix elements in SU(n) for application to complex spectroscopy*, Phys. Rev. A (in press).

[41] P. W. Payne, *Matrix element factorization in the unitary group approach for configuration interaction calculations*, Int. J. Quantum Chem. 22 (1982), pp. 1085–1152.

[42] I. Shavitt, *New methods in computational quantum chemistry and their application on modern super-computers*, Annual Report to the National Aeronautics and Space Administration, Battelle Columbus Laboratories, Columbus, Ohio, June 29, 1979.

[43] R. Ahlrichs, H.-J. Böhm, C. Ehrhardt, P. Scharf, H. Schiffer, H. Lischka, and M. Schindler, *Implementation of an electronic structure program system on the CYBER 205*, J. Comput. Chem. 6 (1985), pp. 200–208.

[44] M. D. Gould and J. Paldus, *Unitary group approach to general system partitioning. I. Calculation of* $U(n = n_1 + n_2)$: $U(n_1) \times U(n_2)$ *reduced matrix elements and reduced Wigner coefficients*, Int. J. Quantum Chem. 30 (1986), pp. 327–363.

[45] M. D. Gould and G. S. Chandler, A spin-dependent unitary group approach to many-electron systems, Int. J. Quantum Chem. 26 (1984), pp. 441–455.

[46] R. Shepard, I. Shavitt, and J. Simons, *Comparison of the convergence characteristics of some iterative wave function optimization methods*, J. Chem. Phys. 76 (1982), pp. 543–557.

[47] G. Born, *U(n) operator manifolds for electron propagator applications*, Int. J. Quantum Chem., Symp. No. 16 (1982), pp. 633-639.

[48] G. Born, *Correlated electron propagator theory and applications for open-shell atoms and molecules*, Int. J. Quantum Chem. 28 (1985), pp. 335–348.

[49] I. Shavitt and G. J. Born, *Unitary group methods for the electron correlation problem*, Progress Report to the National Science Foundation, Battelle Columbus Laboratories, Columbus, Ohio, February, 1980.